NASA/TP–2007-214556

I0049914

The Aviation System Monitoring and Modeling (ASMM) Project: A Documentation of its History and Accomplishments: 1999–2005

Editor:
Irving C. Statler
Ames Research Center
Moffett Field, California

National Aeronautics and
Space Administration

Ames Research Center
Moffett Field, California 94035-1000

June 2007

The NASA STI Program Office . . . in Profile

Since its founding, NASA has been dedicated to the advancement of aeronautics and space science. The NASA Scientific and Technical Information (STI) Program Office plays a key part in helping NASA maintain this important role.

The NASA STI Program Office is operated by Langley Research Center, the Lead Center for NASA's scientific and technical information. The NASA STI Program Office provides access to the NASA STI Database, the largest collection of aeronautical and space science STI in the world. The Program Office is also NASA's institutional mechanism for disseminating the results of its research and development activities. These results are published by NASA in the NASA STI Report Series, which includes the following report types:

- TECHNICAL PUBLICATION. Reports of completed research or a major significant phase of research that present the results of NASA programs and include extensive data or theoretical analysis. Includes compilations of significant scientific and technical data and information deemed to be of continuing reference value. NASA's counterpart of peer-reviewed formal professional papers but has less stringent limitations on manuscript length and extent of graphic presentations.

- TECHNICAL MEMORANDUM. Scientific and technical findings that are preliminary or of specialized interest, e.g., quick release reports, working papers, and bibliographies that contain minimal annotation. Does not contain extensive analysis.

- CONTRACTOR REPORT. Scientific and technical findings by NASA-sponsored contractors and grantees.

- CONFERENCE PUBLICATION. Collected papers from scientific and technical conferences, symposia, seminars, or other meetings sponsored or cosponsored by NASA.

- SPECIAL PUBLICATION. Scientific, technical, or historical information from NASA programs, projects, and missions, often concerned with subjects having substantial public interest.

- TECHNICAL TRANSLATION. English-language translations of foreign scientific and technical material pertinent to NASA's mission.

Specialized services that complement the STI Program Office's diverse offerings include creating custom thesauri, building customized databases, organizing and publishing research results . . . even providing videos.

For more information about the NASA STI Program Office, see the following:

- Access the NASA STI Program Home Page at *http://www.sti.nasa.gov*

- E-mail your question via the Internet to help@sti.nasa.gov

- Fax your question to the NASA Access Help Desk at (301) 621-0134

- Telephone the NASA Access Help Desk at (301) 621-0390

- Write to:
  NASA Access Help Desk
  NASA Center for AeroSpace Information
  7115 Standard Drive
  Hanover, MD 21076-1320

TABLE OF CONTENTS

LIST OF FIGURES

FOREWORD

The Aviation System Monitoring and Modeling (ASMM) Project of NASA's Aviation Safety Program was successful, and its contributions to aviation safety, the products it produced, and the lessons learned are worth documenting. To a large degree, the success of ASMM was due to steps taken during the initial phase of project development to identify what information significantly impacts current aviation safety, and the aviation transportation system as it evolves in the future. This process was accomplished by including members with operational experience in aviation safety on the ASMM team at the outset, and by acquiring the perspectives on information requirements through end-user-needs studies. These studies entailed extensive personal interviews with the end users (e.g., at the air carriers and at the air-traffic-control facilities). The interviews were conducted with individuals by three-person teams composed of two members with operational experience in the domain of the interviewee and one representing the perspective of the ASMM Project goals. A key aspect contributing to the success of the project was the effective extraction of information from diverse data sources and recognition that it could be deemed as operationally significant information only by the end user. The first versions of the ASMM concepts and products were based on the initial perceptions of informational requirements gained through these interviews, but that was simply the starting point of the iterative process in collaboration with the end users. This iterative, evolutionary development is reflected throughout this documentation of the products of the ASMM Project. It provided essential feedback to ensure continuously that relevant products were produced, and that the products were easy to transfer for commercialization to the vendors and/or the user community as they matured. The ASMM Project management recommends this approach: "Involve the end user throughout," in developing future capabilities when success ultimately relies on the end users' perception of their operational value and relevance.

Irving C. Statler
ASMM Project Manager

1.0 INTRODUCTION

Air transportation, one of the most important modes of transportation, is also one of the safest. Nevertheless, the public demands that safety levels continuously improve and that the absolute number of aviation accidents continue to decline, even as air-traffic levels increase.

On February 12, 1997, after the tragedy of TWA 800, President William J. Clinton declared,

"We will achieve a national goal of
reducing the fatal aircraft accident rate by 80% within 10 years."

In response to this presidential declaration, the administrator of NASA announced that NASA would undertake a new program in aviation safety in support of this objective. NASA quickly formed the Aviation Safety Investment Strategy Team (ASIST) in collaboration with the Federal Aviation Administration (FAA) and the National Transportation Safety Board (NTSB), which organized a series of five workshops to examine the options and recommend an approach for NASA to develop the enabling technologies that address the president's goal. The exceptional and dedicated participation from all sectors of the aviation industry in the development of NASA's strategy was remarkable.

Subteams were formed to focus on each of the following:

- Consequences of Human Error
- Weather
- Flight Critical Systems and Information Integrity
- Human Survivability
- Aviation Systemwide Monitoring, Modeling, and Simulation

The result of these deliberations was a recommendation for a three-thrust program covering

- *Accident prevention:* Preventive technologies to eliminate accident precursors
- *Accident mitigation:* Stay-alive technologies to decrease fatalities in survivable accidents
- *Aviation System Monitoring and Modeling (ASMM)*: Technologies to identify existing accident precursors in the aviation system, and to forecast and identify potential safety issues to guide the development of future safety technology.

The ASIST recognized that, although the ultimate metric was fatal accidents, this metric was difficult to use as a metric for measuring research progress because so few accidents occur each year and their circumstances vary greatly. Consequently, the ASIST suggested that the challenge to NASA was to develop technologies and methodologies to accurately identify accident precursors from the much larger set of incident reports and from operational data acquired by flight-data recorders (as depicted in figure 1.0.1).

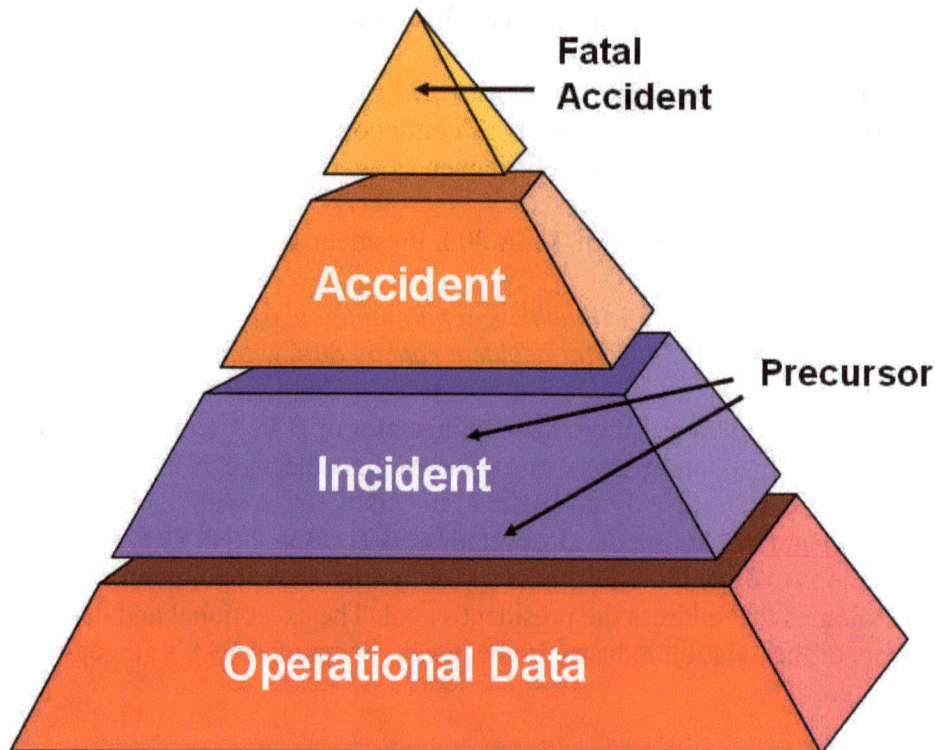

Figure 1.0.1. The Heinrich Pyramid of Data.

Following the ASIST workshops, formative planning and preliminary studies of approaches were conducted during the years of FY98 and FY99. NASA's Aviation Safety Program (AvSP) was initiated in FY00 with the goal of developing technologies that, depending upon implementation, could reduce the aircraft accident rate by a factor of 5 within 10 years, and by a factor of 10 within 20 years.[1]

One of the projects within the AvSP, the ASMM Project, addressed the need to provide decision-makers with systemwide analysis tools for identifying and correcting the predisposing conditions that could lead to accidents. Systemwide analysis is a concept with extraordinary potential benefits and challenges, because of the complex objectives and relationships in a modern aviation transportation system involving a diversity of manufacturers, operators, and regulators. NASA has an important role in modern aviation as a nonpartisan, honest broker, and ASMM is an initial step towards technologies that enable proactive management of safety risk from a systemwide perspective. This report documents the achievements of the ASMM Project during the life of the AvSP from FY00 to FY05.

---

[1] After 9/11 the AvSP became the Aviation Safety and Security Program (AvSSP).

2.0 A PERSPECTIVE ON PROACTIVE MANAGEMENT OF SAFETY RISK

Suppose you are the vice president for safety of a company whose customers enter through the front door, depart through the back door, and in between are exposed to some hazardous processes. How do you know how well your organization is fulfilling its responsibility to protect its customers from harm? Can you be content so long as (almost) every customer leaves your premises reasonably safe and sound? Is it acceptable if your statistical analyses show that, of every hundred customers that enter the front door, only 0.005 fail to make it out the back? Are you willing to wait until one of your customers is unable to make it through the back door before you investigate the safety of the processes in your facility?

This ratio is quite representative of the aviation industry. A great deal of time, money, and effort are invested in forensic analyses of every accident even though, in such (fortunately) rare events, the factors entailed in the next one will, most likely, be totally different. But, how does the vice president for safety of Lower Slobovia Airlines know how well they are operating in between these rare events? Or how about the manager of the air traffic control (ATC) facility at that major hub at Slivovitz, the capital of Lower Slobovia? Or the minister of aeronautic transportation of Lower Slobovia?

As global air transportation continues to expand, the number of aircraft accidents will become unacceptable to society if accidents continue at the current rate. Therefore, efforts are now being invested worldwide to find an approach to reducing still further the already very low rate of aviation accidents. The extremely low frequency of these events presents ASMM with a significant challenge in developing appropriate methods for analysis of these rare events to understand precursors and estimate future impact.

Worldwide, the aviation-safety culture is changing from one of primarily reacting to the last accident to a proactive approach to identification and alleviation of life-threatening conditions and events. This cultural change underlies many new national and international activities, including, for example, the FAA-Industry Flight Operational Quality Assurance (FOQA) program (see appendix A), the UK CAA Operational Flight Data Monitoring (OFDM) program, the Commercial Aviation Safety Team (CAST), the European Aeronautics Vision, the International Civil Aviation Organization (ICAO) initiatives on use of flight data, the Global Aviation Information Network (GAIN), the Eurocontrol Performance Review Commission, positions and initiatives of the unions, including, the Airline Pilots Association (ALPA), the Allied Pilots Association (APA), and the National Air Traffic Controllers Association (NATCA), the JAA (Joint Aviation Authorities) Safety Strategy Initiative (JSSI), etc.

This cultural shift was reflected in the following statement made by the FAA Administrator, Marion C. Blakely, at the North American Safety Conference on February 5, 2003:

"For one, we need to change one of the biggest historical characteristics of aviation safety improvements --- our reactive nature. We must get in front of accidents...anticipate them...and use hard data to detect problems and disturbing trends."

The General Accounting Office (GAO) evaluation report titled "Safer Skies" (2000) said that additional performance measures were needed and precursors associated with past accidents should be used to track safety baseline and improvements from interventions. The FAA Internal Studies, 2004 Strategic Plan said it was necessary to identify risks before they lead to accidents.

All of this recent activity is a reflection of the recognized need for a comprehensive, accurate, and insightful method for continuously monitoring the operational performance of aviation transportation. Aviation policy makers need reliable measures of the frequencies or the trends of aviation-safety incidents, particularly as capacity becomes constrained and the skies are more and more crowded. Technology developers need to know whether new equipment or procedures inserted into the aviation system are producing expected improvements and/or unwanted side effects. Aircraft manufacturers require feedback on their equipment performance and characteristics under the wide range of conditions present in an operational environment to identify and correct precursors, especially now that aircraft life spans are stretching towards 40 years or more. Airline operators juggle to maintain viability in a continuously changing economic environment and need information for situational awareness of global safety issues as well as their own operations. Pilots, air-traffic controllers, mechanics, or anyone working in aviation transportation need a communication channel to report personal observations of safety incidents without intimidation, punishment, or retribution.

It is conducive to the economic growth of the aviation industry and avoidance of the cost and trauma of fatal aviation accidents to provide decision-makers with regular, accurate, and insightful measures of the performance and safety of the National Air Transportation System (NATS). It is conducive to sound decision-making on interventions to provide technology and procedure developers with reliable predictions of the *systemwide* impacts of the changes they are introducing into the aviation system.

The government and the world aviation community continue to routinely amass large quantities of data that could be sources of information relevant to aviation safety. Increasingly, the accumulation of these data outpaces the community's ability to put them to practical use. It is difficult to extract operationally significant information from each of these data sources. It is difficult to combine information related to the same subject when they come from diverse, heterogeneous sources. Often safety data cannot be retrieved after they have been put into computerized storage because of the way that the data were formatted and organized. The ability to monitor continuously, convert the collected data into reliable information, and share that information for collaborative decision-making is the basis for a revolutionary, proactive approach to managing an aviation-transportation system for prevention of accidents. This paper describes the process of proactive management of safety risk, the technical challenges it entails, and the manner in which the NASA Aviation Safety Program attempted to meet those challenges through its ASMM Project.

A proactive approach to identifying and alleviating life-threatening conditions in the aviation system entails a well-defined process of identifying hazards and vulnerabilities, evaluating their causes, assessing risks, and implementing appropriate solutions. This process is not a trivial undertaking. It requires continuous monitoring of system performance in a nonpunitive environment; learning from normal operational experience; comparing actual performance to expected performance; identifying the precursor events and conditions that foreshadow most accidents; understanding the confluence of factors that cause these events; designing appropriate interventions to minimize the likelihood of their occurrence; and having a system in place to monitor the efficacy of the interventions.

Furthermore, performance must be viewed from a total system perspective, including all aspects of the NATS. The aircraft, its subsystems, its flight crew, the air-traffic controller, the dispatcher, the flight mechanic, the operational environment including weather, organizational culture, training, policies, business practices, and procedures are all elements of the complete system, and, invariably, contribute interactively to its performance.

In this complex operating environment of aviation with its many dynamically interactive components, identifying the precursors of the next accident poses a special challenge. A safety incident or an aviation accident is usually the consequence of a confluence of multiple latent and proximate factors. In her book titled "Safeware: System Safety and Computers," (1995), Nancy Leveson says,

"...all hazards are affected by complex interactions among technological, ecological, sociopolitical, and cultural systems."

People are key components of the aviation system, and human error is cited as a contributing factor or cause of 60% to 80% of aviation incidents and accidents. Interestingly, a report from the Aviation Safety Reporting System (ASRS) office shows that the ASRS analysts attributed over 60% of about 121,000 reported incidents to "human performance" (ASRS, 2005). The attribution of "human error" is a social- and psychological-based judgment of human performance made in hindsight that is, in forensic analysis, invariably biased by knowledge of the outcome (Woods et al. (1994)). However, human performance is relied upon to resolve uncertainties, conflicts, and competing demands inherent in large, complex systems and to provide flexible response to the inevitable unexpected event. Human performance is as complex as the domain in which it is exercised and cannot be judged independently of that domain. Human behavior is context-dependent, and little can be understood of the causes of human error without understanding the prevailing as well as the more distal conditions conducive to error. The humans interact with, or are impacted by, all the other elements (including humans) in the aviation system. These interactions influence human performance in ways that may enable or aggravate accidents or incidents. In addition, persons' own actions or inactions can reflexively affect their performance capabilities. Much depends on being able to determine how complex systems have failed and how human(s) contributed to the outcome of such failures. Consequently, the question is, "Why do professional, well-trained, highly motivated operators of the aviation system make mistakes that contributed to an incident or accident?" The focus of the ASMM Project was on uncovering and understanding those conditions that elevate the probability of downstream human errors and that, in turn, may contribute to aviation-safety incidents or accidents. Knowledge about these systemic features that could be the precursors of incidents or accidents assists in the understanding of how they shape human behavior and provides insight into improving the performance of the system.

In this report, the term precursor is used to mean the *symptom* of a systemic problem that is a confluence of causal factors conducive to undesired system behavior (e.g., human fatigue, organizational culture, equipment failure, or procedural discrepancy) that, if left unresolved, has the potential to result in *increased probability* of an accident. A precursor is a measurable deviation from expectations or the norm, and it is important that it not be viewed as being synonymous with causality. The term "causal factors" means both the latent and the proximate factors that constitute the context in which the event occurred and includes:

- Conditions *necessary* for the occurrence of a precursor

and

- Conditions that *increase the probability* of occurrence of that precursor.

Consequently, in the approach to the study reported here, we are searching for the causal factors of the precursor incident or trend and not of the anomalous consequences, per se. It is the problem (i.e., the causal factors of the precursor) that must be treated; not the symptom (i.e., the precursor itself). A process of routinely identifying and minimizing the causal factors that are conducive to a human making a mistake is the essence of the ASMM concept of proactive management of safety risk of the NATS.

Proactive risk management entails a cyclic process of learning continuously from normal operational experience with the following four primary steps (as depicted in figure 2.0.1):

- *Identifying*—collecting data to monitor the system performance continuously and comparing with established standards or expectations to identify potential risks.

Figure 2.0.1. A strategy for improving safety of aviation operations.

- *Evaluating*—diagnosing the factors contributing to or aggravating the potentially hazardous event, estimating the likelihood of future occurrences or confluence of these factors, and assessing the severity and possible impacts of an anomalous consequence.

- *Formulating*—proposing changes in design, training, and/or procedures to mitigate recurrence of the event, assessing the systemwide safety risk of the intervention, estimating benefits and costs, and developing a strategy for implementing the identified change(s).

- *Implementing*—implementing on a small scale or in prototype, evaluating the performance of the intervention, refining, establishing performance standards, and then monitoring the system to assess the efficacy of the intervention and to identify unwanted side effects.

In current aviation transportation operations, the key function that is often missing from this cyclic process is that of having in place a system for monitoring in order to assess the effectiveness of the intervention.

At each of these stages in the cyclic process of proactive management of system-safety risk, airline domain experts, air-traffic managers, and other providers of aviation services make key decisions in a collaborative manner to set the performance standards, gain insight from monitoring, propose the changes, and develop the intervention strategies. The technical challenges are to provide these experts with automated tools that assist them in these collaborative decision-making processes so as to enable the proactive management of safety risk. Decision-makers must be able to focus quickly on those events with the highest likelihood of occurrence and most severe potential consequences. Automated tools, such as those developed by ASMM, can facilitate this work. Computational sciences and information technology can extract useful information from large, heterogeneous, dispersed data sources to help the decision-maker gain insight into operations. However, even the most advanced adaptations of information technologies and computer sciences cannot replace the human expert with automated decisions.

The ASMM Project had the goal to provide the methodologies, the computational tools, and the infrastructure to assist the experts in making the best possible decisions.

3.0 THE ASMM PROJECT: THE GOALS AND OBJECTIVES

The ASMM Project has developed computational tools that enable human experts to focus their attention on the most significant events, and assist them to identify the factors that distinguish unsafe operations from routine operations. One of the goals of the ASMM Project was to provide decision-makers with regular, accurate, insightful information on the performance and safety of the aviation system. The products of the ASMM Project described in this report enable identification of conditions, events, and trends of operational conditions that could compromise safety. Another goal was to provide technology and procedure developers with reliable predictions of the systemwide effects of the changes they are introducing into the aviation system.

The objectives of the ASMM Project were to:

- Convert a bounty of raw aviation operational data drawn from many sources—aircraft flight data recorders, ATC radar tracks, maintenance logs, weather records, and aviation-safety incident reports—into insightful interpretations of the safety of the NATS.

- Develop methodologies and tools to enable efficient monitoring of the NATS by routinely processing large masses of both textual and quantitative data pertaining to all aspects of the performance of the NATS.

- Develop tools to extract and display reliable information from large databases with which experts can gain insight into the performance and safety of the NATS and can identify situations that may indicate changes to levels of safety.

- Assist and encourage stakeholders in the NATS in the use of these tools for their operational evaluation and continuous evolutionary development.

- Develop fast-time simulations that enable reliable predictions of systemwide effects of proposed technological or procedural changes.

Data sources of information about incidents and trends, precursors, and their causal factors are to be found in the extensive data-collection systems that already exist across the aviation industry. Information about precursors resides in a variety of diverse, heterogeneous data sources that range from purely quantitative accident and operational data to textual data in the forms of free text reports and surveys of operational experiences. These sources cover the gamut from totally objective to totally subjective sources of information, and all of these sources are needed to understand not only *what* happened, but *why*. The objective, quantitative data provide information to understand "what" happened, but information must also be extracted from qualitative, textual data sources to help the domain expert understand the subjective aspects of "why" an incident occurred.

As indicated in figure 3.0.1, these data exist in the form of accident reports resulting from forensic analyses; in the form of quantitative operational data, experiential reports, and surveys that are the bases of diagnostic analyses; and in the form of domain expert knowledge that is currently the primary basis of prognostic analysis. The challenge to the ASMM Project team was to develop automated capabilities to extract information reliably from all these sources, integrate that information, and display operationally significant results. These capabilities would enable the domain experts to gain insight into the performance and safety of the NATS, to gain insight into the precursors of the next accident and their causal factors, and to identify situations that may indicate changes to levels of safety requiring mitigating actions.

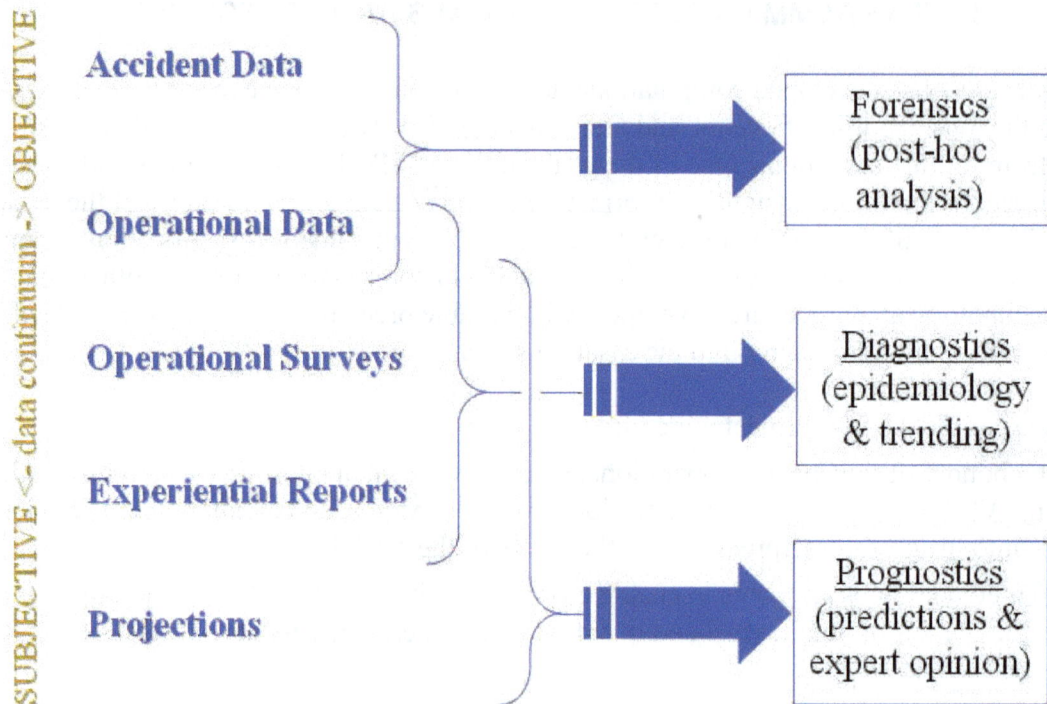

Figure 3.0.1. Where are precursors to be found?

Discovery of precursors requires knowledge of *where* to look, and *how* to look. The sources of information that will enable domain experts to identify incidents as precursors of the next accident are in large heterogeneous, distributed databases collected and owned by various organizations across the entire community. Information must be extracted from all these databases and merged in order to give the decision-maker a complete picture of the systemwide situation. How to look at these sources entails recognition of operational context and multiple operations that identify synchronicity across multiple data sources, and exploit sequential pointers from one dataset to another. Consequently, the ASMM Project has developed automated tools for extracting information from both quantitative numerical data and qualitative textual data, and for recognizing information from any data source that may be relevant to a particular query. Information extracted from quantitative data sources helps the domain expert understand the objective aspects of *what* happened, and from qualitative data sources to understand the subjective aspects of *why* an incident occurred. Each ASMM tool contributes insights into the complete picture of an event, and supports the complementary processes of causal analysis and safety-risk assessment. Causal analysis and safety risk assessment, together with analysis of associated costs and benefits, are all required in order for experts to formulate appropriate interventions.

The ASMM Project addressed a primary technological barrier to proactive management of aviation-safety risk; namely, the capability to extract meaningful and reliable information from *very large*, dispersed, heterogeneous databases efficiently with minimal labor to provide domain experts with the insight for causal analysis and safety-risk assessment.

ASMM pioneered in the application of advanced computational sciences and information technologies to

- mine aviation databases to identify automatically patterns in the data that may be the precursors without advance knowledge of what to look for, and

- present meaningful displays of these events with which to judge their operational significance.

4.0 THE ASMM PROJECT: THE APPROACH

The ASMM Project consisted of four elements that emphasize the exploitation of state-of-the-art information technology to capitalize on aviation data. The elements were called *Data Analysis Tools Development, Intramural Monitoring, Extramural Monitoring,* and *Modeling and Simulations.*

The *Data Analysis Tools Development* element provided for the research that resulted in software to process data into information that presently can be performed only by experts with much time and effort. Capabilities were developed to process textual and numerical aviation data, and to recognize relevant information in diverse databases. The ASMM tools convert these data into displays of meaningful information to help analysts achieve the insight needed to understand the circumstances, focus their attention on operationally significant events, and propose mitigating actions. Within this element, the concepts were developed to prototypical stage, at which point they were transferred to the *Intramural Monitoring and Extramural Monitoring* elements for test and evaluation in operational conditions.

The *Intramural Monitoring* element is intended to provide individual air-service operators (i.e., air carriers and ATC facilities) with the tools needed to monitor their own performance and safety continuously, effectively, and economically within their own organizations. The prototypical tools resulting from *Data Analysis Tools Development* were tested and evaluated under this element of the ASMM Project in collaboration with air carriers, FAA, and vendors of current data analysis software with whom NASA entered into Space Act Agreements (SAAs).

The *Extramural Monitoring* element complements *Intramural Monitoring* and is a comprehensive systemwide statistically sound survey mechanism for monitoring the performance and safety of the overall NATS and for detecting and evaluating the effects of new technologies as they are inserted into the system by seeking the perspectives of the front-line operators (i.e., flight crews, air-traffic controllers, cabin crews, mechanics, etc.).

The *Modeling and Simulations* element of ASMM addresses the need to support predictions and safety-risk assessments by developing and validating systemwide models and simulations. The data collected from all the activities under *Intramural* and *Extramural Monitoring* have been used to support the development and validation of models of the NATS.

Although the products of these four elements of the ASMM Project all have standalone value, they have been planned and developed in concert to be complementary, interdependent, and interrelated in gaining a full picture of system performance. The information extracted by the tools of these four elements must be merged into a systemwide framework enabling aviation policy makers to collaborate in aviation-safety-risk management. This information must be shared while respecting the proprietary rights to some sources of data and sensitivities to potential misuse if they are released outside the owning organization. Although this sharing was not achieved during the life of the ASMM Project because of concerns for potential misuse of proprietary data, it is being addressed in the future activities that are described later in this report.

The following sections describe the activities and products of each of these elements and show how they relate to enabling each of the four steps in the cycle of proactive management of risk. The objective is to assist the human analyst in finding the precursors underlying events or trends that could compromise the safety of the system operations, in identifying potential adverse consequences of technological and procedural changes, and in assessing the safety risks to the NATS.

In another view of the cycle depicted in figure 4.0.1, the specific activities to which the ASMM tools to be described in this report contribute are indicated in orange.

Identifying:

The first step in the cycle of proactive management of risk is to monitor the system continuously by collecting, codifying, and classifying the various data from the system into repositories that can be mined subsequently for safety insights. The databases containing information relevant to aviation safety are very large, heterogeneous (textual and digital), diverse, distributed sources from which information must be extracted and merged to gain a complete picture of a situation. The information (not the data) must be displayed in a way that makes it easy for the domain expert to interpret and compare with expectations or performance standards. The decision-makers must gain sufficient insight to identify those events that present potential risks.

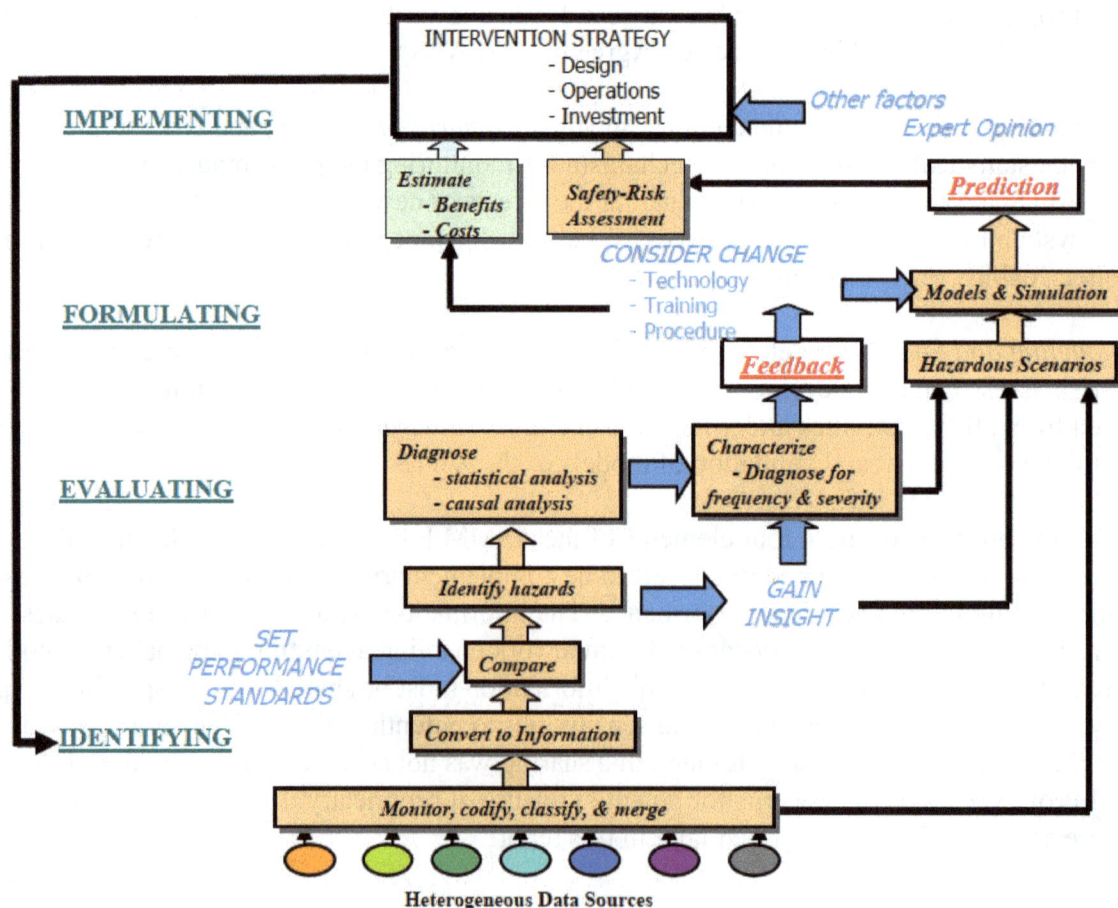

Figure 4.0.1. The elements of proactive management of safety risk.

The research carried out under the *Data Analysis Tools Development* element of the ASMM Project produced advanced software to facilitate efficient, comprehensive, and accurate extraction of information from large, heterogeneous databases throughout the NATS. These data sources consist of qualitative data (textual, categorical, and survey data) and quantitative data (digital flight-recorder data, radar-track data). The ASMM tools and methodologies establish meaningful linkages among the information extracted from these diverse data sources and enable visualization of operationally significant patterns and trends.

Prototypical tools were developed to convert these data into graphical displays of meaningful information to help analysts achieve the insight needed to understand the circumstances and propose a mitigating action. As a result, data can be better classified, more easily prioritized, and more readily combined with data from other sources, and data patterns can be better understood. Data-processing costs decline because the analytic capability of humans are supplemented and enhanced with automation. As they were developed, the conceptual prototypes were tested and evaluated on existing operational databases.

Some of these databases, such as the ASRS (see appendix C) and NTSB databases, reside in the public domain. Others, such as each air carrier's digital flight-recorded data archives, store proprietary information and reside in the private domain. Accordingly, the ASMM approach to monitoring the system entailed a dual strategy of creating two distinct, independent, but complementary, capabilities.

The capability developed under the element called *Intramural Monitoring* entailed a "bottom-up" approach that is intended to provide air-service operators with the tools needed to monitor their performance continuously, effectively, and economically within their own organizations. The primary products of this activity are the Aviation Performance Measuring System (APMS) for processing aircraft flight-recorder data and the Performance Data Analysis and Reporting System (PDARS) for processing ATC data.

The APMS provides an advanced capability for managing, processing, and analyzing digital flight-recorded data. The objective of the APMS research project was to develop the methodologies and tools to demonstrate to U.S. air carriers that very large quantities of flight-recorded data can be monitored, processed, and analyzed routinely, efficiently, and economically. The suite of integrated APMS tools is designed to convert flight-recorded data into information that will be useful feedback to the air carrier for ensuring the quality, reliability, and safety of performance of his/her operations in support of each company's own FOQA programs (see appendix A) and Advanced Qualifications Programs (AQP) (Chidester, (2001)). APMS developed tools that significantly extend the capabilities of the current commercially available analytic methods that are mainly designed to count predefined exceedances. This activity established and maintained collaborations with the air carriers and their FOQA vendors to obtain the perspectives, the operational data, access to other relevant databases, and the users' evaluations in the operational environment that are essential to evolving a suite of data-analysis tools offering maximum benefits at minimal cost.

APMS was a program to test and evaluate software tools for analyses of flight-recorded data.

- Requirements were initially based on user-needs studies.

- The program was built upon exceedance-based FOQA programs.

- The program included enhanced commercial off-the-shelf (COTS) tools for exceedance-based programs.

- The program developed advanced concepts, methods, and tools.

- The program was an engine of technology transfer.

APMS was intended to provide the air-service operators with the tools needed to monitor their own performance and safety continuously, effectively, and economically within their own organizations. An objective was to provide a suite of tools for converting flight data into information customized to each individual user among the diverse operating personnel, and, thereby, to encourage them to share their information for cooperative proactive decision-making. APMS assists an operator in understanding how its aircraft are being operated normally and routinely on the line, and assists the flight-safety analyst to identify atypical, statistically extreme, and safety-related events and trends routinely, empirically, and inductively to support safety and economic decisions.

Intramural Monitoring of ATC was addressed with the PDARS, an ATC radar-track monitoring capability developed collaboratively by NASA and the FAA.

PDARS was a program to test and evaluate software tools for analyses of radar-track data.
- It routinely collects, extracts, processes, and merges ATC data.

- It routinely computes quantitative performance measures.

- It produces and disseminates performance-measurement reports.

- It aids identification of operational problems and causal analysis.

- It accesses system-design and simulation tools for "what-if" studies.

- It archives basic operational data and performance statistics.

This project was carried out in collaboration with the ATC community (FAA and NATCA) to obtain the perspectives, the operational data, access to other relevant databases, and the users' evaluations essential to evolving a suite of data-analysis tools to support the identified informational needs of air-traffic management. When ASMM ended in FY05 and full responsibility for PDARS was assumed by the FAA, all 20 U.S. Air Route Traffic Control Centers (ARTCC) facilities in the continental U.S., 13 major terminal radar approach control facilities (TRACON), 5 regional offices, and the Command Control Center in Herndon, Virginia, were all using PDARS and receiving daily reports customized to each facility. By agreements among the individual facilities, these reports are shared among the facilities. PDARS performance measurements relate to system throughput, delays, system predictability, and other key ATC performance indicators (den Braven and Shade (2003)).

The capability developed under the element called *Extramural Monitoring* entailed a "top-down" approach that is intended to provide a systemwide view of the performance of the NATS based on data considered in the public domain. The primary product of this element was a comprehensive systemwide survey mechanism called the National Aviation System Operational Monitoring Service (NAOMS) for monitoring the overall performance and safety of the NATS with provision for detecting and evaluating the effects of new technologies as they are inserted into the system. NAOMS achieved scientific integrity by using meticulously designed survey instruments and a carefully crafted statistical sampling design customized to each constituency.

The objective was the permanent field implementation of a NAOMS responsible for developing and maintaining a comprehensive and coherent survey of the operators of the aviation system (i.e., its pilots, controllers, mechanics, dispatchers, flight attendants, and others) on a regular basis. That objective was not achieved largely because of the cost of designing and implementing the statistically sound survey for each new constituency and the lack of adequate funds.

A crucial, practical measure of NAOMS' success was its ability to support the aviation community in its assessment of safety risks and the efficacy of government/industry interventions. Accordingly, NAOMS cultivated a close working association with the aviation industry and organized labor, including the CAST. CAST views NAOMS as being a valuable tool for measuring the systemwide impacts of the interventions that their Joint Safety Analysis Teams (JSAT) developed. NAOMS actively supported this process by incorporating core safety event questions that addressed CAST priorities and by developing topical questions that addressed focused safety concerns. Consequently, the value in viewing the aviation system through the eyes of its operators was clearly demonstrated on the basis of the survey conducted with nearly 28,000 interviews of Part 121 commercial pilots.

Currently, aviation databases capture information about specific parts of the NATS, but no existing database addresses its performance and safety *as a whole*. NAOMS established a new national capability that, if implemented as a continuing service, would

- *quantitatively* track aviation safety trends;
- acquire information about, and pointers to, precursor events through statistically valid survey techniques for measuring incident rates;
- monitor the impacts of technological and procedural changes to the NATS; and
- contribute to the development of a data-driven basis for safety decisions.

The concepts and capabilities of the NAOMS, APMS, and PDARS will continue to evolve independently, but information derived from each will complement the others in the process of continuously monitoring, assessing, and projecting the health of the overall system when all the data sources can be accessed. Until then, the ASMM Project provided a comprehensive survey system for monitoring safety performance on a national scale, while helping air-services organizations build their intramural capability for safety monitoring and for establishing a potential database for the Aviation Safety Program.

In addition to the data sources represented by APMS, PDARS, and NAOMS, the ASMM Project also drew on ancillary data sources such as meteorological records to further develop its understanding of contextual factors contributing to safety events. An important product of this activity was the Aviation Data Integration System (ADIS), which provided the safety analyst easy access to weather information, Jeppeson charts, and air traffic relevant to a specific event. In particular, the ADIS team developed an archive of weather information (either the Airport Terminal Information

Service (ATIS) or the METAR)[2] at all the major airports in the world. At the end of the ASMM Project, this capability was transferred to the FAA, which made it accessible to all air carriers with FOQA programs.

Another database used as a resource for the ASMM Project was the ASRS which NASA has managed on behalf of the FAA for nearly 30 years (see appendix C). The ASRS is a well-known repository of experiential reports of incidents having safety implications gathered from the operators of the aviation system. Although the ASRS was not formally an activity of the Aviation Safety Program, the authors' experience with ASRS and their access to its analysts stimulated and informed many ASMM research and development activities. The ASMM Project supported the development of another product called Incident Reporting Enhancement Tools under the Extramural Monitoring element. While these were designed to improve our accessibility to, and extraction of information from, the ASRS database, they have applicability to other similar textual databases such as the Aviation Safety Action Program (ASAP). The ASMM Project benefited from a more convenient access to this unique repository of safety reports to test new capabilities that were needed for the Ames Safety Accountability Program (ASAP), NAOMS, and other sources of textual data. The ASRS benefited from upgrading its legacy systems to state-of-the-art capabilities for managing, processing, and accessing textual incident reports. The aviation community benefited from having ASRS data more accessible and better connected to other aviation-safety resources.

The ASRS repository of over 150,000 reports offered a database for testing and evaluating some of the tools for processing and analyzing textual data that were developed under the *Data Analysis Tools Development* element of the ASMM Project. The ASRS database was used as a representative resource of textual reports to develop and evaluate new capabilities for their automated analyses. The first generation of an intramural text-processing capability applied to a database of ASAP reports was developed in collaboration with an air carrier under a SAA.

Evaluating:
The second step in the cycle of proactive management of risk is to evaluate the operational significance of the event that was identified in the monitoring phase. The decision-makers must be able to quickly focus on those events with the highest potential severity and likelihood of reoccurring. This focus requires an understanding of the contextual factors and conditions that were conducive to the identified incident so that the domain expert can ascertain the likelihood of future occurrences and assess the severity of potential consequences. It also requires statistical analyses of the databases to estimate the frequencies of occurrence of the identified event or of the confluence of factors.

Information must be extracted from qualitative data sources to help the domain expert understand the subjective aspects of "why" an incident occurred, and from quantitative data sources to understand the objective aspects of "what" happened. Therefore, automated capabilities were developed to process both textual and numerical aviation data, and to recognize relevant information in diverse databases. The results of the searches of heterogeneous databases are presented in displays for visualization of information to help the analysts achieve the insight needed to assess the operational significance, understand the circumstances, and formulate a mitigating action.

---

[2] METAR is the international standard code format for hourly surface weather information. The acronym is French and translates, roughly, into Aviation Routine Weather Report.

Each of the ASMM tools contributes to such an insight into the picture of a safety event or trend, and can be used to support a complementary and synergistic process of hazard evaluation, causal analysis, and safety-risk assessment. The objective is to minimize the demands on the human expert to identify the operational events deserving of further attention.

A demonstration of how the several ASMM products would be used in such an evaluation process preparatory to formulating an intervention was an application to a study of In-Close Approach Changes (ICACs) that is described in appendix B.

Formulating:

Having identified an operationally significant event and understood its contextual factors, the next step in the process of proactive management of risk is to formulate an intervention. For the most part, it is up to the experts in industry and the FAA to *formulate* and to *implement* the interventions. However, an objective of the element of ASMM called *Modeling and Simulations* is to aid the decision-makers in these two steps of the process.

There can be many contextual factors important to an event that is identified to be a precursor of an anomalous condition and, therefore, there are many possible interventions to minimize the probability of recurrence of those factors or of their confluence. Such an intervention can reside *at any element* of the total system in its

- *Design* (including man-machine interface and quality of information),

- *Procedures* (including organization, requirements, and documentation), or

- *Training* (including education and checking).

The suggested interventions may be in any one of these but, usually, a change in one will necessitate a change in the others as well. Frequently, all three types of faults enter into the confluence of factors that result in an accident—but they do not necessarily all occur in the same element of the system. In order to decide on which of these many choices and combinations to implement and decide on a strategy for implementing the changes, the decision-maker needs to be able to forecast the systemwide effects of the proposed changes.

The contribution of *Modeling and Simulations* to the goal of ASMM is to support reliable prediction of the systemwide effects of new technologies and procedures on operations and communications with emphasis on incorporating appropriate human behavioral models. This element provides for the development of tools for modeling the NATS at a level of detail sufficient to track key safety characteristics to support prediction and decision-making. Techniques were developed for representing the interactions of multiple human and nonhuman agents in complex dynamic scenarios. Models were developed to explicitly incorporate human performance into existing NATS modeling tools and were validated with data obtained from *Intramural* and *Extramural Monitoring*.

The elements of the NATS, including the contributions of individual multiple operators, individual elements of the system, the interactions among multiple (human and nonhuman) agents, technologies, and large-scale system flow and control issues, were modeled. This modeling served as a computational testbed for simulating and analyzing system performance. Simulations of the system operations in non-real time (i.e., in contrast to realtime human-in-the loop simulations) enable the analyst to efficiently explore NATS operations from various perspectives by exercising models of the system. Fast-time systemwide simulations developed in the ASMM Project can be used to answer questions such as, *"Does the solution have any secondary, propagated, or side effects?"* and *"Does the solution provide for graceful degradation in unanticipated operation anomalies?"* and *"Does the proposed intervention address the right question and in the right way based on an understanding of the joint cognitive system?"* Analytical tools were developed in parallel with the fast-time simulations to assist the analysts in identifying the causal factors of an event and in assessing the safety risks.

Implementing:
Implementation of an intervention for an identified problem is accomplished via prototypes, their effectiveness is evaluated, refinements are implemented, and then full-scale deployments are facilitated. The decision on the appropriate intervention to mitigate the occurrence of some event entails considerations by the FAA and the industry of costs versus benefits in addition to the safety risks. Therefore, the products of the ASMM Project play only a supportive role to the decision-makers in this step of proactive management of safety risk.

The step that is often missing from the full cycle of proactive management of risk as it is currently pursued in the aviation industry is that of having in place a system for monitoring in order to assess the effectiveness of the intervention measured against expectations. This step is comparable to the first step in the process (i.e., identifying) and closes the loop in the cycle. This step requires that metrics be established and those data that are needed to evaluate the intervention be appropriately collected, codified, and classified for retrospective search. The monitoring system should have been in place before the intervention to gather the baseline data for comparison of the before to the after. Once again, as described in the initial step, the sources of relevant information will be in large heterogeneous, distributed databases, and will need to be merged to gain a complete picture of the systemwide situation. Once again, all of the ASMM tools are applicable to facilitate efficient and insightful analyses of all relevant data.

Figure 4.0.2 shows the relationships of the products of ASMM to the elements of the cycle of proactive management of risk as discussed previously. NASA's role is largely the development of technologies and methodologies to aid the decision-makers in identifying events and trends that could compromise safety, in evaluating their operational significance, and in identifying their causal factors, particularly those latent and proximate systemic features conducive to human error. Although our fast-time simulation capability aids in predicting the systemwide impact of interventions, the phases of formulation and implementation of these interventions are the responsibilities of the industry and the FAA. Moreover, they entail other considerations, such as economics, besides the safety risk.

	ASMM PRODUCTS			
	DATA ANALYSIS TOOLS	APMS PDARS	NAOMS INCIDENT REPORTING UPGRADES	FAST-TIME SIMULATION OF SYSTEM-WIDE RISKS
IDENTIFY				
Monitor and Compare with Expectations	●	●	●	●
Uncover Potential Risks	●	●	●	●
Recognize Improvements	●	●	●	●
EVALUATE				
Diagnose Causation	●	●	●	●
Quantify Frequency	●	●	●	●
Assess Severity	●	●	●	●
FORMULATE				
Consider Change	●			●
Assess Safety Benefits & Risks				●
Estimate Economic Benefits & Costs				
IMPLEMENT				
Implement Locally				
Evaluate Intervention	●	●	●	●
Refine				●
Implement Full-Scale				

NASA ROLE

FAA-INDUSTRY ROLE

● Denotes an area where ASMM tools, methods and datasets will enable proactive risk management

Figure 4.0.2. ASMM products vs. the proactive management cycle.

5.0 THE ASMM PROJECT: THE PRODUCTS

In the following sections, each of the ASMM products mentioned briefly in the discussion of the approach will be described in detail within the element in which it was produced.

5.1 Data Analysis Tools Development

The recorded data of a single flight or the content of a single experiential report can entail values of hundreds or thousands of parameters. It is impossible for a human to perceive the relationships among flights or reports that are described in so many dimensions. Therefore, the objectives of the activities under *Data Analysis Tools Development* were to efficiently and reliably convert numerical and textual data into meaningful and actionable information.

- Develop methods for automatically identifying the unexpected from numerical data,

- develop new methods of automatic text understanding, and

- create mechanisms to link information extracted from disparate textual and numerical data sources.

5.1.1 Numerical Data Analysis

For analyses of digital (numerical) data (such as flight-recorded data and radar-track data), the PROFILER was developed by Battelle's Pacific Northwest Division as the underlying data-mining tool. The PROFILER is a tool to conduct atypicality analysis—to identify multivariate, statistically extreme flights. PROFILER extracts a "signature" for each parameter in each phase of flight, uses multivariate cluster analysis to group flights by similarity along flight signatures, searches for differences among clusters of flights, calculates an atypicality score for each flight, and provides a plain-language description of what makes targeted flights atypical. The distribution of atypicality scores— a function of the Mahalanobis distance from the population multivariate data centroid and multivariate cluster results—is used to identify flights for examination. This calculation is illustrated in figure 5.1.1 in a two-dimensional representation.

PROFILER

- Automatically produces a list of atypical flights

- Produces baseline operational envelope on each parameter

- Identifies the specific parameters that are atypical

- Provides summary information for each atypical flight

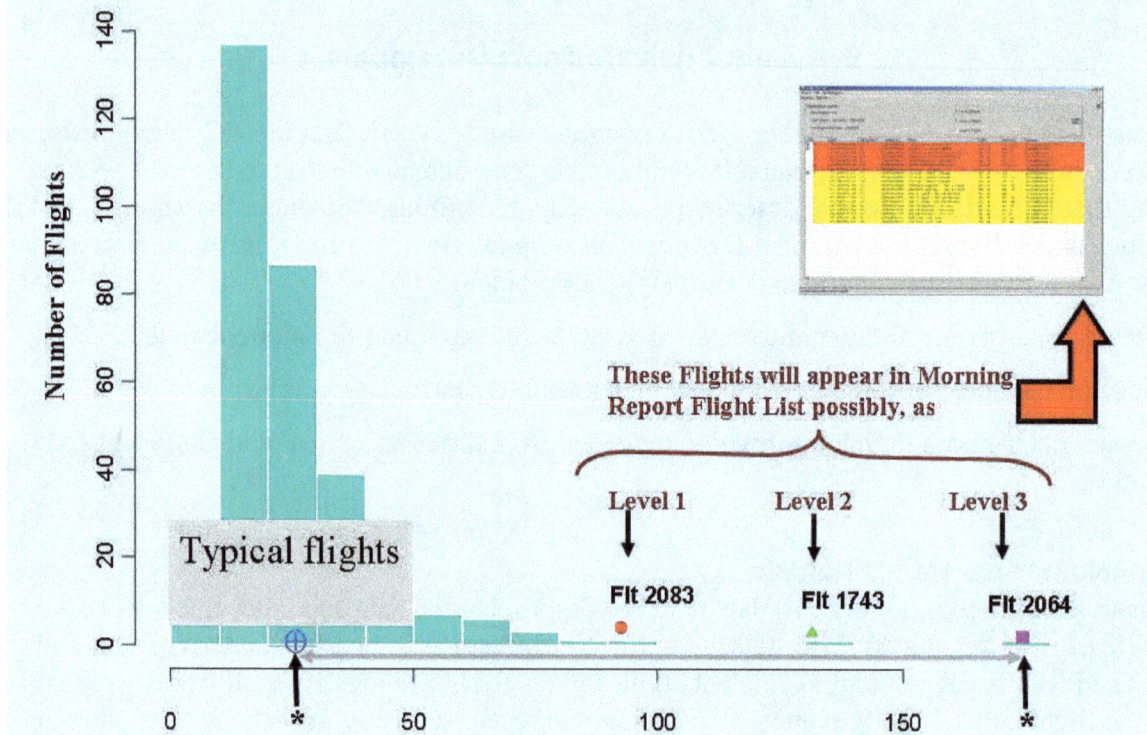

*Atypicality score = Mahalanobis distance as measured from the center of the n-dimensional population data centroid

Figure 5.1.1. Calculation of atypicality score.

PROFILER conducts the mathematical analyses transparent to the user. The results of these analyses were implemented into a user's display called The Morning Report of Atypical Flights because the analyses of the previous day's flights are performed overnight. The implementation and the use of The Morning Report tool is discussed in section 5.3, "*Intramural Monitoring.*"

The Morning Report is a tool to search for the unexpected. It differs from and complements current FOQA tools that focus on exceedances, which are deviations from a predefinition of the appropriate performance envelope. No such predefinition is needed for The Morning Report. It automatically finds the typical operations and identifies statistically significant deviations from those typical operations. (See section 5.3 and Amidan and Ferryman (2005) for additional description of The Morning Report.)

The Morning Report displays a summary page as soon as processing is completed. It shows when the report was completed, the number of flights analyzed, and the date range of those flights. It summarizes the number of atypical flights identified and their level of atypicality. It is similar in its presentation to the FOQA exceedance analysis in that it identifies individual flights worthy of attention. Clicking the flight list button of The Morning Report displays the most extreme 5% of flights (level 2 and 3 atypicalities with the option to display level 1 flights) in much the same way that the FOQA tools identify and display the levels of the exceedances. Atypical flights within normal operations may, or may not, point to unsafe conditions lurking in the aviation system (Amidan, Cooley, et al. (2002); Amidan, Swickard, et al. (2002); Ferryman (2001)). The analyst's access to displays of the specific parameters that were identified as atypical for a flight and of comparisons to the typical baseline for each helps the analyst determine the operational significance of the event. The results of the operational tests of the prototype of The Morning Report conducted with APMS and then adapted to PDARS are reported under "*Intramural Monitoring*" in section 5.3.

The Morning Report of Atypical Flights was a recipient of the R&D 100 Award as one of the most innovative new products of 2005.

Complementary to The Morning Report is the Pattern Search tool, an aid to retrospective search of flight-recorded or radar-track data that enables the user to define a pattern of multiple parameters, and search for that pattern in a large database (Chidester (2001)). When The Morning Report identifies an atypical pattern in a subset of data, the Pattern Search tool is then used to search the entire database for that new pattern. The Pattern Search tool is discussed further in section 5.3.

The focus of the data analysis that resulted in The Morning Report was on the performance of the aircraft and, therefore, on the continuous flight-data parameters such as altitude, airspeed, accelerations, and rate of descent. In this process, it was useful to ignore the potential atypicalities in the sequences of the discrete data, except as they were used to define "snapshots" of a flight. However, the extraction of information from the flight-recorded data of the discrete parameters related to activities in the cockpit was always of interest. Therefore, during the final year of the program, a study was undertaken that focused on the discrete sensors, specifically, sensors recording pilot actions or switches. The purpose was to extract information about the sequences in which the values of these parameters change during the course of a flight and to find anomalies in the sequences and in the flight behavior based on this information. The conceptual approach was similar to that of The Morning Report in that this analysis entails two tasks as part of detecting atypical events in flights: first, the identification of atypical flights, and second, finding the events during the course of those flights that are anomalous or atypical. The underlying statistical processes that were developed to accomplish these tasks are described in Budalakoti et al. (2006).

The first step in the process is to merge the data from all of the discrete sensors into a single sequence for each flight that records a sensor only when it makes a transition. The Longest Common Sequence (LCS) (see Budalakoti et al. (2006)) was used as the similarity measure for clustering flight data. New algorithms were developed that use Bayesian Networks to efficiently identify anomalous events during the flights within each cluster. These new algorithms ran up to five times faster than the current commonly used methods for clustering on LCS. The information that the process provides to the safety analyst is classified into:

- A sequence of switching that was expected at a given stage did *not* occur.
- A sequence of switching that was *not* expected did occur.
- A sequence of switching occurred in an unexpected order.

The performance of these algorithms was demonstrated using operational data during the approach and landing phase of 6400 flights for which the base (typical) clusters were developed and anomalous events in the flights were identified in the atypical clusters (Srivastava (2005)) and Srivastava et al. (2006)). There were about 700 individual parameters. The 6400 distinct data sequences varied in length from 600 to over 9000 switch changes, with the average sequence length of about 2500. A further preprocessing was performed that removed parameters believed to be secondary. These parameters are recordings of the reaction of the aircraft to the pilot's command. For example, when the pilot changes the flap position, not only is the command action recorded, but also the reaction of a wing sensor that indicates that the position changed. After this reduction, the number of switch changes during the approach and landing phase of a single flight in these 6400 flights ranged from about 200 to over 1600.

Figure 5.1.2 displays a histogram showing the distribution of sequence lengths. As is clear from the figure, around 4000 of the 6400 sequences have lengths less than 500 switches. Another 1500 have lengths around 1000, and around 500 sequences have lengths in the 1000 to 1600 range.

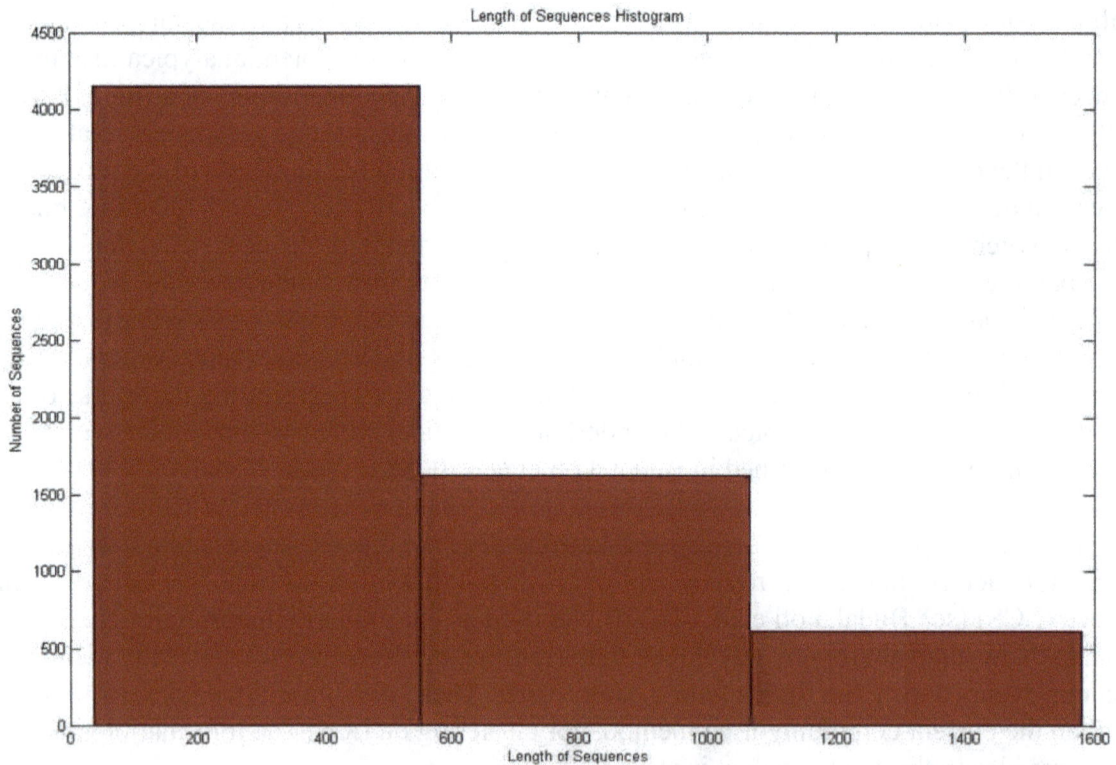

Figure 5.1.2. Histogram of sequence lengths in 6400 flights.

Current algorithms for calculating the LCS become very slow in practice as the sequence length increases. So a new algorithm was developed to calculate the LCS, which is much faster and scales better to long sequence lengths. Figure 5.1.3 shows the time taken by two current standard algorithms for calculating the LCS compared with the time using the new hybrid algorithm that was developed. The two current algorithms to which the comparison was made are the dynamic programming (DP) algorithm (Cormen et al. (1977)) and the Hunt-Szymanski (H-S) algorithm (Hunt and Szymanski (1977)). The horizontal axis gives the lengths of the sequences being compared, and the vertical axis gives the time taken. As is clear from the graph, the new hybrid algorithm is about a factor of five faster than the conventional DP and H-S algorithms currently used for clustering the sequences dataset.

The process of clustering on the similarities of normalized (by length of sequence) LCS for the 6400 flights produced a large cluster of about 3300 flights with an average similarity score of about 0.75. It was also found that about 75% of the switch changes occur in the same order for most flights. The smallest cluster contained 846 flights with an average similarity score of 0.55. The similarity score of a cluster is defined as the average normalized LCS between the elements of the cluster and the cluster centroid. The anomalous sequences were those in the clusters that were statistically dissimilar from those in the dominant cluster of 3300 flights.

Not only were the anomalous sequences identified, but, unlike many anomaly-detection systems, the newly developed algorithms explained why these chosen sequences were considered to be anomalous. There was only time to demonstrate a statistically viable concept before the AvSP ended, and there was not time to evaluate its uncertainty or the operational validity of the results of the process.

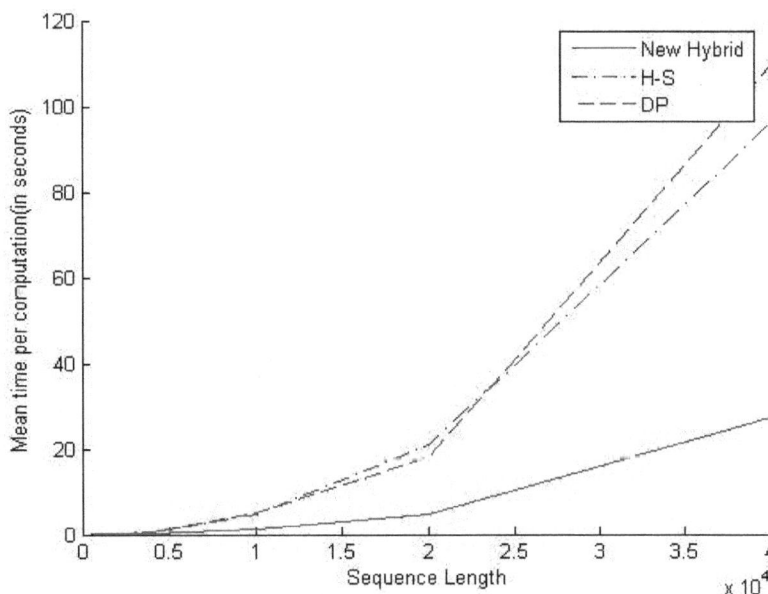

Figure 5.1.3. Performance comparison of current algorithms with the new hybrid algorithm.

Figure 5.1.4 represents how the algorithms present the results in an example based on the landing phase of an actual flight. The horizontal axis, read from left to right, represents the time remaining until the plane touches the runway. The vertical axis indicates the confidence with which the algorithm predicts a switch should or should not have been activated to make the sequence closer to the norm. The positive direction of the vertical axis represents the suggested insertions (i.e., switches) the algorithm thinks the pilot should have pressed at that stage but did not. The negative direction represents the suggested deletions, (i.e., switches) the algorithm believes the pilot should not have pressed at that stage, but did. Thus, a graph with no bars would be representative of a normal flight. The colors of the bars indicate the altitude at the time. (Note: Actual altitude data were not used to generate this graph.)

As an example, in one application of sequenceMiner, it was discovered that the landing gear had been activated in an unusual manner on flight 1147. The flight path of flight 1147 in terms of airspeed and altitude versus time to touchdown is shown in figure 5.1.5. Upon viewing this figure, a domain expert's comments were "Landing gear goes up and down more than once. This is probably a go-around. Unable to determine whether pilot or ATC-related. Low descent after bringing landing gear up. Needs to be investigated (with flap information)." This identification of a go-around was confirmed when viewed with the perspective shown in figure 5.1.6.

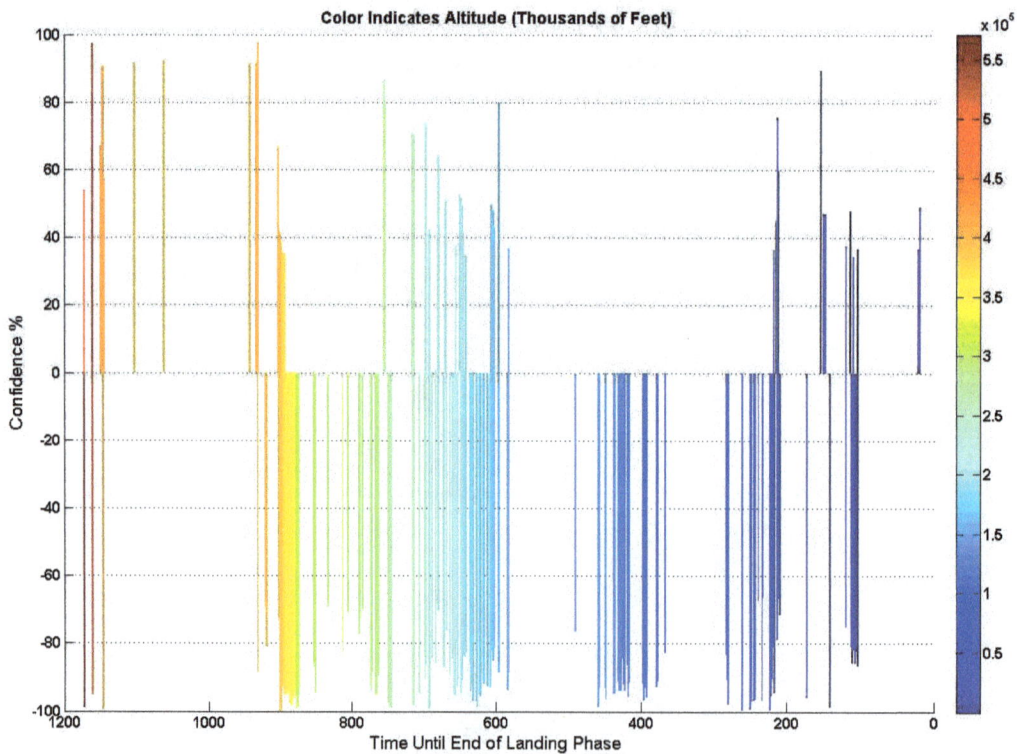

Figure 5.1.4. Switching anomalies during landing phase of 6400 flights.

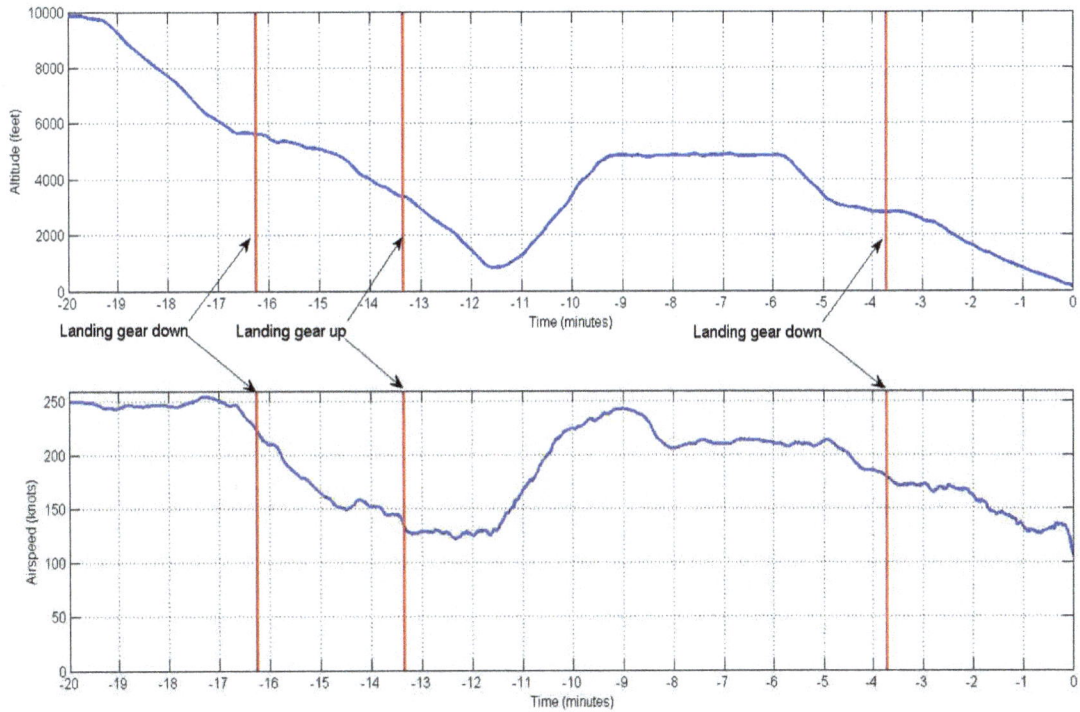

Figure 5.1.5. Switching anomalies during landing phase of flight 1147.

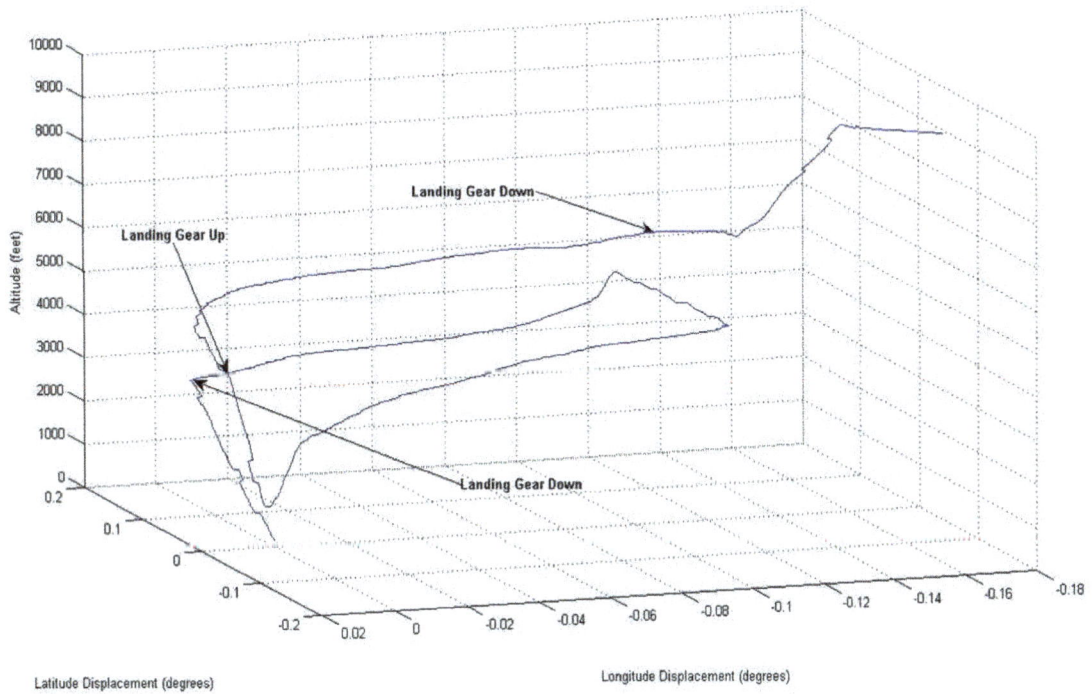

Figure 5.1.6. A 3-D view of the landing phase of flight 1147.

It is certainly true that a go-around such as this one would have been found by any analysis of the continuous flight data. However, it is significant that it was discovered solely from the data on the sequence of switches. The sequenceMiner tool opens the possibility of searching for unknown anomalous situations in the discrete data in conjunction with the continuous data.

5.1.2 Textual Data Analysis

Analyses of numerical data with tools like The Morning Report and Pattern Search provide information about the characteristics of an event describing what happened, but the question remains: Why do professional, highly trained, motivated operators make mistakes?

The formulation of the appropriate intervention in the process of proactive risk management requires:

- Understanding *why* it happened.

- What were the causal, contributing, and aggravating factors?

- Which of these factors are systemic?

Consequently, the next challenge was to automatically extract information from databases of experiential reports to give the safety analyst insight into the *why* of incidents. This exploitation of incident reports requires sophisticated tools to analyze free text and to classify reports in a way that is meaningful to experts.

In order to extract useful information (whether quantitative or textual) from large databases, it is necessary to identify global patterns and relationships to aid decision-making. A model was needed with which to guide the automated analyses of textual data. The rationale for such a model is described in detail in Maille et al. (2006) and is summarized here.

An aviation incident report is the story of the evolution of the reporter's world from a safe state, through a sequence of events and states to a compromised or an anomalous "unsafe" state. Report narratives embody naturally occurring chains of events and the transitions between events as depicted in figure 5.1.7(a). The following three states can define the "world" of our aviation transportation system:

- **Safe:** All the aircraft and people (crews, traffic controllers…) and all the key systems (aircraft systems, instrument landing system (ILS)…) are in a state that approximates normalcy.

- **Compromised:** An element of the aviation transportation system is in an undesirable state; a person is involved in a situation, an aircraft is experiencing compromised performance, or undesirable environmental factors are impinging upon the aircraft. However, the required properties of all the involved aircraft are still nominal and the separation between aircraft complies with norms.

- **Anomalous:** One or more of the required properties of an aircraft or an involved person is observably not in compliance with pertinent norms.

30

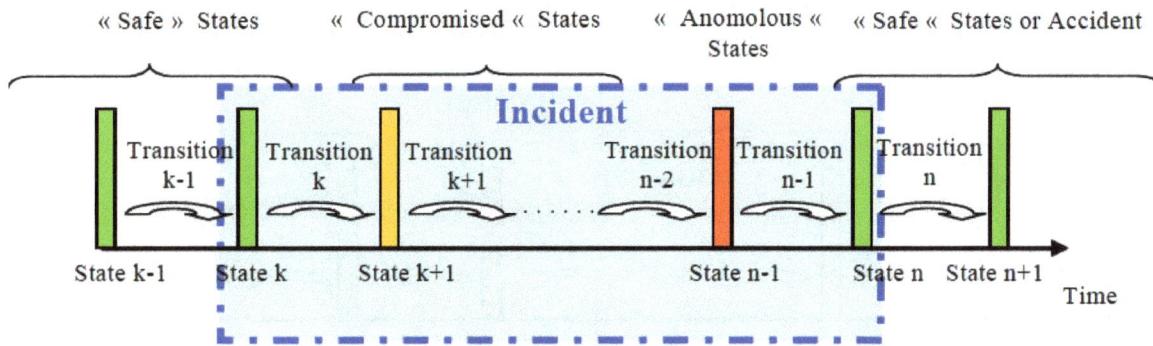

Figure 5.1.7(a). The experiential report.

Further, what is referred to in this analysis of textual reports as an aviation incident will be characterized as follows:

- Incident: An incident is a finite sequence of states and transitions such that:
  - The first state is safe,
  - The last state is safe,
  - All the other states are either compromised or anomalous, and
  - At least one state is anomalous.

If the state does not return to a safe one, the story is not related to an incident, but to an accident. If no anomalous state is reached, the story is not considered to be an incident.

Figure 5.1.7(a) may well be the model of an incident report needed to achieve the objectives of this study. However, it is much too complex to be used as a basis for identifying similarities with automated clustering tools, at least for our initial attempt. Therefore, for the sake of simplification and justified on the basis of the experience with ASRS reports, it was decided to emphasize three parts of the generic incident model represented in figure 5.1.7(a) for the concept of Scenario represented in figure 5.1.7(b); namely, the first (safe) state (the beginning of the story); the sequence of states and transitions that lead the world to the anomalous state; and the anomalous state. Thus, the high-level definition of the scenario that was adopted for the present model was:

$$\text{Scenario} = \{\text{Context} + \text{Behavior} \rightarrow \text{Outcome}\}$$

With this simplified definition of scenario:

- The Context fits the exact description of the situation in the last safe state.

- The Behavior contains all the problematic events that occur during the transition from the last safe state to the anomalous state.

- The Outcome describes why the anomalous state is considered anomalous. It does not necessarily contain all the parameters used to describe the state.

31

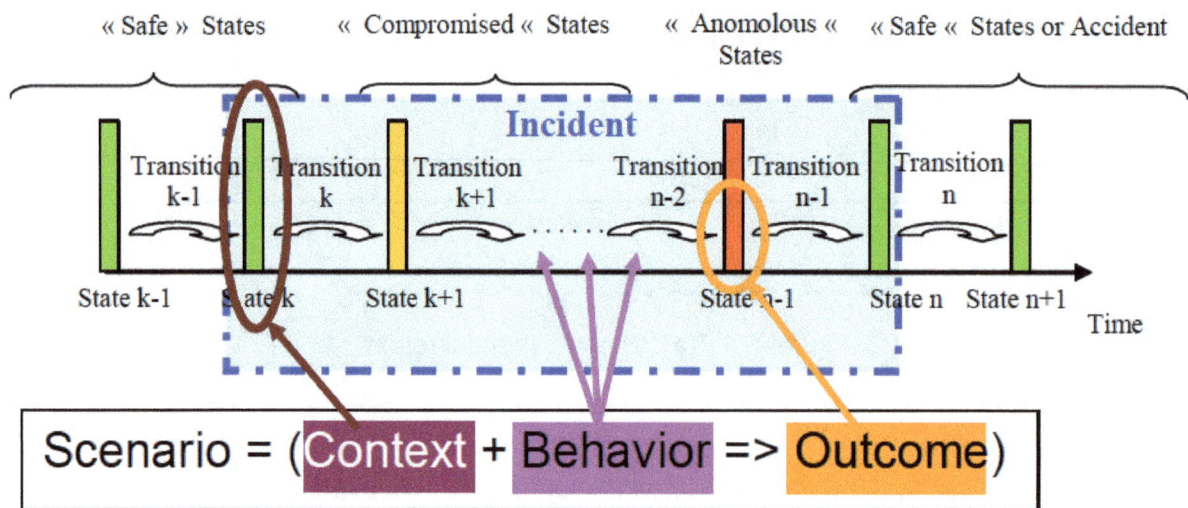

Figure 5.1 7(b). The scenario model of the experiential report.

This simplified model may not apply to all systems or even for the aviation transportation system; other Scenarios could be stated and will, perhaps, have to be explored in future work. For example, another Scenario might highlight the recovery to the final safe state to study which parameters influenced the recovery process (and so prevented an accident). Also, this model does not provide for the possible changes in context across the three states. The context of the last safe state is assumed to prevail throughout the incident. Nevertheless, there is merit in starting with the simplest possible model.

The notion of the Scenario was useful to guiding the automated analysis and extraction of clusters of similar incident reports from large databases. Indeed, when the search is not based on a predefined specific issue or a preselected example report, a meaningful way to build clusters is to group reports with similar Scenarios. Identifying the main Scenarios in a database (i.e., their profiles) should help experts identify major safety issues and obtain clues for designing intervention strategies.

The model that was established to guide automated analyses established, also, the capabilities needed to perform the analyses. The state-of-the-art of text-analysis tools that could meet these needs was reviewed. Figure 5.1.8 diagrams the state-of-the-art of Natural Language Processing (NLP) showing the stages of expression detection with examples of the expressions that would be identified with a particular concept (e.g., the concept of "familiarity").

Mature

NLP Depth

String Matching

Part-of-Speech/ Entity Extraction

Syntactic Parsing

Event Extraction

Event Co reference

Scenario Template

research

Example: familiarity

"Unfamiliar", "Unacquainted"

"This setting was * new to me"
(* = "somewhat", "quite", "entirely"
but ≠ "not", or "not quite")

"The regulations had * changed *
since…."

"I had not performed this procedure
for years….."

"I (did something) … we went too
high because on this plane the
(tool) is more sensitive than on the
737 " (suggesting that the pilot is
familiar with the B737, but not the
plane he is flying right now)

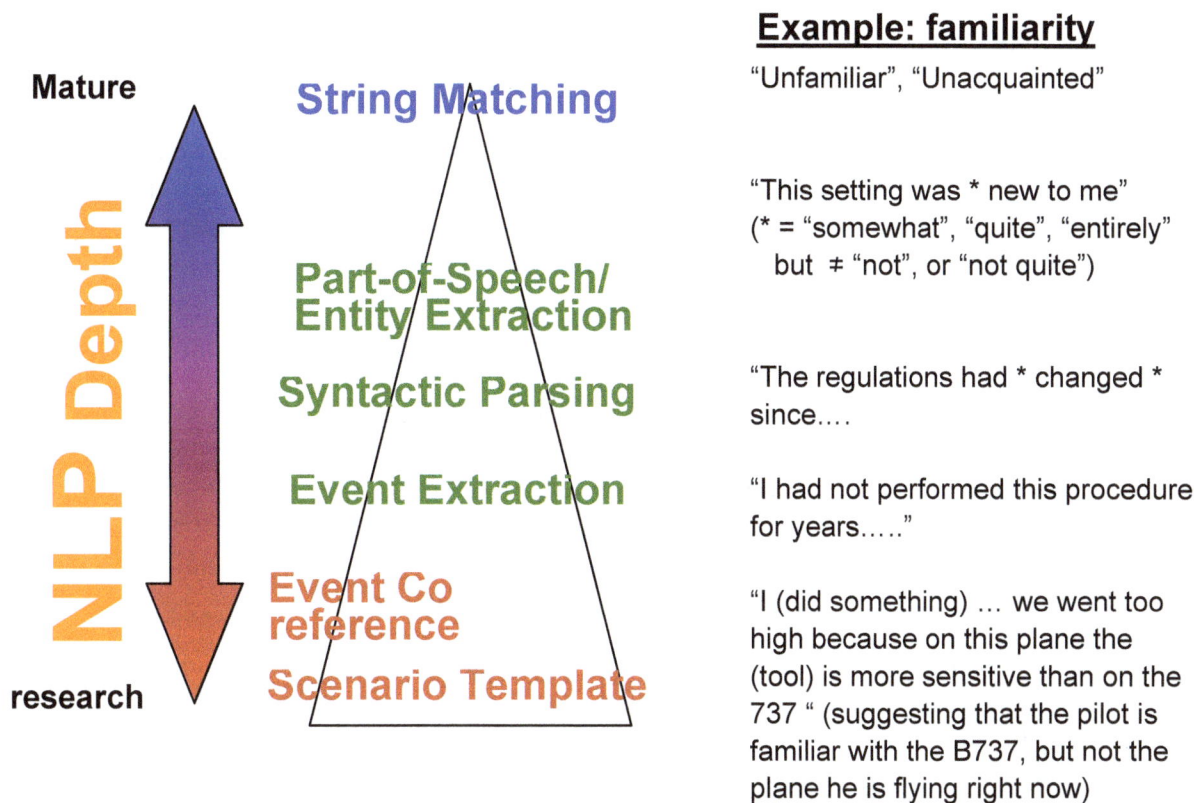

Figure 5.1.8. Stages of expression detection.

Many techniques based on statistical correlation (commonly known as "bag of words" classification) cope very well with string matching and identification of similar expressions and have been used in a variety of successful search engines. However, capabilities for extracting meaning are extremely limited and even syntactic parsing or event extraction push the state-of-the-art of NLP. The Scenario model of the incident reports requires some of the capabilities of both bag of words and NLP for the desired analyses.

The Scenario is defined by the subsets of parameters that describe the Context, the Behavior, and the Outcome of the incident model that are specific to the "story" of a particular incident report. These descriptors constitute the taxonomic structure[3] for our world of aviation-safety-incident reports. Although a wide variety of taxonomic structures is used in the different accident/incident databases, the taxonomic structure of the ASRS database was chosen because this database was used to develop the ASMM's text-analysis capabilities and because it is representative of reports of incidents in our world of aviation (Maille et al. (2006)).

---

[3] By taxonomic structure, we mean a structured set of terms that describe some domain or topic. The taxonomic structure provides a skeletal structure for a knowledge base.

A report to the ASRS comprises a set of attributes (the fixed fields) and the values of those attributes (entered by the reporter) plus the reporter's narrative of the incident. A structured set of "descriptors" is currently used to describe the incident and store it in the database.

The study started with a prescribed subset of incident reports (aggregated by a selected phase of flight and aircraft type). Then, domain experts identified all the possible anomalous states (i.e., Outcomes) for the selected aircraft type and phase of flight. In fact, the set of anomalous states that were used were those that the ASRS had identified based on its nearly 30 years of experience. With the help of the ASRS analysts, the specific attributes in the fixed fields of the ASRS reports that describe each of the possible anomalous state were defined. Further, the domain experts identified the descriptive parameters (i.e., attributes in the fixed fields) of the Context important to the selected aircraft type and phase of flight. Then, in the first stage of automated filtering, clustering was performed based on similarities among the objective parameters that define the Context of the Scenario combined with the objective parameters that define each possible Outcome of the Scenario in the subset of reports. Because all of these objective parameters were contained within the fixed fields of the ASRS reports, relatively conventional statistical techniques were adequate. Figure 5.1.9 may help in the understanding of this perspective on Context and Outcome.

Identification of all the objective parameters of the Context and the Outcome that exist in a particular Scenario is an adequate definition of *what* happened, and available statistical techniques are adequate for analyzing the *Contest* and the *Outcome* from the fixed fields of an ASRS report. The Automatic Language Analysis Navigator (ALAN), a text comprehension tool developed by Battelle's Pacific Northwest Division that clusters textual data, was used. ALAN uses well-defined objective factors in the fixed fields or in the narrative to identify aviation-safety reports that have similar topics, or to identify clusters of reports that are similar to a given exemplar (Willse, et al. (2002)).

- **Context**
  - Aircraft make/model
  - Airport equipment
  - Weather (snow, ice, fog, …)
  - Light (dark, dawn/dusk)
  - Metrological conditions (IMC, borderline)
  - Phase of flight (takeoff, cruise, approach, landing…)
  - Etc.

- **Outcome**
  - Altitude deviation
  - Conflict (airborne or ground)
  - Ground incursion
  - In-flight Wx encounter
  - Maintenance problem
  - Nonadherence
  - Airspace violation
  - Etc.

Note: In ASRS reports, these are usually identifiable from the fixed fields.

Figure 5.1.9. The parameters of Context and Outcome.

In fact, clustering on the *what* may prove to be pragmatically sufficient for an effective retrospective search for similar incidents, even though the explanation of *why* may be quite different for the clusters of incidents. However, other incident report forms may not have all the information needed to define the Context and the Outcome and would require analysis of the free text to extract this information.

All this, so far, has been preparatory to pursuing the objective of automatically defining the *why*. The understanding of *why* the incident happened relies on exploitation of the free text and the extraction of subjective parameters.[4] Natural language understanding is needed to extract information about *Behavior* from free narratives.

Experience with reading a great many ASRS reports has shown that reporters tend to tell why the incident occurred. Frequently, reporters provide "advice" in their reports to the ASRS, which implies the reporter's perception of causal factors and potential interventions. The advice is often associated with the words "should" or "ought," as in the following few exemplary excerpts taken from ASRS reports:

I FEEL THAT THESE DEER SHOULD BE EITHER RELOCATED OR INSTALL A FENCE AROUND THE ARPT

THIS APCH SHOULD BE OTS IF SUCH A RESTR IS REQUIRED, AND WINDS ALOFT ARE FROM THE N.

I FEEL THAT THERE SHOULD HAVE BEEN A HOLD SHORT LINE BEFORE RWY 1.

A SE TXWY SHOULD BE INSTALLED.

THE INCORRECT PAPERWORK SHOULD HAVE BEEN DISCOVERED WELL BEFORE DEPARTING, BUT PRESSURE TO KEEP THINGS MOVING PUT US IN A 'GO' MODE

I THINK THE SIGN SHOULD BE TO THE L (S) OF TXWY G AND PARALLEL TO RWY 35L/17R.

I FREQUENTLY HAVE TO ASK FOR WIND INFO FOR TKOFS AND LNDGS WHICH SHOULD ALWAYS BE GIVEN.

BECAUSE OF TIME PRESSURE TO GET FLT OUT, THE REQUIREMENT TO CHK OIL QUANTITY WAS OVERLOOKED AND NO ONE FROM MAINT SHOWED UP TO CHK. AN ENTRY SHOULD HAVE BEEN MADE AND SIGNED OFF IN LOGBOOK BUT WAS NOT.

PERHAPS A TAXI INSTRUCTION OF A DIFFERENT TYPE, IN REGARDS TO THE TXWY MERGER, OUGHT TO BE GIVEN SOME CONSIDERATION.

---

[4] The experiential narratives of ASRS reports are a rich source of information regarding the behaviors of pilots, air-traffic controllers, other persons, and automated agents during the course of the reported events. However, the unstructured nature of these data creates an analytical challenge.

IT OUGHT TO HAVE A WARNING HORN AT 300 FT FROM ALT

TWR OUGHT NOT TO SEQUENCE RELATIVELY FAST ACFT SUCH AS MY M20R BEHIND TRAINERS

WE OUGHT TO GO TO SCHOOL ON THIS TO PREVENT PROBS THAT COULD RESULT AT NIGHT OR IN THE WX.

HE OUGHT TO FIND A NEW GATE FOR THAT INBOUND FLT

Notice that in these examples and many others (though not all others), the causal factors and recommended interventions are linked to contextual factors rather than the identification of cognitive behavioral factors such as loss of situational awareness (which was, in fact, the first direction that this study of Behavior took).

It was decided that this argument was valid, that it would be fruitless to try to separate contextual factors from behavior, and that the two are going to be closely connected in the experiential report. Consequently, the focus shifted to capturing fully from the free text the contextual factors that the experiential report says influenced the reporter's behavior. To avoid confusion with the objective parameters of the Context (from the fixed fields) that were addressed previously, these factors derived from the narrative that influenced the behavior are referred to as "Shaping Factors." Consequently, the causal factors of an event as modeled by the Scenario consists of the factors that described the context of the initial safe state (i.e., the Contextual Factors) and the factors that influenced the reporter's behavior (i.e., the Shaping Factors).

With the assistance of colleagues in human-factors research and operational personnel, and based on the experience of the ASRS reports, a set of 14 factors that influence human performance in aviation operations was formulated. Table 5.0 lists the set of 14 Shaping Factors with brief definitions and exemplary expressions taken from ASRS reports that were used to evaluate the ability to identify factors such as these automatically from the free narrative of a set of ASRS reports (Posse et al. (2004)).

TABLE 5.0. SHAPING FACTORS – A SUBSET OF CONTEXT FACTORS DESCRIBING EXPERIENTIAL TEXTUAL DATA

Factor	Definition	Example
Attitude	Any indication of unprofessional or antagonistic attitude by a controller or flightcrew member, complacency, or get-home-itis (being in a hurry to get home).	*"I believe a contributing factor was complacency flying a very familiar approach, also it was our last leg get-there-itis."*
Communication environment	Interferences with communications in the cockpit such as noise, auditory interference, radio frequency congestion, or language barrier.	*"We were unable to hear because traffic alert and collision avoidance system were very loud."*

TABLE 5.0. SHAPING FACTORS (Concluded)

Factor	Definition	Example
Duty cycle	A strong indication of an unusual working period, including a long day, flying very late at night, exceeding duty-time regulations, and having short and inadequate rest periods.	*"Flight had previously been delayed and we had minimum rest period coming up, less than 9 hours."*
Familiarity	Any indication of a lack of factual knowledge, such as new to or unfamiliar with the company, airport, or aircraft.	*"Both pilots were unfamiliar with the area."*
Illusion	Illusions include bright lights that cause something to blend in, a black hole, white-out confusion, or sloping terrain.	*"I was flying and was experiencing a black hole effect."*
Physical environment	Unusual physical conditions that could impair flying or make things more difficult, such as unusually hot or cold temperatures inside the cockpit, cluttered workspace, visual interference, bad weather, or turbulence.	*"This occurred because of the intense glare of the sun."*
Physical factors	Pilot ailment that could impair flying or make things more difficult, such as being tired, fatigued, drugged, incapacitated, influenced by alcohol, suffering from vertigo, illness, dizziness, hypoxia, nausea, loss of sight, or loss of hearing.	*"I allowed fatigue and stress to cloud my judgment."*
Preoccupation	A preoccupation, distraction, or division of attention that creates a deficit in performance, such as being preoccupied, busy (doing something else), or distracted.	*"My attention was divided inappropriately."*
Pressure	Psychological pressure, such as feeling intimidated, pressured, pressed for time, or being low on fuel.	*"I felt rushed to complete the checklist in time."*
Proficiency	A general deficit in capabilities such as inexperience, lack of training, not qualified, not current, or lack of proficiency.	*"The biggest safety factor here is the lack of adequate training in the newer autopilot system."*
Resource deficiency	Absence, insufficient number, or poor quality of a resource, such as overworked or unavailable controller, insufficient or out-of-date chart, equipment malfunction, or inoperative, deferred, or missing equipment.	*"Later I learned the minimum equipment list was wrong."*
Task load	Indicators of a heavy workload or many tasks at once, such as short-handed crew.	*"Due to high workload, I forgot to switch to tower."*
Unexpected	Something sudden and surprising that is not expected.	*"Had we known of him prior to takeoff we would have made adjustments."*
Other	Anything else that could be a shaper, such as shift change, passenger discomfort, or disorientation.	*"This happened during shift change." "Ill passenger on board."*

These Shaping Factors are not mutually exclusive, and some may be more difficult to discriminate than others. Nor are these factors suggested as the "full and complete" set of contextual factors that can influence human performance. At this stage, they were selected simply to test the concept on their automated identification in a subset of ASRS reports.

The next step was to test the ability to use these definitions and exemplary phrases to discriminate reliably the Shaping Factors from the free narratives of the ASRS reports that pertained to each Scenario. However, it is not sufficient to simply find such expressions in the narratives, because the context in which they are used must be accounted for. Most importantly, the narratives must be checked for the presence of negative modifiers in close proximity to a given expression that changes the meaning of the expression. For example, the expressions "I have not had an exhausting day," "If I had felt I was tired I would...," "...fatigue was not an issue," indicate that fatigue was, in fact, not a significant factor.

Consequently, it was recognized that the traditional keyword-based and bag-of-words classification approaches that worked so well on the fixed fields for analyzing the Context and the Outcome of each Scenario would be of limited value for analyzing the Shaping Factors of Behavior. More complex pattern extraction methods were required, and an understanding of the nature of the narratives to be analyzed was needed in order to identify the most appropriate available methods.

The narratives of incident reports are, generally, highly informal and replete with the acronyms, idioms, multiple spellings, misspellings, and ambiguous abbreviations that are specific to each domain, and even to tasks. The informal reporting setting tends to produce poor grammar, spelling variants, very domain- or task-specific expressions, as well as jargon. Stream of consciousness permeates the reports, which often exhibit feelings of anger, guilt, or defense. Styles and even languages vary dramatically, from telegraphic to very detailed, and they depend on the narrator, whether or not he/she is a pilot, an air-traffic controller, a flight attendant, a mechanic, or ground personnel. The following ASRS report is representative of incident reports and illustrates some of these characteristics:

"I RETURNED FROM A LCL TRNING FLT. PER TWR, WE LANDED ON RWY 4R AT MDW. SHORTLY AFTER OUR T/D, THE ACFT EXPERIENCED SEVERE NOSE SHIMMY VIBRATION. WE SLOWED DOWN AND TURNED LEFT AS PER TWR INSTRUCTIONS OFF OF RWY 4R. WE SWITCHED TO GND AND THEY TOLD US TO TAXI TO THE N RAMP. WE PROCEEDED TO DO SO. FIVE SECS LATER, GND ASKED US HOW DO WE HEAR. WE TOLD THEM LOUD AND CLEAR. AFTER 15 SECS GND TOLD US TO CALL THE TWR. WE DID AND TWR TOLD US WE XED AN ACTIVE RWY (4L) W/O THEIR PERMISSION. TWR SAID AS WE WERE LNDG HE TOLD US TO TURN LEFT AND HOLD SHORT OF RWY 4L AND REMAIN WITH HIM. WE OBVIOUSLY DID NOT HEAR HIM DUE TO THE EXTREME NOISE CAUSED BY THE NOSE SHIMMY. GND SHOULD NOT HAVE TOLD US TO TAXI TO THE N RAMP. TWR TOLD US ON THE PHONE THAT THERE WAS LNDG TFC ROLLING ON RWY 4L AT THE TIME WE XED RWY 4L. "

Consequently, the automation effort combined language normalization and knowledge infusion from domain experts. Normalization produces readable and correct English text, whereas knowledge infusion elicits the peculiar expressions used by personnel involved in aviation operations. A way was needed to standardize the language of our world of flight operations before an automated analysis of this sort of text could be expected.

Under the ASMM Project, the Battelle's Pacific Northwest Division developed a preprocessing tool called PLADS to standardize the language of unstructured text so as to facilitate its reliable automated analysis. PLADS uses a combination of standard dictionary English and domain-specific concepts to improve text-mining effectiveness. PLADS is composed of software (Java, Matlab, and Perl) and lexicons. The development of the lexicons when PLADS is adapted to a new domain for the first time requires an expert in PLADS working with an expert in the reporting language of that domain (Posse et al. (2005)).

PLADS is an acronym of the names of the following five stages of filtering performed on each report prior to its automated analysis.

- **P**hrases identified and concatenated. Identify phrases in the unstructured text by statistical means by identifying 2-, 3-, 4-, or 5-word strings that occur more often than one would expect based solely on the individual word frequency. Then concatenate the phrase together into what would be identified as a single word to subsequent software: e.g., ClassCAirspace, UnitedStatesOfAmerica.

- **L**eave some words unprocessed.

- **A**ugment some words to make the meaning more useful for computer analysis by subsequent software. Some words have "too much" information; that is, they may be abbreviations for instruments and/or concepts with make/model/series, or numeric values of selected concepts. Examples include:
  - "B-757-300" might be augmented with the word "airplane."
  - "FL28" ("FL26," "FL30") means "flight level at approximately 28,000 (26,000, 30,000) feet." Augmenting with "FlightLevel" enables subsequent software to identify these 3 (and others) as related to a flight-level concept leaving the refinements of which specific flight level to finer grain analysis.
  - "24L," "24R," "25L," and "25R" all relate to runways. Augmenting with the word "runway" enables the software to capture that concept.
  - Proper names are often augmented with the more general concept; e.g. "Dallas" augmented with "city."
  - Airport abbreviations are often augmented with the word "airport"; e.g. "LAX," "ORD," and "DFW."

- **D**elete some words to simplify the analysis. These words are often called "stop" words. Examples include: "the," "a," and "an." Some times numbers are deleted.

- **S**ubstitute some words for others. Often there are many ways to express the same concept, including synonyms, abbreviations, jargon, and slang. For example, "pilot" might be substituted for these words: "pilot," "pilots," "co-pilot," "captain," "co-captain," "left seater," "PIC," "Pilot-in-Charge," "plt," and "plts." Standard abbreviations can be checked and full meanings substituted. Numbers that are spelled out can be replaced by the numeral.

PLADS does not directly assess anything. It is used as a preprocessor in conjunction with other text-analysis tools, whether they entail statistical (i.e.; bag-of-words classification) or NLP tools, to facilitate analysis of free text.

Following is an example of an ASRS report shown previously after it has been processed through PLADS:

I RETURNED FROM A LOCAL TURNING[5] FLIGHT. PER TOWER, WE LANDED ON RUNWAY 4R AT MDW. SHORTLY AFTER OUR TOUCHDOWN, THE AIRCRAFT EXPERIENCED SEVERE NOSE SHIMMY VIBRATION. WE SLOWED DOWN AND TURNED LEFT AS PER TOWER INSTRUCTIONS OFF OF RUNWAY 4R. WE SWITCHED TO GROUND AND THEY TOLD US TO TAXI TO THE NORTH RAMP. WE PROCEEDED TO DO SO. FIVE SECONDS LATER, GROUND ASKED US HOW DO WE HEAR. WE TOLD THEM LOUD AND CLEAR. AFTER 15 SECONDS GROUND TOLD US TO CALL THE TOWER. WE DID AND TOWER TOLD US WE CROSSED AN ACTIVE RUNWAY WITHOUT THEIR PERMISSION. TOWER SAID AS WE WERE LANDING HE TOLD US TO TURN LEFT AND HOLD SHORT OF RUNWAY 4L AND REMAIN WITH HIM. WE OBVIOUSLY DID NOT HEAR HIM DUE TO THE EXTREME NOISE CAUSED BY THE NOSE SHIMMY. GROUND SHOULD NOT HAVE TOLD US TO TAXI TO THE NORTH RAMP. TOWER TOLD US ON THE PHONE THAT THERE WAS LANDING TRAFFIC ROLLING ON RUNWAY 4L AT THE TIME WE CROSSED RUNWAY 4L.

The work described previously to capture the Context and Outcome of ASRS reports was based on the fixed-field (objective) data for which statistical (i.e., bag of words classification) tools worked very well. However, something more was needed to capture the significant information that the reporter conveys in the free text. The information in the free text does not lend itself to statistical classification because the requirement is to extract meaning about Shaping Factors that tends to be more subjective than objective (as is evidenced by the exemplary statements from ASRS reports used in table 5.0).

After an investigation into available tools, the General Architecture for Text Engineering (GATE) tool was selected to capture NLP of the ASRS narratives after they had been processed through PLADS. GATE, a software tool created by University of Sheffield (http://nlp.shef.ac.uk/), is one of the most widely used human-language processing systems in the world. GATE has had great success at TREC (Text Retrieval Conference series) cosponsored by the National Institute of Standards and Technology, the Information Technology Laboratory's Retrieval Group of the Information Access Division, and the Advanced R&D Activity of the Department of Defense (DOD) in competition with numerous other techniques.

GATE comprises an architecture, framework, and graphical development environment to identify evidence of specific concepts contained in unstructured text. The concepts may be vaguely defined, and phrased in a way as to require subtle insight to identify their existence (Manning and Schültze (1999) and Cunningham et al. (2002)). GATE provides framework for applying customizable tools for data-mining. A GATE gazetteer is a list of expressions that map to a specific set of concepts. There may be many specific expressions enumerated that will be associated to one specific concept.

---

[5] Notice that "TRING" was incorrectly converted into "TURNING" rather than "TRAINING."

Gazetteers may produce more false positives, but less false negatives because exact matches are determined out of context. Therefore, gazetteers are a good way to start encoding simple expressions of the Shaping Factors to control the false-negative rate. GATE includes, for example, a pattern specification language called Java Annotation Patterns Engine (JAPE), which describes patterns to match that will be associated with specific concepts. In the particular domain of this study, reports can be tagged with respect to the Shaping Factors, where the tags correspond to the patterns coded in JAPE for matching to the text. JAPE rules can be refined very rapidly without having to rewrite large pieces of code and can easily handle matching exceptions. The tools available within GATE enable identification of specific concepts using synonyms and Boolean expressions to identify natural language phrases that were not envisioned.

Following is an example of the application of JAPE to Pilot Fatigue, which is one of the Physical Factors of the 14 Shaping Factors defined in table 5.0. The guidance was that a hint of the possibility of Pilot Fatigue being a factor would be when a pilot mentions that an incident occurred during a certain leg of a several-day/-leg trip. Mining the ASRS database revealed all of the following variants:

LAST LEG OF A 2 DAY TRIP	LAST FLIGHT OF THAT DAY
LAST LEG OF A TWO DAY TRIP	LAST FLIGHT OF A FOUR DAY TRIP
LAST LEG OF A 3 DAY TRIP	LAST FLIGHT OF A 5 LEG
FINAL LEG OF A 3 DAY TRIP	LAST FLIGHT OF A LONG BYT UNPROFITABLE DAY
LAST LEG OF A 3-DAY TRIP	LAST SEGMENT OF A 3-DAY
FINAL LEG OF A 3-DAY TRIP	FINAL SEGMENT OF A LEG WAY
LAST LEG OF A LONG 3-DAY TRIP	NEXT TO LAST LEG OF TRIP
LAST LEG OF A LONG 5-LEG DAY	DAY 2 ON A 4 DAY TRIP
LAST LEG OF 3 DAY TRIP	THIRD DAY OF A 3 DAY TRIP
FINAL LEG OF 3 DAY TRIP	THIRD DAY OF A 4 DAY TRIP
LAST LEG OF 3-DAY TRIP	LAST LEG OF DAY #3
LAST LEG OF A FOUR DAY TRIP	3RD LEG OF A 4 LEG
LAST LEG OF A 4 DAY TRIP	3RD DAY OF A 4-DAY TRIP
FINAL LEG OF A 4 DAY TRIP	THIRD LEG OF A 4 LEG, 1 DAY TRIP
LAST LEG OF A 4-DAY TRIP	THIRD DAY OF A 3 DAY TRIP
FINAL LEG OF A 4-DAY TRIP	THIRD DAY ON A 4 DAY TRIP
LAST LEG OF 4 DAY TRIP	THIRD LEG OF A 2 DAY TRIP
LAST LEG OF AN ALL DAY TRIP	4TH AND LAST LEG OF THE TRIP
LAST LEG OF OUR 3 DAY TRIP	4TH LEG OF A LONG DAY
LAST LEG OF A SIX LEG WAY	FIFTH LEG OF THE DAY
LAST LEG OF THE DAY	SIXTH LEG OF A SEVEN LEG DAY
FINAL LEG OF THE DAY	10TH LEG OF AN 11 LEG DAY ON DAY 3 OF A 3 DAY TRIP
LAST LEG OF THE TRIP	11TH LEG OF THE DAY
LAST LEG OF A LONG TRIP	10 HRS INTO A 12 HR DUTY DAY
FINAL LEG OF THE TRIP	NEARING THE END OF A LONG 3-DAY TRIP
LAST LEG OF A 4 LEG	END OF A 2-DAY TRIP
LAST LEG OF 4 LEGS	END OF A 3 DAY TRIP
LAST LEG OF A LONG DAY	DAY THREE OF A THREE DAY TRIP
LAST LEG OF A VERY LONG DAY	DAY 3 OF A THREE DAY TRIP
LAST LEG OF THE NIGHT	DAY 4 OF 4 DAYS
LAST FLIGHT OF THE DAY	DAY 4 OF A 4 DAY TRIP
LAST FLIGHT OF A 2 DAY TRIP	THIRD DAY OF A THREE DAY TRIP
LAST FLIGHT OF THE TRIP	SECOND DAY OF A THREE DAY TRIP
LAST FLIGHT OF THAT DAY	LAST NIGHT OF A 6 NIGHT TRIP

JAPE enables the provision of rules to cope with all such possible variations.

Although there are numerous other methods of doing NLP, this study found NLP via GATE to be the most effective at identifying specific concepts in the free text of ASRS reports. GATE/JAPE was used to identify the Shaping Factors associated with a subset of 17,155 ASRS reports identified by a search of the ASRS database for 10 specific anomalous Outcomes (Ferryman et al. (2006)). Figure 5.1.10 is an example of some of the results of these analyses.

This example is for the Scenario with the Outcome of Landing without a Clearance and it is associated with the Approach and Landing phase of flight. The green asterisks * are the probabilities of the occurrences of each of the 13 Shaping Factors (the Other category was omitted in this discussion) during an Approach and Landing phase of flight. The red dots • are the probabilities of the occurrences of the Shaping Factors during the Scenario of Landing without a Clearance in this phase of flight. The hash marks on either side of the red dots are the 80% and 99% uncertainty intervals. A Shaping Factor is significantly associated with this Scenario when the ratio of the latter probability (i.e., the red dot) to the former (the green asterisk) is greater than 1.0, as in the cases of those indicated by the blue arrows ⊆ for Communication Environment, Duty Cycle, Physical Environment, Physical Factors, Preoccupation, Proficiency, and Task Load. When both probabilities are small, the numerical stability of their ratio is questionable as in the cases of Attitude, Familiarity, Illusion, Pressure, and Resource Deficiency.

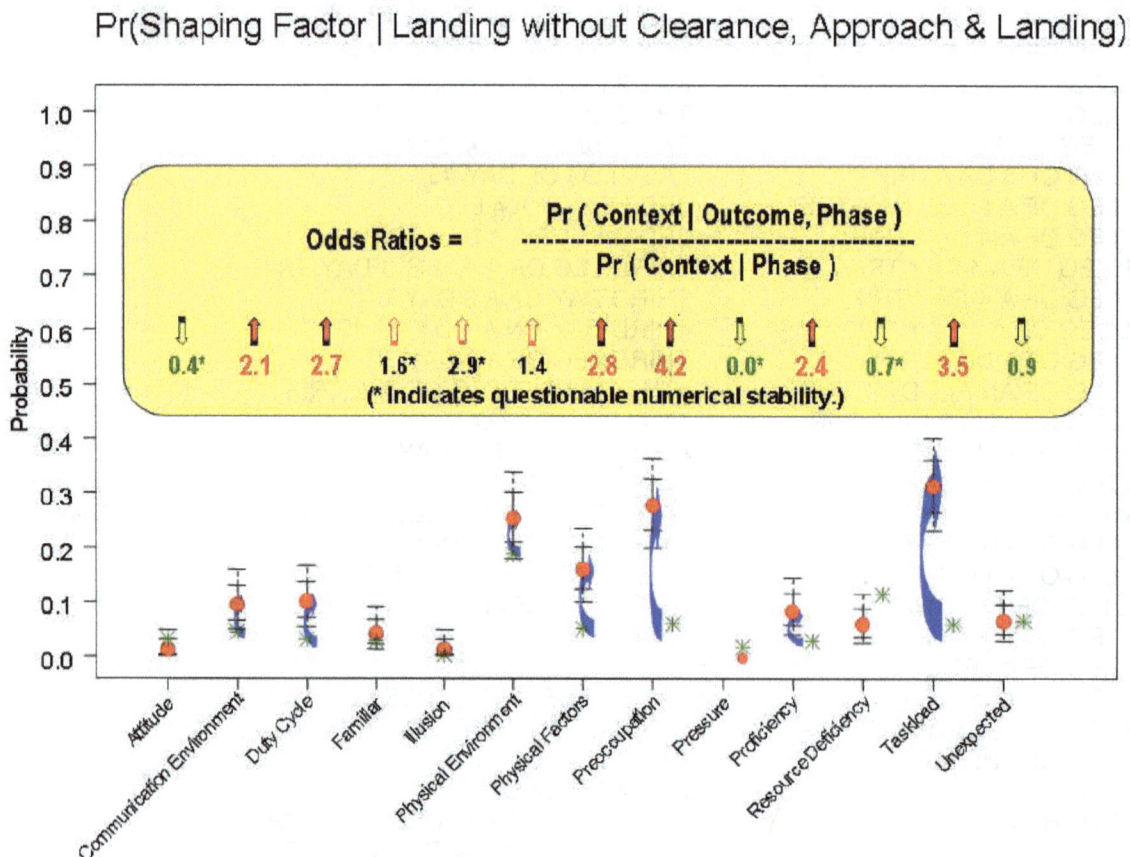

Figure 5.1.10. An example of the probability of Shaping Factors.

In the results of this experiment shown in table 5.1, the significant Shaping Factors are identified and ranked (by odds ratio) for each of the 10 anomalous Outcomes in its associated phase of flight.

Validation of results of the study was based solely on expert opinion. When the results of both the probably important (shown in table 5.1) and the probably unimportant Shaping Factors that had been identified in associations with anomalies were presented to a group of experts in aviation operations, they unanimously agreed that, in each of the 10 Scenarios, the results seemed entirely reasonable and plausible. This outcome demonstrated a capability to analyze the narratives of experiential reports for identification of the Shaping Factors of the human behavior in a group of reports of the same anomalous incident that is consistent with expert opinion. This validation was considered adequate for purposes of defining why the event occurred, considering the reliability of the incident reports.

The experiments used the ASRS database as a resource for developing the analysis methodology. Nevertheless, the approach is expected to be applicable to any incident-reporting system because the ASRS is representative of all such databases. The methodology is sufficiently generic to be used with any database of textual reports of aviation incidents. This fact was demonstrated by applying the process to a set of ASAP reports that were in a format quite different from those in the ASRS database, showing that it was possible to reliably categorize the reports by their Context and Outcomes into the anomalies that had been defined by the ASRS Office. The process could be adapted to any other set of anomalies or to new ones given their definitions and an exemplary report. A capability was demonstrated to extract information with adequate credibility on the Shaping Factors associated with each identified anomaly found in a large database of ASAP reports (that could have come from a single air carrier or de-identified from multiple air carriers).

There remained one final objective to be achieved by the Data Analysis Tools Development team: to create mechanisms to link information extracted from disparate numerical and textual data sources, as, for example, FOQA and ASAP data. The systems for analyzing FOQA and ASAP databases were developed independently, they are heterogeneous databases, and they have different classifications. Nevertheless, there are potential synergies because the FOQA data describe *what* happened, whereas the ASAP data have information about *why* it happened. A solution was sought to link the complementary information extracted from each with concordance based on the same anomalous event within common constraints of flight phase, aircraft type, airport of origin or destination, and month(s) of the year, A process was developed for accomplishing this objective, and the capability was demonstrated by linking information about Shaping Factors from ASAP reports with information about the other contextual features of the same anomalous incidents extracted from FOQA data. The underlying process by which the linkage is performed is diagramed in figure 5.1.11. The user's interface for using this process is shown in figure 5.1.12.

TABLE 5.1. SHAPING FACTORS IDENTIFIED WITH SELECTED ANOMALIES

Ground Flight Phase
- *Maintenance problem* is associated with:
  - Pressure (Relative importance = 2.1)
- *Ground incursion* is associated with:
  - Familiarity (Relative importance = 2.0)
  - Communication environment (Relative importance = 1.9)
  - Pre-occupation (Relative importance = 1.8)
  - Task load (Relative importance = 1.8)

Take-off Flight Phase
- *Air conflict* is associated with:
  - Task load (Relative Importance = 1.9)
  - Communication environment (Relative importance = 1.7)
- *Nonadherence* is associated with:
  - Preoccupation (Relative importance = 1.6)

Ascent Flight Phase
- *Altitude deviation* is associated with:
  - Preoccupation (Relative importance = 2.7)
  - Proficiency (Relative importance = 2.1)
  - Physical factors (Relative importance = 2.1)
  - Duty cycle (Relative importance = 1.9)

Cruise Flight Phase
- *Altitude deviation* is associated with:
  - Preoccupation (Relative importance = 3.3)
  - Physical factors (Relative importance = 2.0)
- *Air conflict* is associated with:
  - Communication environment (Relative importance = 1.6)

Descent Flight Phase
- *In-flight Wx encounter* is associated with:
  - Pressure (Relative importance = 3.1)
- *Altitude deviation* is associated with:
  - Preoccupation (Relative importance = 1.8)
  - Proficiency (Relative importance = 1.6)
  - Familiarity (Relative importance = 1.5)

Approach and Landing Flight Phase
- *Landing without clearance* is associated with:
  - Preoccupation (Relative importance = 4.2)
  - Physical factors (Relative importance = 2.8)
  - Duty cycle (Relative importance = 2.7)
  - Proficiency (Relative importance = 2.4)

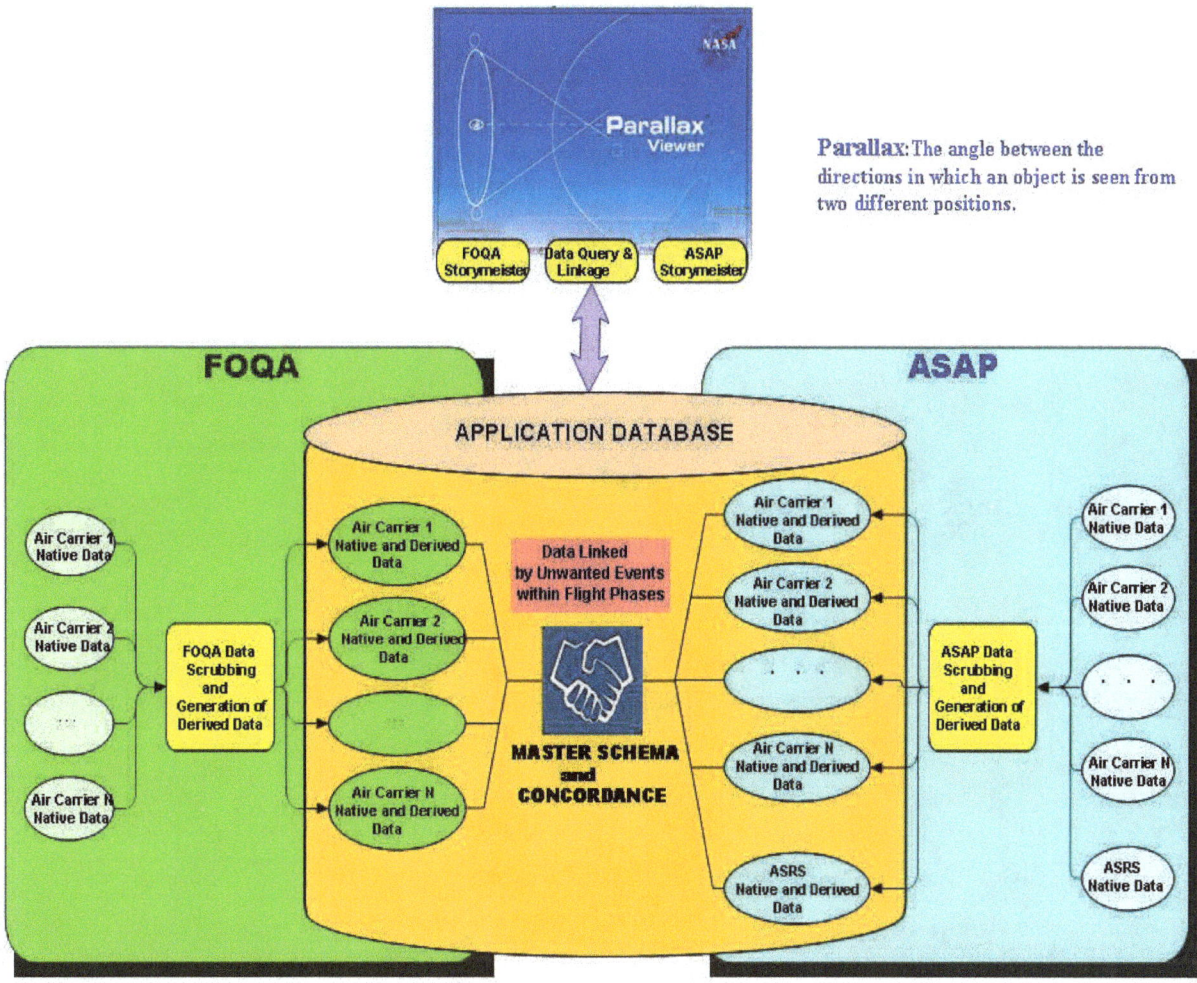

Figure 5.1.11. The Process for FOQA – ASAP linkage.

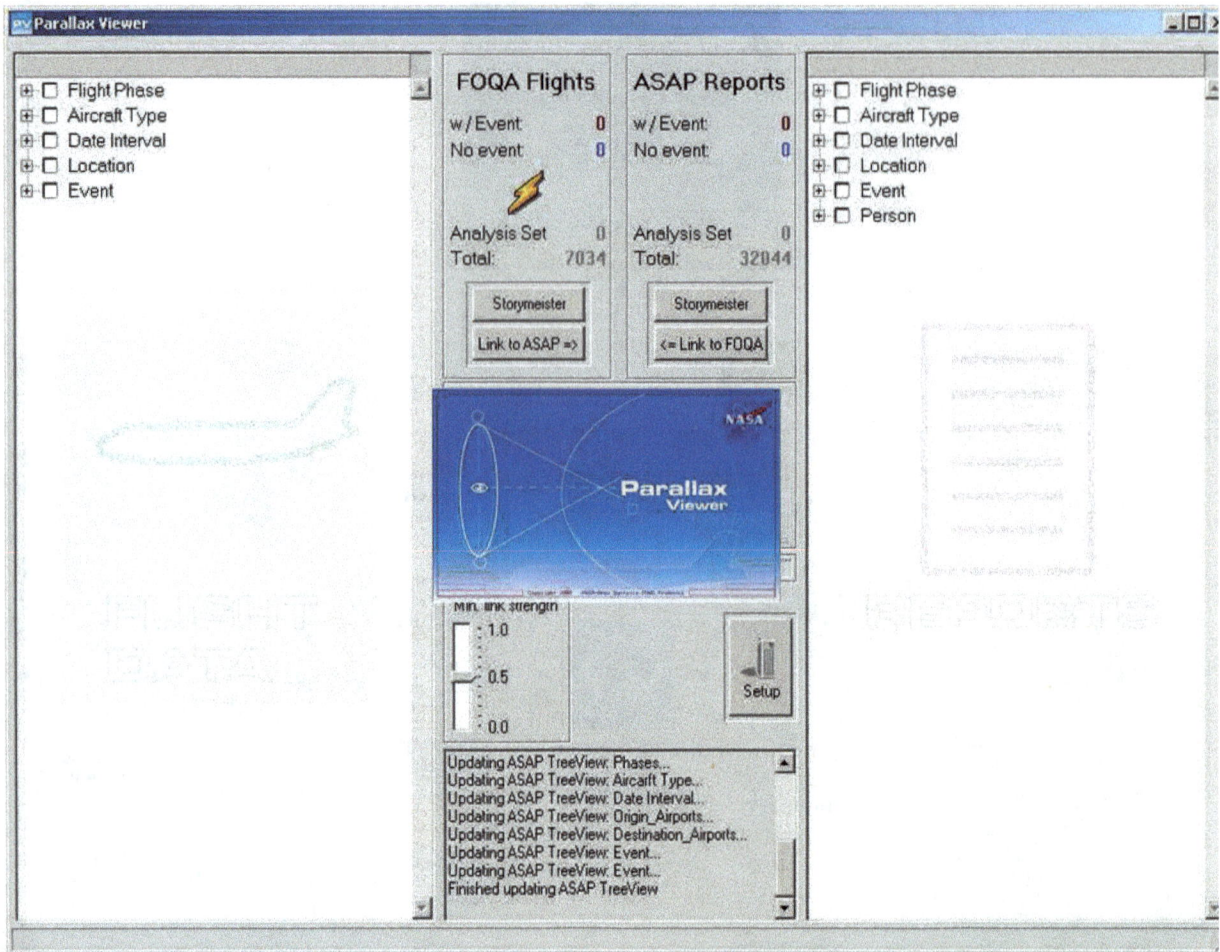

Figure 5.1.12. Parallax: The interface to FOQA-ASAP linkage.

The user's interface, called Parallax (shown in figure 5.1.12), allows the user to initiate a query starting from either the ASAP databases or the FOQA databases by specifying the constraints that define the data set (i.e., anomalous event, flight phase, aircraft type, airport, and time period). For example, to use Parallax, the user selects (from the drop-down menus shown at the top left of the figure) the phase of flight, the aircraft make/model, the time interval of the search, possibly the airport, and the event type. A single click initiates the search of the FOQA data.

The criteria for the Pattern Search tool to search flight-recorded data for each of the anomalous events are predefined and stored. These criteria have been defined in terms of the flight-recorded parameters by domain experts who are knowledgeable about flight data and familiar with these anomalies.[6] The Pattern Search tool is used on the FOQA database of a single air carrier or multiple air carriers to categorize flights with respect to each of the anomalous Outcomes (as depicted on the left side of the figure).

---

[6] Not all of the ASRS anomalies (for example, cabin events such as passenger misconduct or passenger illness) are identifiable from flight-recorded data.

The portion of the screen just below "FOQA Flights" automatically displays the number of flights in that subset with and without the selected event. Below that are displayed the number of flights in the selected subset and the total number of flights available in the database.

When the user clicks on Storymeister under FOQA Flights, the Flight-Data Storymeister is displayed, as illustrated by figure 5.1.13. The Flight-Data Storymeister, developed by Battelle's Pacific Northwest Division, identifies the Contextual Factors and describes the event in natural language. The Flight-Data Storymeister provides a high-level preliminary analysis of the search results in the format and language familiar to the user. It presents the discriminating features of the selected set of flight data. Discriminating features are parameter values with unexpectedly high or low frequencies or discrete parameters that show unexpected transition patterns. (In the example of figure 5.1.13, these discriminating features are the parameters shown at the left and ranked by frequency of occurrence.) In the application to flight data, the key discriminating features are the parameter averages. Storymeister contrasts flights having *one or more* of the selected anomalous events with flights that have none of the selected events within the constraints (i.e. flight phase, month(s), aircraft type, and airport) of the subset of reports being analyzed. An example of a display from the Flight-Data Storymeister is shown in figure 5.1.13. To the right are shown the location of the event and below that, the secondary combined events.

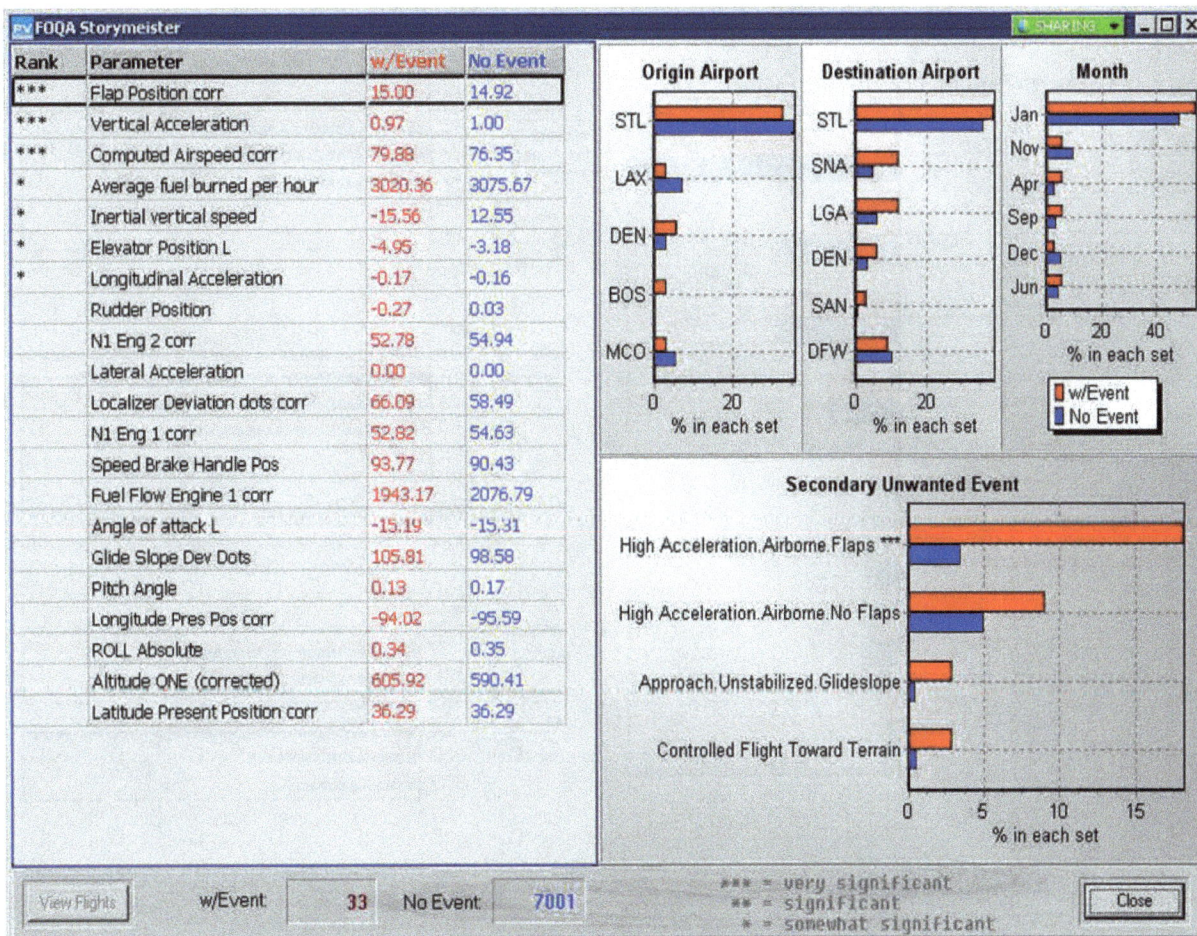

Figure 5.1.13. An example of a display from the Flight-Data Storymeister.

A click on Link to ASAP on Parallax (figure 5.1.12) initiates the comparable analyses of the ASAP database. As indicated in figure 5.1.11, concordance of the incident-report data with the flight data is based on the same anomalous event under the common constraints of flight phase, aircraft type, airport, and time of year. From any single database of incident reports, such as a set of ASAP reports from a single air carrier or from multiple sources of incident reports (perhaps including ASRS) (depicted on the right side of figure 5.1.11), the reports are automatically categorized with respect to the anomalous Outcomes as defined by the ASRS (or any modification of those). For each anomaly, the subset of reports is analyzed for the Shaping Factors that are identified and the results are described to the user in natural language by the Incident-Report Storymeister developed by Battelle's Pacific Northwest Division. An example of a display from the Incident-Report Storymeister is shown in figure 5.1.14.

Just like the Flight-Data Storymeister, the Incident-Report Storymeister provides a high-level pre-liminary analysis of the results in the format and language familiar to the user. It presents the discriminating features of the selected set of reports. Discriminating features are values in the fixed fields of the reports and textual terms in the narrative with unexpectedly high or low frequencies. In the application to ASAP reports, the key discriminating factors are the Shaping Factors. Storymeister contrasts flights having one or more of the selected anomalous events with flights that

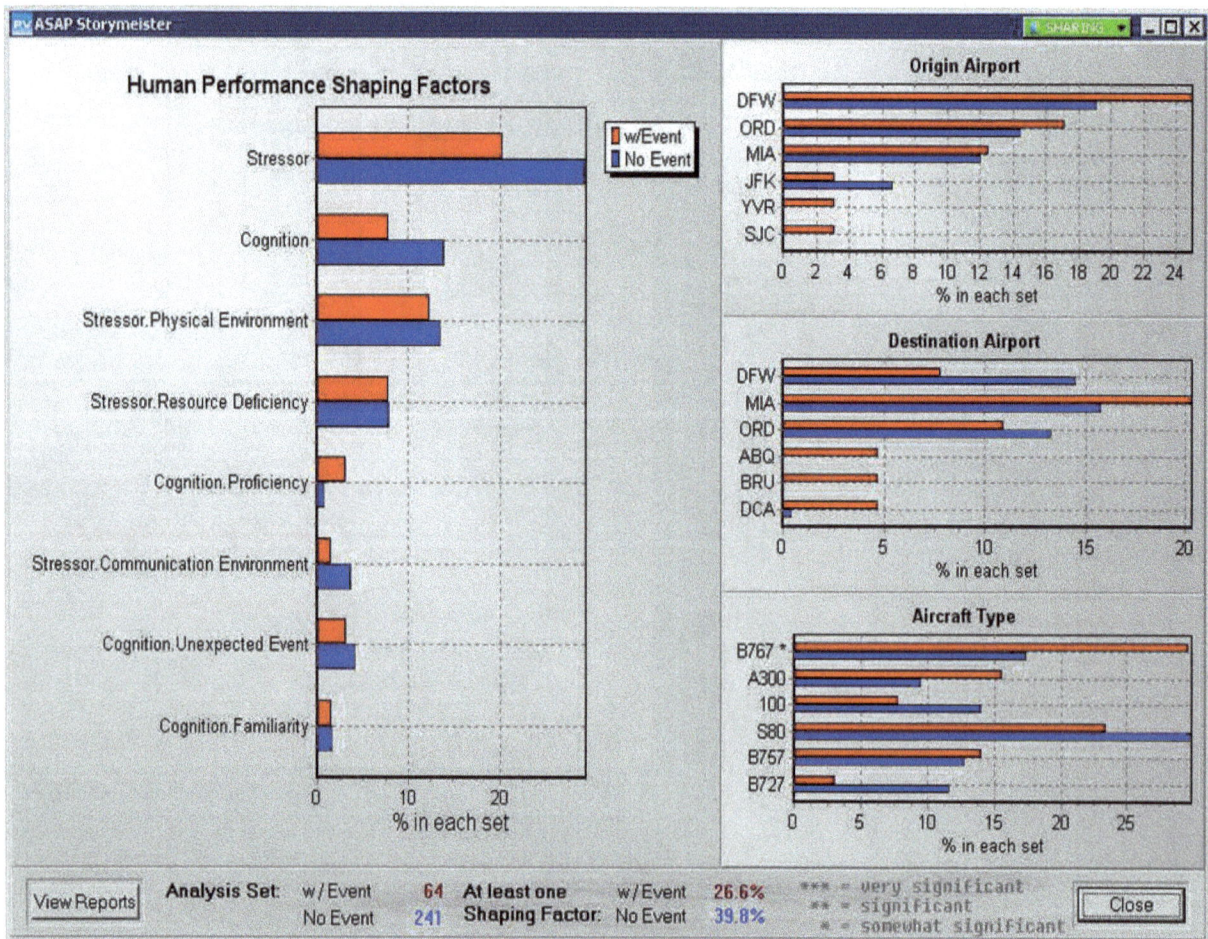

Figure 5.1.14. An example of the display from the Incident-Report Storymeister.

have none of the selected events within the constraints (i.e., flight phase, month(s), aircraft type, and airport) of the subset of reports being analyzed. For each such anomalous event, Storymeister states the relative importance of the specified Shaping Factors.

The information from the FOQA data about the Contextual Factors together with the information from the ASAP reports about Shaping Factors gives the safety analyst understanding and insight into the causal factors of an incident and enables decisions defining the most appropriate intervention.

In summary, the *Data Analysis Tools Development* element of the ASMM Project produced the following capabilities:

- The PROFILER – the statistical analysis of digital data that underlies the displays of The Morning Report of Atypical Flights.

- The Pattern Search tool – a convenient way for safety experts to state the criteria for any pattern of numerical parameters and conduct a search of a large database.

- PLADS – a preprocessing tool to standardize the language of unstructured text so as to facilitate its automated analysis.

- ALAN – a text comprehension tool that clusters textual data.

- Adaptation of GATE/JAPE – an NLP capability that analyzes free text and identifies the Shaping Factors of human behavior in an incident report.

- Parallax, a process for FOQA-ASAP linkage – a way for a safety analyst to combine the information from the FOQA data about the Contextual Factors with the information from the ASAP reports about Shaping Factors for insight into the causal factors of an incident.

5.2 Extramural Monitoring

The activities under the *Extramural Monitoring* element of the ASMM Project resulted in two products: Incident-Reporting Enhancement Tools and the NAOMS.

5.2.1 Incident-Reporting Enhancements

The Incident-Reporting Enhancements were designed to improve the accessibility to, and extraction of information from, the ASRS database. However, for the most part, these tools were general and are equally applicable to other similar textual databases such as the ASAP. The ASRS database was used as a representative resource of textual reports to develop and evaluate new capabilities for their automated analyses such as those described in the previous section. (See appendix C for a description of the ASRS.)

The motivation for modernizing the ASRS processing was to make it more convenient to access this unique repository of safety reports in order to test new capabilities in text mining that were needed for the ASAP, NAOMS, and other sources of textual data. However, the ASRS benefited from upgrading its legacy systems to state-of-the-art capabilities to improve the efficiency of report processing and database search, increase the rate of volume-handling capability, and improve the quality of the processing and analyses. The specific aspects of the Incident-Reporting Enhancements included ASRS database migration, electronic report submission, the Analyst Workbench, and improved ASRS search and analysis capability. The ASRS legacy database that had been housed on a VAX at the NASA contractor's headquarters was migrated to a state-of-the-art Oracle 8 system.

The capability for electronic report submission has as its goal to maximize access to digitized data for submitted reports through coordination with 1) airline ASAP programs, and 2) direct ASRS electronic submission from the Website with realization that some form of paper submissions will continue into the future. ASRS is currently receiving ASAP reports from 15 pilot groups, 3 mechanics groups, 1 flight-attendant group, and 5 dispatch groups. ASRS supports 24 programs with 15 airlines, and several more are awaiting coordination with ASRS. In 8 programs with 6 airlines, reports are being submitted through secure, electronic data transmission at various intervals (daily 7 days per week, once per week, etc.). When it is completed, the electronic report submission will provide a secure end-to-end Web-based ASRS system that:

- enables ASAP reporters to securely submit reports with a Web browser without leaving traces to protect confidentiality provisions of ASRS,

- enables ASAP reporters to obtain an ASRS report receipt electronically without requiring an email address, and

- enables ASAP reports to be automatically entered into the ASRS processing stream without human intervention and without threatening the integrity of the ASRS database.

The Analyst Workbench is a customized, browser-based, cross-platform software application used for report processing. It tracks reports electronically through the report-production process, it automates analyst coding of over 1200 values in the current 12-page hardcopy coding form, and it enables 100% of those data needed to identify the type of event to be captured, thereby providing statistical indication of types of events being reported to ASRS. Following are the modules of the Analyst Workbench that have been implemented and are operational in the ASRS:

- Online Resources

- Preliminary Screening

- Initial Screening

- Secondary Screening

- Decap ID Strip

- Multiple Report Matching

- Final Classification

- Full Form Coding Prototype

- Query Wizard as Basis for "Public Access"

Improvements have also been realized in the capability of the ASRS for data search and analysis with new database query tools; namely, BRIO (a COTS query tool) and Query Wizard (a browser-based tool). Database mining has benefited from the installation in the ASRS of Perilog (a tool developed at NASA Ames Research Center) that provides relative ranking of narrative text as well as keyword search, phrase search, phrase discovery, and exemplar matching.

5.2.2 National Aviation System Operational Monitoring Service

The other product of the *Extramural Monitoring* element was the NAOMS, which enables a scientifically designed survey of all the constituents of the aviation system to obtain the "front-line" operators' experiences with system performance and a statistically sound basis for evaluating frequencies of incidents and trends that might compromise safety (figure 5.2.1). This approach complements and enhances information gathered currently through voluntary reporting systems, such as the ASRS and the ASAP, that cannot be used for statistical analyses. NAOMS may provide the first indication of a developing new situation or trend deserving of further investigation.

The value of receiving experiential reports on performance of the system from the perspectives of its operators has been well established in the 30 years of ASRS. It is important to have such systems that allow the communication of safety issues among the stakeholders in the aviation community. The ASAP programs that have been recently implemented internally by over 20 U.S. air carriers have recognized the value of voluntary experiential reporting. However, voluntary reports cannot be used as the basis for deriving frequencies of events or for any other statistical analyses of system performance because they do not necessarily represent the entire population. They can only provide information on an event important to the reporter that may or may not be operationally significant from a system perspective. The NAOMS addresses this limitation and enhances the value of front-line reporting by the addition of statistical validity, allowing results to be broadly and reliably interpreted.

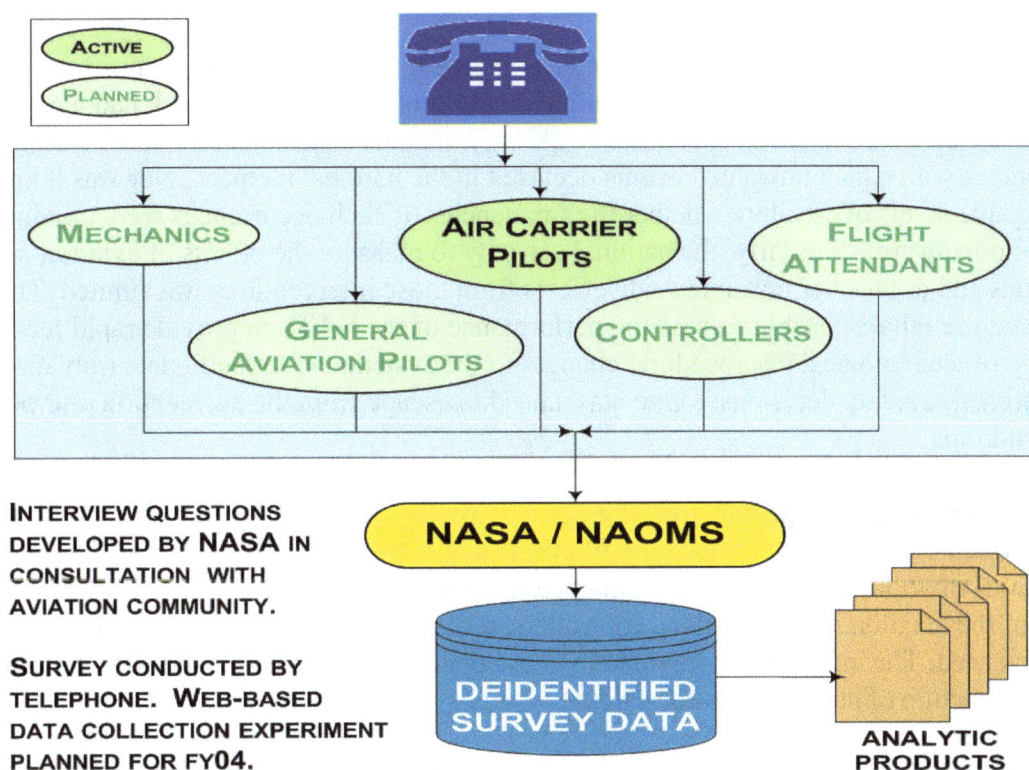

Figure 5.2.1. The NAOMS concept.

The need for this kind of information has been documented. For example, the following appears on page 13 of the report of the White House Commission on Aviation Safety and Security (1998): "The most effective way to identify incidents and problems in aviation is for the *people who operate the system* (pilots, mechanics, controllers, dispatchers, etc.) to self-disclose the information." The General Accounting Office's (GAO's) Safer Skies Review (2000) emphasized the need for additional performance measures. The NTSB's Safety Report on Transportation Safety Databases (2002) said that there was a "Need to address the problem of *under-reporting in current aviation safety data systems*." Numerous databases attempt to capture safety-related information concerning the NATS; e.g.:

- NTSB Accident/Incident Database
- FAA National Airspace Incident Monitoring System (NAIMS)
- FAA Accident/Incident Data System (AIDS)
- Aviation Safety Reporting System (ASRS)

Numerous databases also attempt to capture safety-related information concerning specific parts of the NATS; e.g.:

- FOQA
- PDARS
- ASAP

However, prior to NAOMS, there had not been a database that addressed the performance and safety of the NATS as a whole in a quantitatively defensible fashion. Although the U.S. had many aviation safety-data collection efforts, no program gave decision-makers statistically defensible estimates of the frequencies with which unwanted events occurred in the national airspace. Nor was it known with acceptable levels of certainty whether the frequencies of such occurrences were trending upwards or downwards. Similarly, the national capacity to measure the effects of aviation-safety interventions and to uncover unwanted side effects from those interventions was limited. There was a need to acquire reliable, stable data on the performance of the NATS to provide rapid feedback on the efficacy of technological or procedural changes to the system and to facilitate a truly data-driven basis for proactive safety decisions. There was a need to escape from the accident du jour policy-making syndrome.

NAOMS was expressly designed to meet these needs by way of a comprehensive, statistically sound survey on which evaluations of frequencies of events and trends can be reliably based to overcome the limitations of voluntary reporting systems described previously. Information is solicited from the operators of the aviation system—pilots, controllers, mechanics, and others—about the things they have experienced. The information provided by these operator groups can be fused to develop a system-level picture of national aviation safety.

There is no other way to get a top-down view of the system performance on a sound statistical basis until, perhaps, flight and radar data are available from all sources from which information is extracted and integrated routinely. NAOMS is primarily directed to acquiring event information (what happened), but, it can also shed some light on precursors to events (why it happened.) Further,

NAOMS provides information about frequencies of events and trends that may trigger further investigation using the other data sources. The "bottom-up" approach of FOQA/APMS and PDARS and the "top-down" approach of NAOMS supplement each other in some aspects, and complement each other in others.

The objectives of NAOMS were to:

- provide decision-makers in air carriers, air-traffic management, and other air services providers with regular, accurate, and insightful measures of the health, performance, and safety of the NATS;

- provide decision-makers with reliable information to determine that changes of technology or procedures introduced into the system are producing expected improvements without producing unwanted side effects;

- establish the baseline of operational performance against which to measure the systemwide impact of changes;

- contribute to the information extracted from other data sources to support investigations of systemwide safety issues; and

- provide a quick look or "snapshot" of an issue of interest to the aviation community at a particular time.

NAOMS achieved scientific integrity by using well-crafted survey instruments and rigorous analytic methods based on established and proven methodologies. Surveys have been used to shape national policy for many decades, especially in areas such as public health policy and economics. Following are a few examples of Federal Government surveys and their sponsoring agencies:

- Survey of Income and Program Participation (Census Bureau) 1984 –

- Consumer Expenditure Surveys (Census Bureau) 1968 –

- Annual Housing Surveys (Census Bureau) 1973 –

- Survey of Consumer Attitudes National Science Foundation (NSF) 1953 –

- Health and Nutrition Examination Surveys, National Center for Health Services (NCHS) 1959 –

- National Health Interview Surveys (NCHS) 1970 –

- American National Election Studies (NSF) 1948 –

- Panel Study of Income Dynamics (NSF) 1968 –

- National Longitudinal Surveys, Bureau of Labor Statistics (BLS) 1964 –

- Behavioral Risk Factor Surveillance System, Centers for Disease Control (CDC) 1984 –

- Monitoring the Future, National Institute on Drug Abuse (NIDA) 1975 –

Survey methods are mature and well understood, and aviation safety is a natural topic for survey data collection. NAOMS was built on the best survey methodological practices with deliberate and thorough instrument development and robust statistical methods. Methodological decisions were made with regard to events to address, question structure, question grouping and order, length of recall period, sample source and size, and data-collection mode. These decisions entailed briefings to aviation-safety decision-makers; consultations with industry and government safety groups; workshops; reviews of the aviation-safety databases; focus groups with active professional participants; trades among accuracy, completeness, quality, and cost; and field trials with active line pilots to determine how pilots organized memories of safety events and to select the optimum data-collection mode.

During the late 1990s, a series of workshops and briefings to the FAA and industry were conducted to familiarize the aviation community with the concept of NAOMS and to engage their active participation in the process by which the survey questionnaires were developed. The survey questionnaire design for each constituency is based on consultations with representative industry and government working groups and ASRS analysts. NAOMS needed "buy-in" from the community to achieve its goals. Further, the required approvals of the survey by the Office of Management and Budget (OMB) entailed comprehensive and rigorous reviews.

The structure of the survey instrument is shown in figure 5.2.2. Section A provides the fundamental information on the interviewee's activity (that constitutes the denominator of the statistical analysis). Section B provides information to reveal trends (the numerator of the statistical analyses). Section C addresses special topics as, for example, information about effects of a change. Section D asks for an evaluation of the survey.

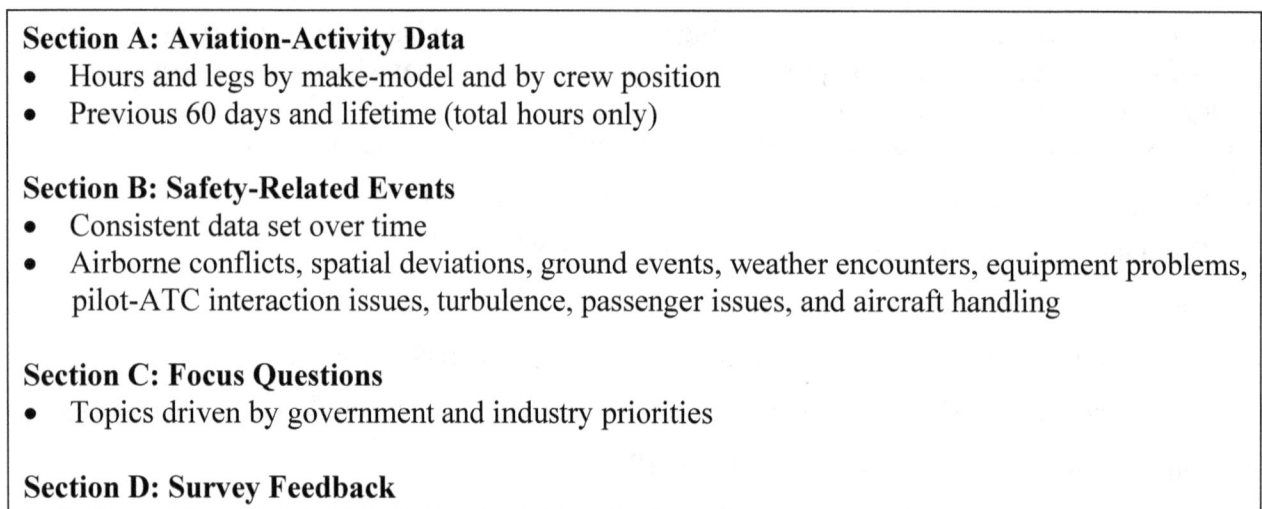

Section A: Aviation-Activity Data
- Hours and legs by make-model and by crew position
- Previous 60 days and lifetime (total hours only)

Section B: Safety-Related Events
- Consistent data set over time
- Airborne conflicts, spatial deviations, ground events, weather encounters, equipment problems, pilot-ATC interaction issues, turbulence, passenger issues, and aircraft handling

Section C: Focus Questions
- Topics driven by government and industry priorities

Section D: Survey Feedback

Figure 5.2.2. The structure of the NAOMS survey instrument.

A field trial of NAOMS was conducted in 2000 followed by a successful program launch in 2001. The initial respondent group was air-carrier pilots who were chosen randomly from the FAA Airman's Medical Database.[7] The aviation community's response to NAOMS was enthusiastic. During about 4 years of test and evaluation, almost 30,000 air-carrier-pilot interviews were completed

---

[7] The survey was later extended to General Aviation (GA) pilots and a brief experiment was conducted in which 4200 interviews were completed. Sixty-nine percent of the GA pilots contacted completed the interview. Resource limitations caused the suspension of the GA-survey effort about six months after it was begun.

before interviewing stopped in early FY05. NAOMS achieved exceptional response rates—83% of those contacted completed their interview, which is well above what is considered the "gold standard" for responses to surveys. This level of response was crucial to achieving statistical validity.

Figures 5.2.3 through 5.2.9 are examples of a few of the results after 2 years of the survey of commercial airline pilots and about 15,000 interviews. Figures 5.2.3 through 5.2.6 are examples of responses to the types of questions asked under section A: Demographics, the results of which are frequently used in calculating rates of reported events.

Respondent Flight Experience	Mean Value
Total Life-Time Flight Hours	10,094 hours
Last 60 Days Flight Hours	97.8 hours
Last 60 Days Departures	37 Departures

Figure 5.2.3. Section A: Demographics – Flight experience.

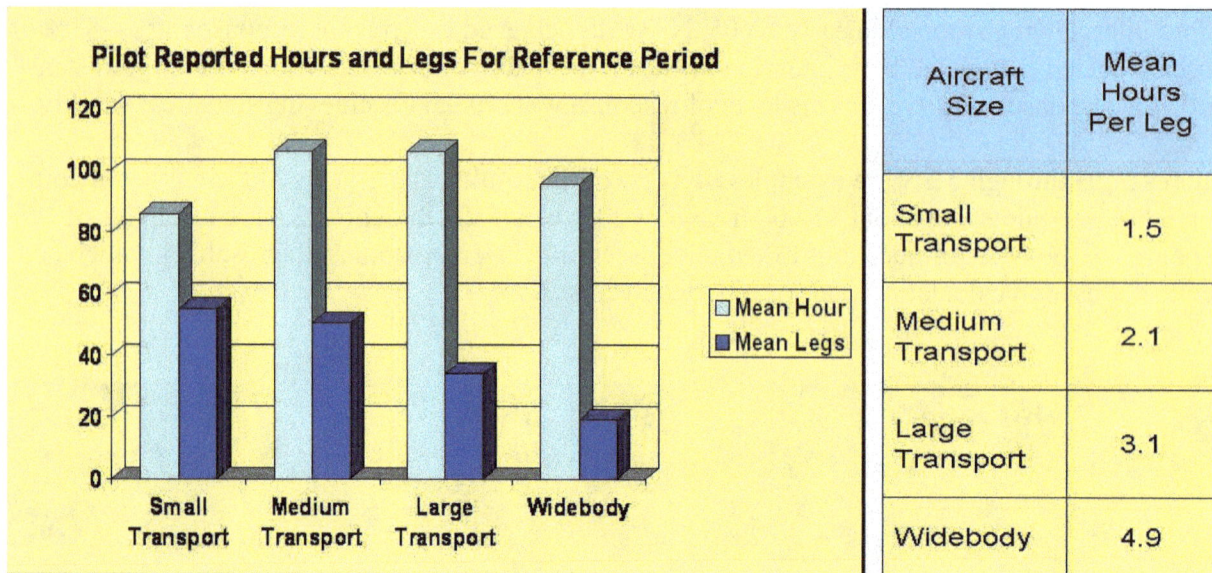

Pilot Reported Hours and Legs For Reference Period

Aircraft Size	Mean Hours Per Leg
Small Transport	1.5
Medium Transport	2.1
Large Transport	3.1
Widebody	4.9

- Small Transport < 100 k lbs GTOW
- Medium Transport ≥ 100 k lbs and < 200 k lbs GTOW
- Large Transport > 200 k lbs GTOW with single aisle
- Widebody > 300k lbs GTOW with two aisles

Figure 5.2.4. Section A: Demographics – Hours and legs by aircraft size.

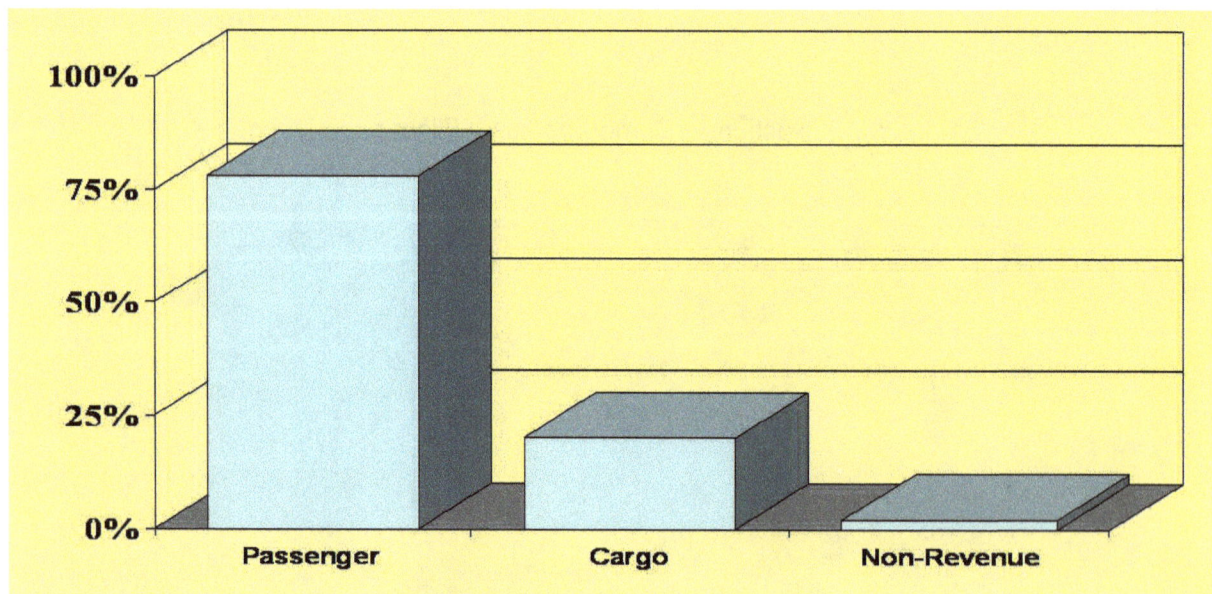

Figure 5.2.5. Section A: Demographics – Type of flight.

Figure 5.2.6 shows a comparison between the estimates of flight time per leg based on the NAOMS survey with the estimates reported at that time by the Bureau of Transportation Statistics (BST).

Figure 5.2.7 is a sample of a result from section B queries and compares pre- to post-9/11.

Aircraft Category	Estimate Source	Mean Hours Per Leg
Small Transport	NAOMS	1.5
	BTS	1.3
Medium Transport	NAOMS	2.1
	BTS	1.9
Large Transport	NAOMS	3.1
	BTS	2.9
Widebody	NAOMS	4.9
	BTS	5.3

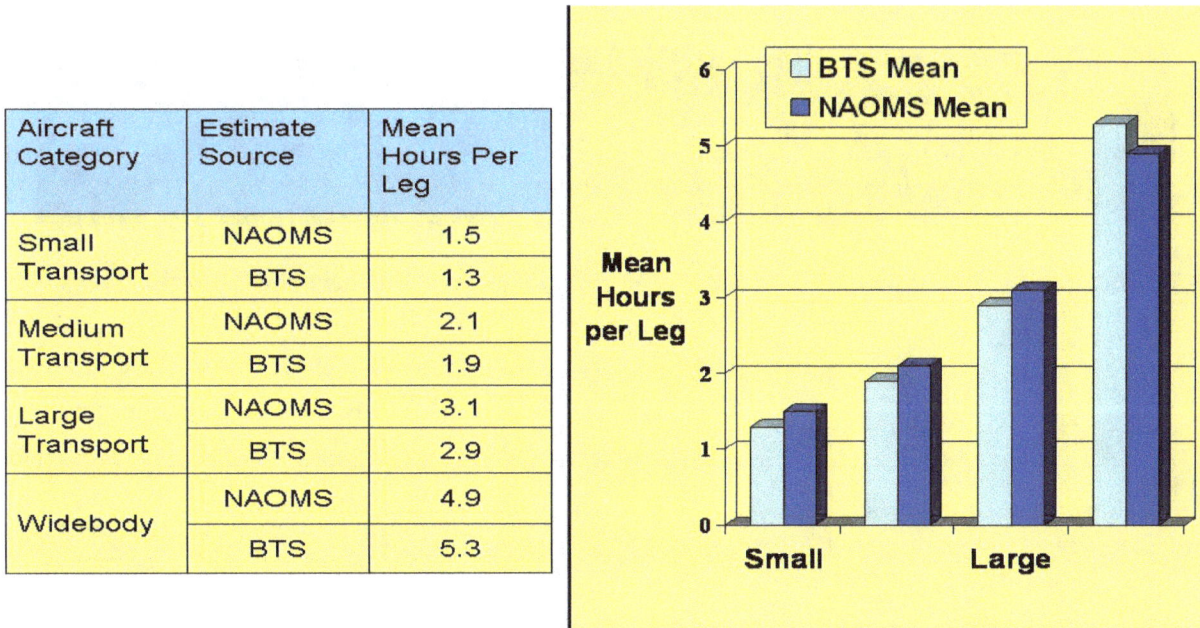

Figure 5.2.6. Flight-time-per-leg estimates – NAOMS compared with BTS data.

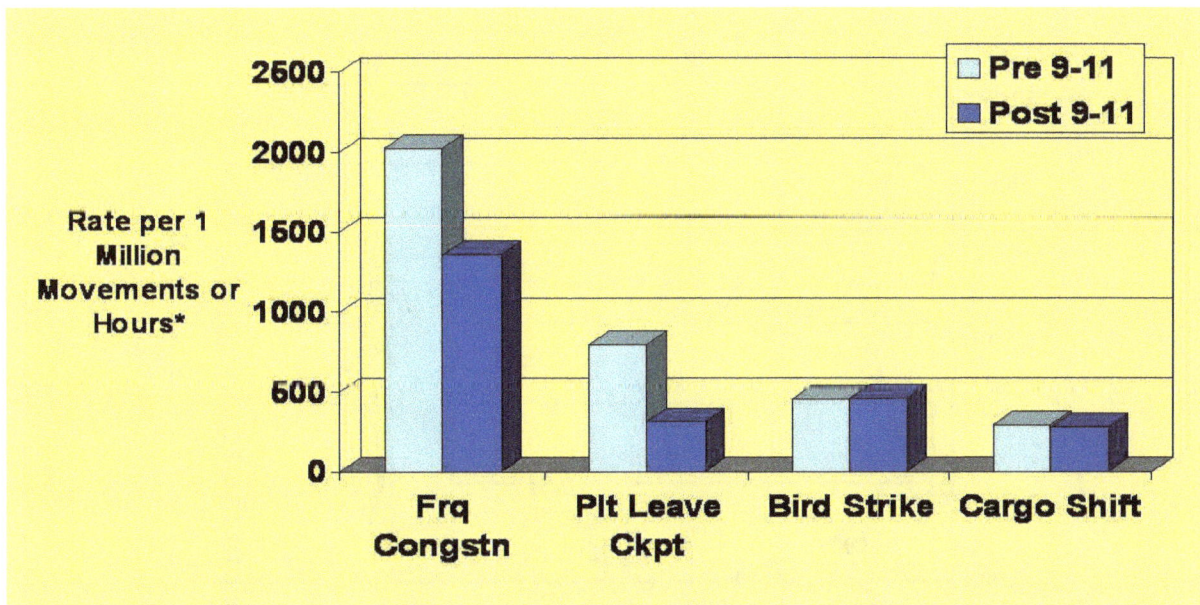

* Rate for Bird strikes is calculated for each departure.

Figure 5.2.7. Section B: Safety-related events – Evaluation of sample events.

Figure 5.2.8 shows some of the data obtained by NAOMS in support of the special study of In-Close Approach Changes (ICACs) that is reported in appendix B.

Figure 5.2.9 is an example of the response to a question in section D in which the respondents showed they had a high degree of confidence in over 90% of their answers.

	Approaches Flown	Percentage of Approaches Flown	Extrapolated Annual Events	Comment
Total Approaches Flown	296,165	100.00	8,000,000	Estimated
Total Number of ICAC Requested by ATC	17,943	6.0	484,675	Estimated
Total Number Accepted by Pilots	16,802	5.7	453,855	Estimated
Total Number of ICAC Approaches with Issues	1,083	0.4	29,254	Estimated

Figure 5.2.8. Section C: Number of ICACs requested by ATC.

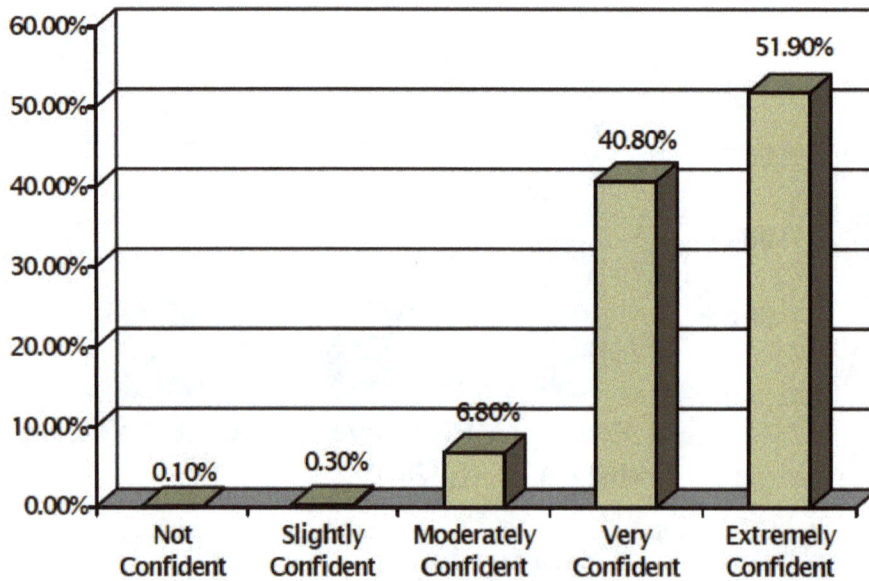

Figure 5.2.9. Section D: Confidence in accuracy of answers.

The results of the survey on air-carrier pilots demonstrated the potential value of NAOMS if implemented as a permanent service. Although it remained work in progress, NAOMS had demonstrated the potential ability to develop statistically defensible estimates of a wide range of safety occurrences and to track their long-term trends. This potential was demonstrated, for example, by the contributions from NAOMS to the ASMM study of ICACs mentioned in section 4.0 and described in appendix B. However, public release (and, therefore, presentation in this report) of all the results of the surveys that clearly show this potential ability was limited because their interpretation and operational significance required domain expertise. NASA turned to the CAST for this help, and the results were offered to that body for use in their deliberations.

The CAST has a critical need for safety metrics to validate that the industry's safety investment in the interventions recommended by the CAST has been well spent and that the risk-reduction assumptions are valid. The NAOMS survey is one of the few indicators currently available to provide insight of whether they are on track or that adjustments are needed. The NAOMS team worked with the CAST in response to a request from the CAST's Joint Implementation Data Analysis Team (JIMDAT), and showed that NAOMS can be used to assess the effectiveness of safety interventions.

The CAST determined that section C of the NAOMS survey could be used to access information about several of the safety enhancements that were particularly related to the pilot community. The JIMDAT selected Procedures and Training as the desired focus for the initial use of NAOMS. A subteam of JIMDAT members worked in collaboration with the NAOMS team to construct a questionnaire for section C that would address the specific issues identified by the JIMDAT. NAOMS conducted interviews for 2.5 months of air-carrier pilots. The completion rate was a phenomenal 92%, with 1150 completed interviews using 56 main questions covering a variety of issues, including air-carrier reactions to go-arounds, unstabilized approaches, and ASAP reporting. The results were generally encouraging of the CAST's interventions to improve safety.

Figures 5.2.10 (a) and (b), showing the responses to some of the questions pertaining to go-arounds, are examples of the many results of the study that was conducted in collaboration with a team from the JIMDAT.

JIMDAT SECTION C RESULTS: Safety Reporting				
Question Number	Question	Response	Response Value	Response %
JD25	Does your airline have a no-fault missed approach or go-around policy?	Yes	1076	94%
		No	55	5%
		Refused	0	0%
		Don't know	17	1%
		Total	1148	100%
If Yes, Refused or Don't Know, Skip to JD26				
→»JD25a	Would you favor the institution of such a policy, oppose it, or neither favor nor oppose it?	Favor	35	64%
		Oppose	6	11%
		Neither	14	25%
		Refused	0	0%
		Don't know	0	0%
		Total	55	100%
JD26	During the last 60 days did you perform a missed approach or go around?	Yes	193	17%
		No	955	83%
		Refused	0	0%
		Don't know	0	0%
		Total	1148	100%
If No, Refused or Don't Know, Skip to JD27				

Figure 5.2.10(a). Example of responses to NAOMS survey in support of the JIMDAT.

Question Number	Question	Response	Response Value	Response %
→»JD26a	Did you receive any feedback from your airline regarding this missed approach or go around?	Yes	8	4%
		No	185	96%
		Refused	0	0%
		Don't know	0	0%
		Total	193	100%
If No, Refused or Don't Know, Skip to JD27				
→»JD26b	Was that feedback positive, negative, or both positive and negative?	Positive	7	87%
		Negative	0	0%
		Neither	1	13%
		Refused	0	0%
		Don't know	0	0%
		Total	8	100%
JD27	Does your airline participate in the safety reporting program called ASAP also known as the Aviation Safety Action Program?	Yes	992	86%
		No	96	9%
		Refused	0	0%
		Don't know	60	5%
		Total	1148	100%
If No, Refused or Don't Know, Skip to JD28				

Figure 5.2.10(b). Continuation of example of responses to NAOMS survey in support of the JIMDAT.

The pilot community is aware of and supportive of company procedures and policies, they are generally positive about the training they are receiving, and there is general acceptance of innovations such as the FOQA and ASAP. NAOMS was able to identify several specific areas where there was room for improvement, and indications were that pilots from airlines with large fleet sizes responded more positively than pilots from airlines with smaller fleet sizes. NAOMS has demonstrated its standalone capability to provide insight into the effectiveness of safety enhancements in a system-wide and timely manner. As ASMM ended, the NAOMS team was working with CAST and with the ALPA to continue the NAOMS service for air-carrier pilots by transferring responsibility of the survey of Part 121 pilots to ALPA.

The goal of NAOMS was to create a new national permanent service. The structure and methodologies developed have proven their potential as a basis for such a service to:

- routinely measure the safety of the NATS in a quantitatively precise way,

- assess trends in the performance of the NATS that may have implications of compromised safety and to identify the factors driving those trends, and

- identify safety and efficiency effects of new technologies and/or procedures as they are inserted into the operating environment.

NAOMS established and demonstrated a quantitative statistically defensible systemwide safety assessment tool that is the basis for such a national service. The statistical validity of the results obtained were ensured by the careful design of the methodology and demonstrated by comparison with the few other sources of similar data such as the one shown in figure 5.2.6. However, it was possible during the life of the ASMM Project to apply NAOMS only to the pilot community that was represented by results such as the examples shown in figures 5.2.3 through 5.2.10. The approach needs to be adapted to the other constituents, including, in particular, the air-traffic controllers as well as mechanics and cabin crews. Until these constituents have been included, NAOMS will not be able to present a full and unbiased perspective of the system performance.

In summary, the *Extramural Monitoring* element of the ASMM Project produced the following capabilities:

- The Incident-Reporting Enhancement tools – a suite of new capabilities designed primarily for the ASRS, but applicable to any incident-reporting database that includes:

  - ASRS database migration to ORACLE 8,

  - Electronic report submission,

  - The Analyst Workbench, and

  - Improved ASRS search and analysis capability.

- NAOMS – a quantitative statistically defensible, systemwide safety assessment tool that is the basis for a permanent national service.

5.3 Intramural Monitoring

The *Intramural Monitoring* element of the ASMM Project entailed a "bottom-up" approach that was intended to provide air-service operators with the tools needed to monitor their performance continuously, effectively, and economically within their own organizations. The primary products of this activity were the APMS for processing aircraft flight-recorder data, and the PDARS for processing ATC data.

5.3.1 Aviation-Performance Measuring System

APMS was a research and development program to support and enhance the industry's program for analyzing aircraft flight data called the FOQA program (see appendix A). Advanced tools would reduce the labor necessary to extract information from the very large volume of data generated by these programs and increase the information that could be extracted relative to then-existing commercially available software. NASA would also take on developmental risks of advanced capabilities that, without a demonstrated market, had too much risk for commercial vendor development. Under these guidelines, government research focused on the development of advanced statistical techniques, the limits of data storage, the limits of processing speed, and the limits of the level of data-quality filtering available for FOQA at the beginning of APMS.

APMS was initiated in 1993 in response to a request from the FAA to provide capabilities that would minimize the costs of operating a FOQA program and, thereby, encourage U.S. air carriers to undertake the investment. The value of continuously monitoring and analyzing flight-recorded data had been demonstrated by several European air carriers. In 1997, the FAA withdrew its funding of the APMS, and NASA supported the continuing work until FY00, when it became part of the ASMM Project of the AvSP. Throughout the entire time, APMS was an activity to both develop advanced software tools and provide a toolbox of developed capabilities to minimize the time of human experts. The APMS team built tools responsive to user needs and pushed the limits of processing technology. The prototypical tools resulting from *Data Analysis Tools Development* were operationally tested and evaluated under *Intramural Monitoring* in collaboration with air carriers, the FAA, and vendors of current data-analysis software with whom NASA entered into SAAs.

The tools developed by APMS fit two general categories, those developed early in the program that were *pulled* by stated airline-user needs and those that grew from a NASA *push* for advanced technology. Following are the tools in the first category, developed between 1994 and 2001, that have become industry standards.

The Event Processing System (EPS) (figure 5.3.1) was designed to solve the workflow management problem inherent to FOQA data processing that had been identified during user-needs studies with air-carrier personnel. As flights are downloaded each day, commercial FOQA software searches for and identifies predefined exceedances. As APMS began, managing the review, validation, tracking, and reporting of exceedances, rates, locations, etc., was a significant, labor-intensive challenge, especially because the analyses were performed by a team of line pilots working individually as they had free time. The EPS assists the FOQA analyst in analysis, decision-making, decision-tracking, reporting, communicating, and recording for future reference information related to flight exceedances. The EPS gives the analysts an organized interactive to-do list of each flight and its exceedances and enables them to review each flight and validate, invalidate, defer, and/or comment upon the exceedance. It provides record-keeping tools for evaluations and recommended actions for each event, and enters the processed events into a database for subsequent reporting.

The Graphics Viewer (figure 5.3.1) assists the FOQA analyst in interpreting and verifying exceedances or special events by displaying the data of a specific flight in great detail for any selected period of time surrounding the event, from a few minutes to the entire flight. Though any commercial FOQA software provides a viewer, APMS pioneered enhancements such as single-click links to flight animation and parameter baselines representing the population of flights operating in or out of the same airport, automatic display of parameter sets relevant to each exceedance, and automatic location of exceedances and detected patterns within a flight. As APMS evolved, this tool became the focal point for advances related to examining individual flights because, inevitably, understanding causal factors of any event entails examination of individual flights. Advanced tools link back to the viewer, and from the viewer an analyst can access weather information or display data in an air-traffic format.

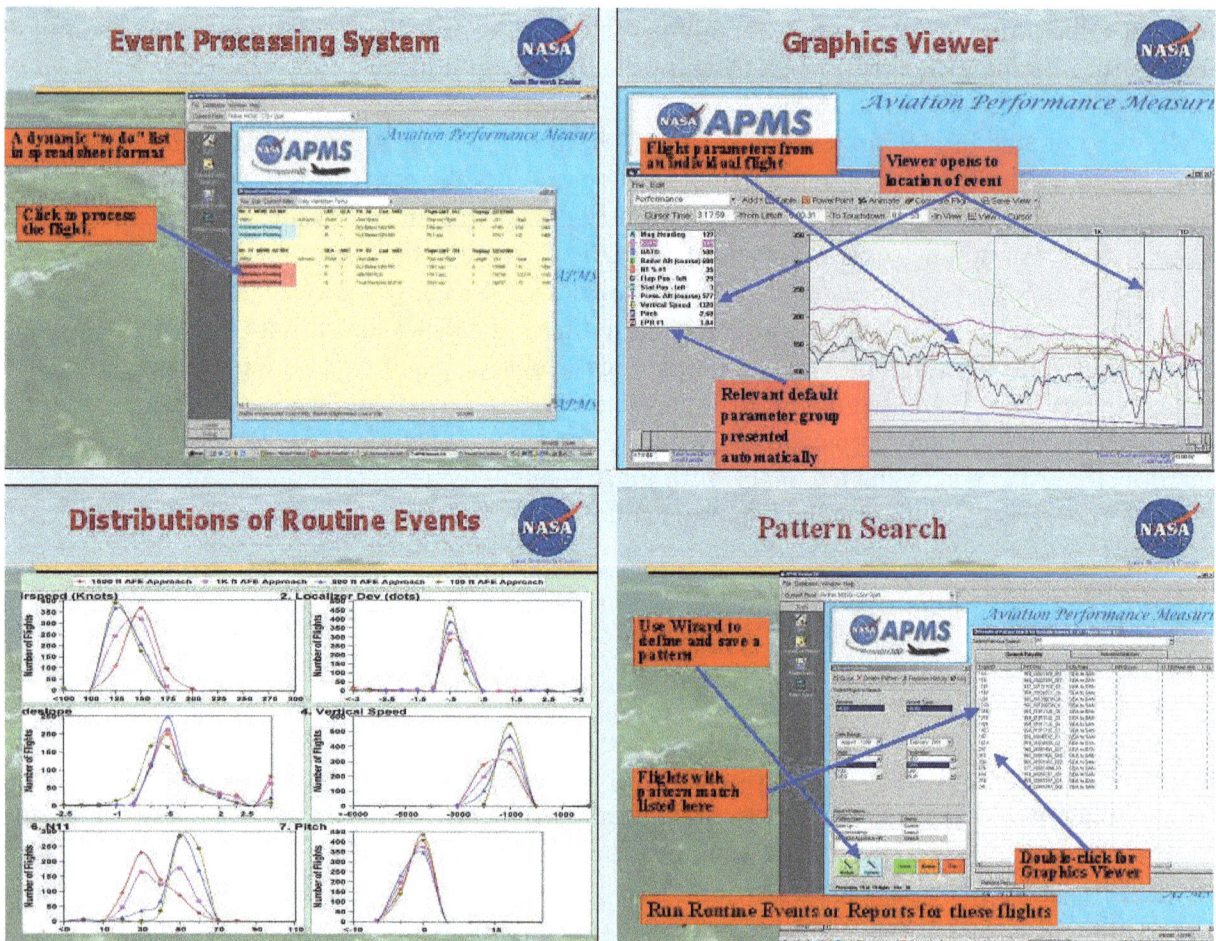

Figure 5.3.1. APMS concepts that have become industry standards.

The Routine Events tool (figure 5.3.1) presents the distribution of flight parameters around selected snapshots in time at "check points" that occur within each flight and are tied to standard operating procedures, giving the FOQA analyst an immediate and obvious point of comparison of performance observed in line operations to the standards specified in operator procedures. The analyst or team selects any set of standard events, for example 1000 feet above touchdown on approach or a flap setting during approach, and the tool displays the distribution of any selected set of parameters for any selected subset of flights at that selected "check point." For example, the tool enables monitoring for compliance with stabilized approach procedures by fleet or destination by displaying airspeed (relative to Vref), vertical speed, engine power settings, flap and gear position, and localizer and glide-slope deviations at 10,000 feet above touchdown. Where compliance is good, the distributions would be tight and centered on reference speed plus additives, a descent rate of about 800 feet per minute, approach thrust setting, fully configured flaps, gear down, and on localizer and glide slope. Where compliance is poor, the variances would be greater, central tendencies skewed, and extreme values more populated. This tool has become a backbone for studies of specific issues, and has been generalized for new derived parameters, such as energy indexing.

The Pattern Search tool (figure 5.3.1) was developed as a way of reviewing any sample or population of flight-recorded data for the occurrence of newly defined events. The user specifies the criteria defining an event, such as an unstable approach, and the search of the selected dataset provides a list of flights fitting the pattern. In a sense, this scenario is similar to finding exceedances, except for the ability to specify any pattern of any number of parameters or any sequence of parameter events. The tool displays all flights fitting the pattern, and makes them available to other tools, such as the viewer, routine events, animation, and airspace display. Pattern Search requires the storage of, and access to, all relevant data in order to provide the flexibility of searching throughout the dataset, whereas exceedance-based FOQA tools typically limit the data storage to about two minutes on either side of an exceedance. A key contribution of APMS was the push to retain all the downloaded data, not just scan for exceedances and discard all but a few minutes of data.

The tools described in the following paragraphs were developed between 2001 and 2005 and grew from a NASA *push* for advanced technology, including the ADIS, The Morning Report of Atypical Flights, and the Aircraft Energy-State Indexing and Display tools.

The ADIS (figure 5.3.2) was developed to allow access to other sources of information about a flight of interest without re-identifying the flight to an analyst. FOQA data are de-identified data by negotiated agreements among pilots and their representatives, airlines, and the FAA. Without a de-identified link to these other informational sources, they would be accessed very rarely, and only after a monitoring team agreed it was necessary. ADIS allows the linkage of a flight to the weather or air-traffic data at any time within the flight, while screening from the display the time and date information that would re-identify the flight to the analyst.

ADIS enables the analyst to routinely access any information the airline and other entities are willing to make available for analysis. Industry has already implemented easy access to flight animation and to approach charts, so NASA pursued the more difficult ones: weather and air-traffic radar data. A linkage of a de-identified flight to the related ASAP report has not been demonstrated but is viable.

Integration of information has been pursued to facilitate causal analysis and risk assessment of a flight identified as atypical or having an exceedance by providing an analyst with the capability to examine conditions under which events occur, particularly the effects of weather and air traffic. For example, a deviation from glide slope in daylight visual conditions may be trivial (automated guidance may or may not be in use while a pilot navigates by sight to the runway) or the deviation may be purposeful to avoid wake turbulence from a preceding aircraft. A similar deviation in instrument conditions, when runway and terrain features are not visible, may represent a significant risk. Applications of ADIS will facilitate understanding the consequences of weather and traffic situations on flight operations.

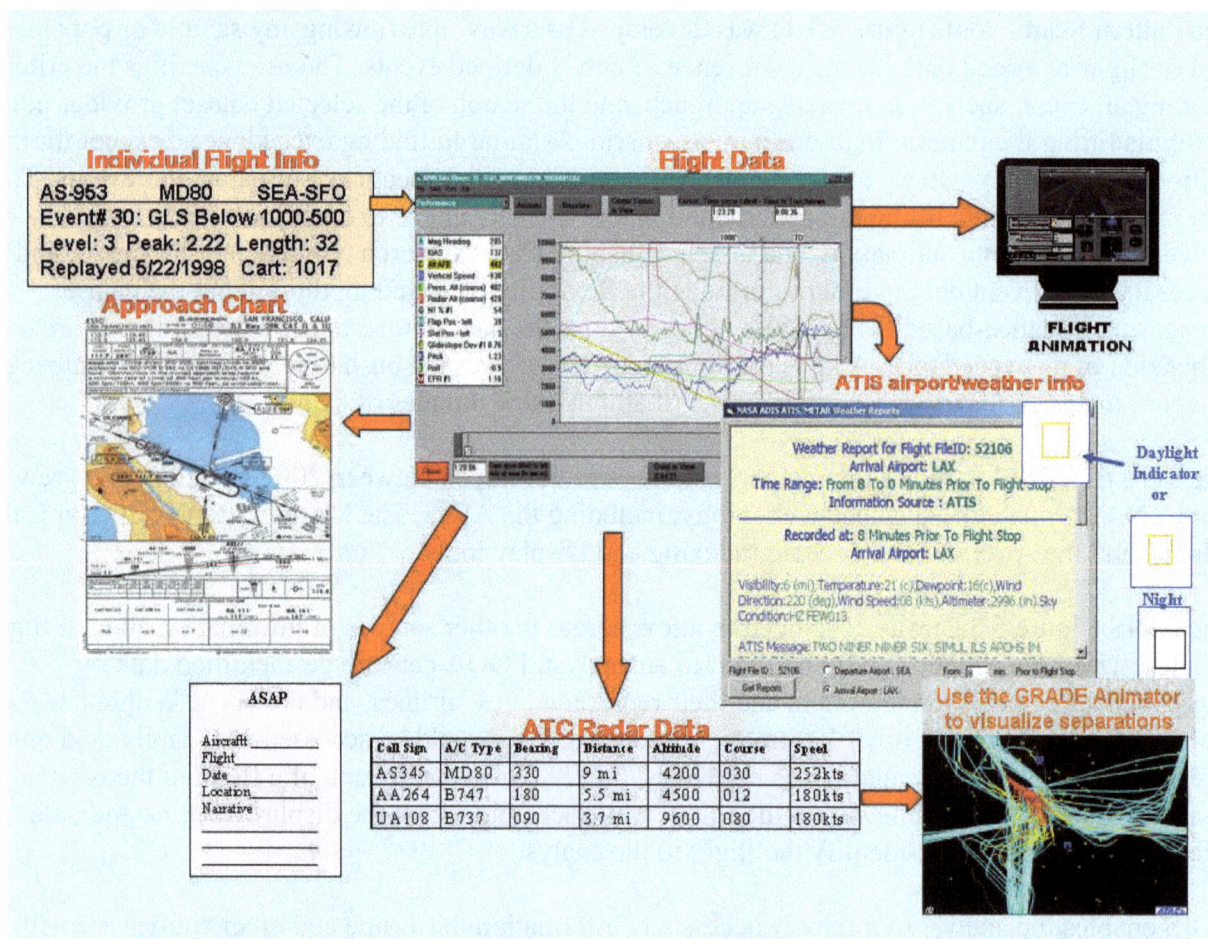

Figure 5.3.2. The concept of the Aviation Data Integration System.

ADIS started an archive of weather information in the form of digital Air-Traffic Information Server (ATIS), Meteorological Terminal Aviation Routine Weather Report (METAR), Runway Visual Range (RVR), and light conditions for use in retrospective analyses. Figure 5.3.3 shows screenshots of the ADIS tool for accessing weather information to relate to flight-recorded (FOQA) data. The analyst makes selections on the left and receives the information on the right. ATIS is presented for those airports where digital ATIS is used and includes NOTAMS. METAR information is displayed for all other airports. At the conclusion of the ASMM Project, the weather-information archive was transferred to the FAA, which established a permanent archive service to all air carriers with FOQA programs. Also, the technology for linking to the weather information while maintaining the de-identification of the flight was demonstrated in operational evaluations and was licensed to one FOQA vendor.

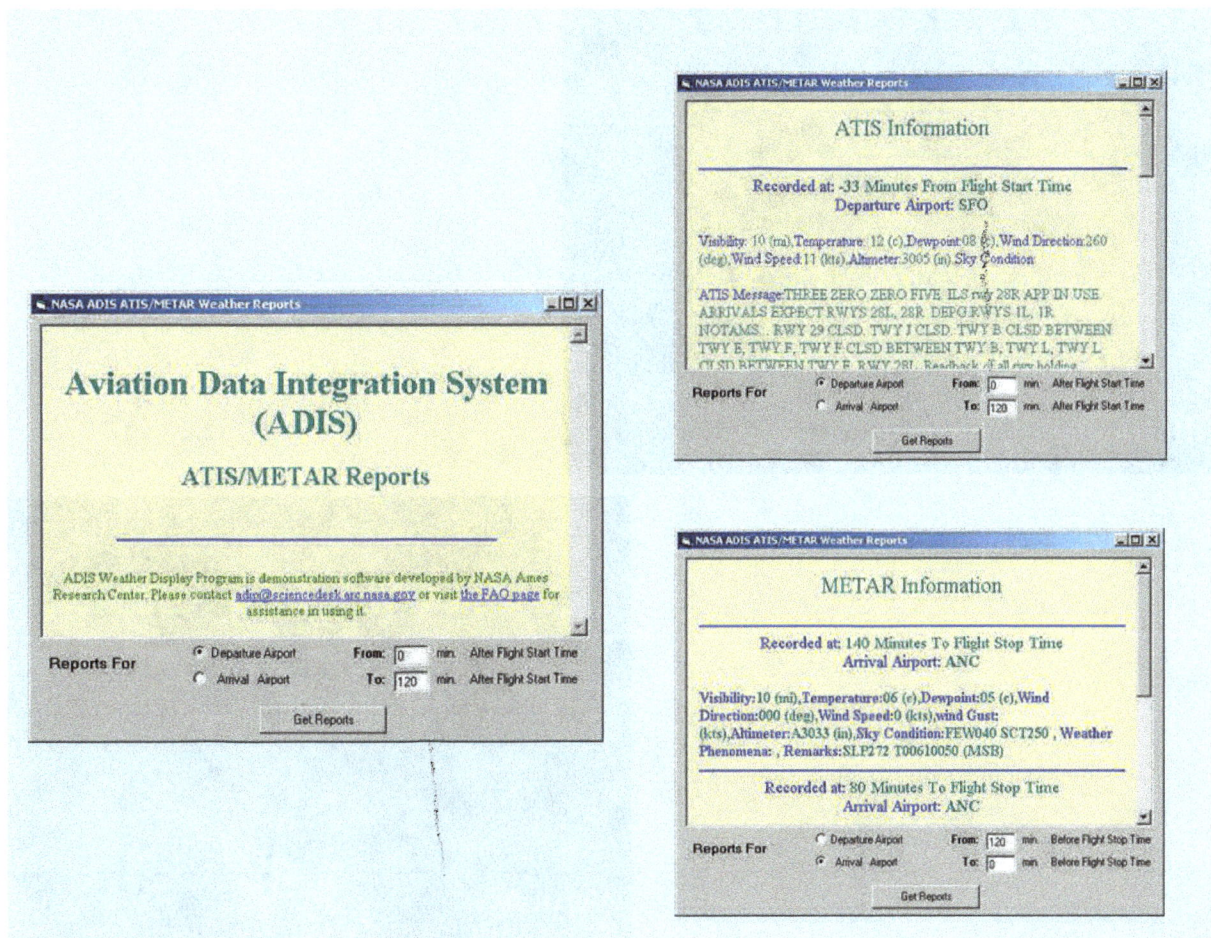

Figure 5.3.3. Weather data (implemented).

Automated linkage of flight-recorded data to air-traffic radar-track data generated by the PDARS program was demonstrated using ADIS and the Graphical Airspace Design Environment (GRADE) tool, under license from ATAC Corporation, the FAA's primary contractor for PDARS. Figure 5.3.4 shows how flight data can be displayed along with associated traffic while maintaining dc-identification of all data. The track in white is the flight of interest (in this case because of high energy state). The green tracks are of all other flights arriving at the airport in the same hour. The upper left display shows the track laterally, the lower right vertically, illustrating the high energy state of the flight relative to other arrivals. The use of this capability will be described with regard to a specific study on high-energy arrivals later in this section.

The operational evaluation of air-traffic integration was more limited than the weather integration because access to and use of ATC data was more constrained. Unlike the weather information, this capability was not implemented as an ongoing service.

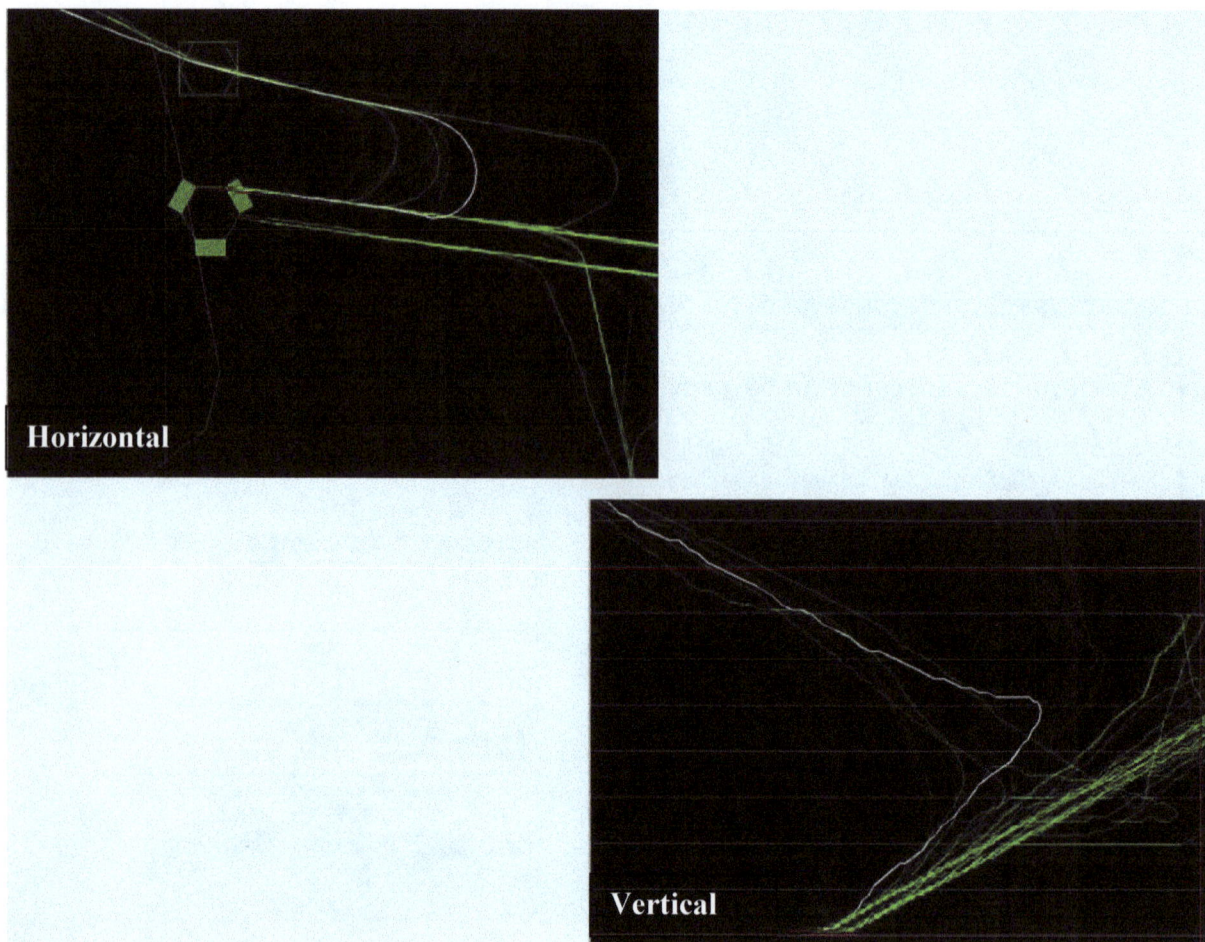

Figure 5.3.4. Example of display of air-traffic data in GRADE.

The Morning Report of Atypical Flights discovers unenvisioned circumstances by searching for flights that are multivariate-statistically extreme within a flight phase, compared to other flights in similar circumstances. It is based on the PROFILER that was described in section 5.1, *Data Analysis Tools Development*. The PROFILER uses multivariate cluster analysis to group flights by similarity along flight signatures derived from parameter values, calculates an atypicality score (figure 5.1.1) for each flight, and provides a plain-language description of what makes targeted flights atypical; these data are then displayed in The Morning Report of Atypical Flights. Atypical flights are different from exceedances where flights have violated limits that have been predefined by airlines based on standard operating procedure (SOP), and the supposition is that airlines know what to look for and have correctly identified all relevant parameters. The Morning Report is a tool for searching for unknown unknowns. The implementation of The Morning Report display is similar to that of exceedance-based FOQA analysis in that it identifies individual flights worthy of attention (Chidester (2003)).

Analysis using The Morning Report tool starts with two simplifying provisions. First, below 18,000 feet, flights are compared only to other flights (of the same make/model aircraft) departing from or arriving at the same airport. Above 18,000 feet, flights are compared to all other flights

68

(actually the most recent 1000 flights of the same make/model aircraft). Second, any and all recorded parameters can be used in the multivariate analysis and in calculating an atypicality score. However, only those parameters that are likely to be operationally significant are selected for analysis so as to minimize the possibility of discovering trivial problems. For example, use of anti-ice is so rare in summer months that its operation will make a flight highly atypical, but not worth examining, so this entry would be dropped from the parameter list for that time of year.

The Morning Report calculates the flight signature for each phase of every flight being analyzed. The phases of flight are takeoff, low-speed climb, high-speed climb, cruise, high-speed descent, low-speed descent, final approach, and landing. Flight signatures for each phase are defined by compressing the data for each parameter into 18 variables, including initial and final values of the parameter at the start and at the end of the phase, the mean, standard deviation, minimum, and maximum of the value of the parameter, its slope, its acceleration, and the (residual) noise calculated over the entire phase.

The flight signature is a characterization of each phase of flight that satisfies the needs of the analysis for The Morning Report and reduces the computational time and complexity greatly because of the high degree of data compression. However, the temporal relationships among the parameters are lost in this characterization. That information is retained in the APMS in its full-flight data (.ffd) file for use in analysis in which those temporal aspects are significant.

The flight signatures in each phase (18 values per selected parameter) are submitted to a principal component analysis to derive a reduced number of uncorrelated dimensions. The flights are clustered to find similar flights. Each flight is compared to the available similarly situated population (of up to the most recent 1000 flights). The atypicality score is then calculated using the Mahalanobis distance of each flight from the multivariate data centroid, and a weighting using a cluster analysis of the reduced principal components. The larger the Mahalanobis distance from the global centroid and the fewer the number of flights in a cluster the greater the atypicality.

As currently implemented, The Morning Report contrasts the most extreme 20% of flights (defined as atypical) to the least extreme 80% (defined as typical). The "typical" 80% is used for a display of a typical performance envelope for each parameter. Atypical flights are divided into three levels; the most extreme 1% is called level 3, the next most extreme 4% is called level 2, and the next most extreme 15% is called level 1.[8] The Morning Report display identifies the most extreme 5% of flights to the safety analyst, who judges their operational significance. The selection of these bounds is arbitrary. They were chosen as a starting point because they correspond to the current definitions of the levels of significance that are typically assigned to exceedances.

As new flights are uploaded into the APMS each day, The Morning Report display automatically alerts the monitoring team to those flights that had level-2 or -3 events. Plain-language text describes the parameters and aspects of those parameters that caused each of those flights to be identified as atypical. It is important to recognize that by searching for multivariate-statistical extremity, The Morning Report will not detect unsafe states that occur often and, therefore, are not atypical.

---

[8] These definitions of the levels are expected to evolve with experience to minimize uninteresting alerts.

The APMS team began testing the first version of The Morning Report in 2003 with three airlines under SAAs. By the end of the ASMM Project in 2005, these airlines had used The Morning Report in processing over 16,000 flights. The APMS team reviewed those flights identified by The Morning Report as having at least one phase in level 3, i.e., the most extreme 1%. The key question was whether The Morning Report alerts were useful and interesting, whether they were redundant to the exceedance-based analysis, or whether they were unique but of no consequence. Interesting results were those flights that the domain experts thought could be operationally significant and deserved further investigation. Uninteresting flights were those that the domain experts could explain easily and were deemed not operationally significant or were the result of bad data. Initially, about half of the alerted flights were found to be interesting by the first two airlines.

Improvements in data-quality filtering were introduced and parameters that were frequently causing trivial alerts were eliminated from analysis. One example of such a trivial alert was caused by engine parameters, each of which might be only slightly atypical but would outweigh all the other parameters by their sheer number in the multivariate statistical analysis. By the time the evaluation was conducted with the third airline with all these changes implemented, 74% of The Morning Report alerts were deemed to be operationally interesting by the airline representatives and deserving of further study.

The Morning Report displays the page shown in figure 5.3.5 when processing is completed. It shows when the report was completed, the number of flights analyzed, and the date range of those flights. It summarizes the number of atypical flights identified and their levels of atypicality. A click on the Go To Flight List button displays the most extreme 5% of flights (level 2 and 3 atypicalities).

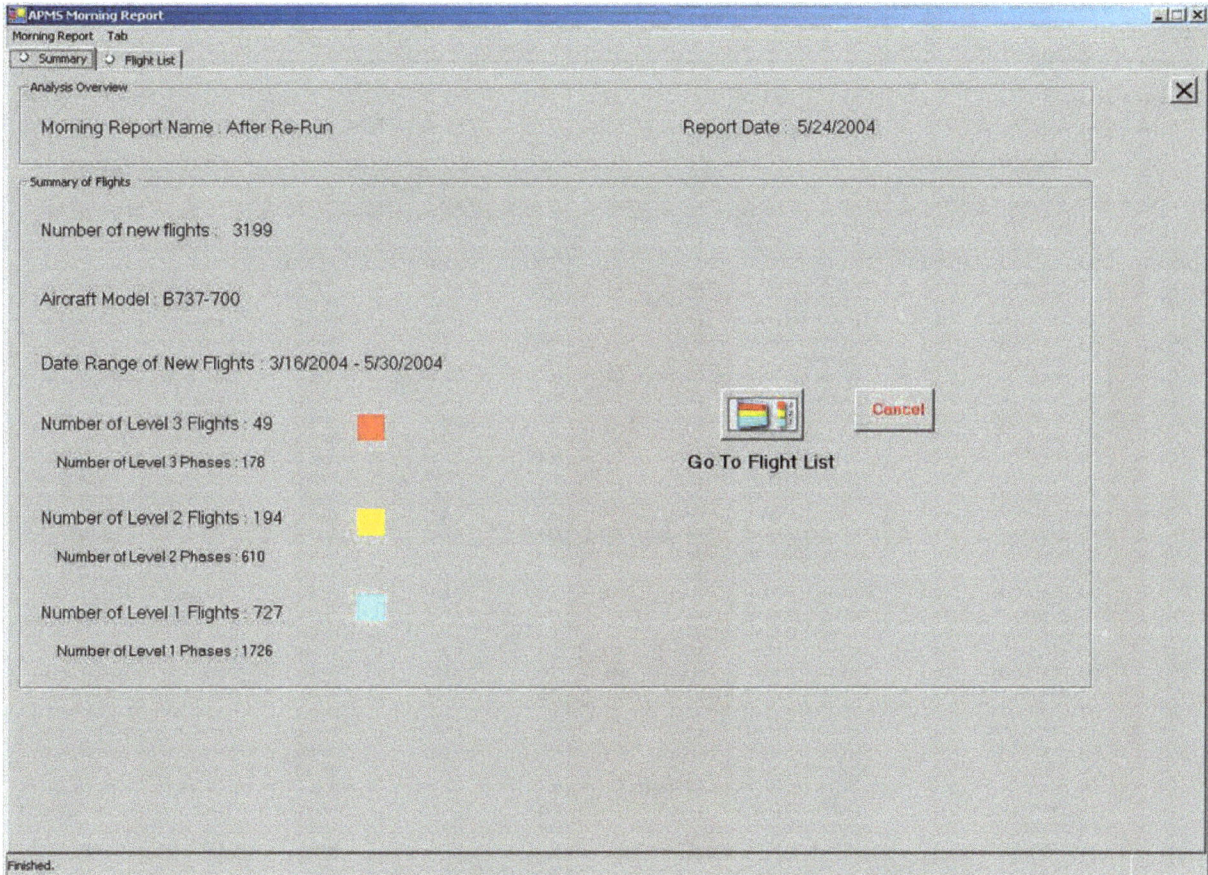

Figure 5.3.5. The opening (executive-level) screen of The Morning Report.

The page that displays the list of flights that had level 2 and 3 atypicalities is shown in figure 5.3.6. It includes information for finding the flight in the database for further analysis, along with the (encrypted) aircraft tail number to assist the analyst in determining whether the atypicality of any aircraft is in performance or in the data recording of that particular aircraft. By selecting a single flight on this screen and clicking on the "Explore Flight" button, a screen like figure 5.3.7 appears that shows the trace of the atypical flight parameter for the atypical phase and compares it to a typical performance envelope of that parameter. There is an option to display level 1 atypicalities.

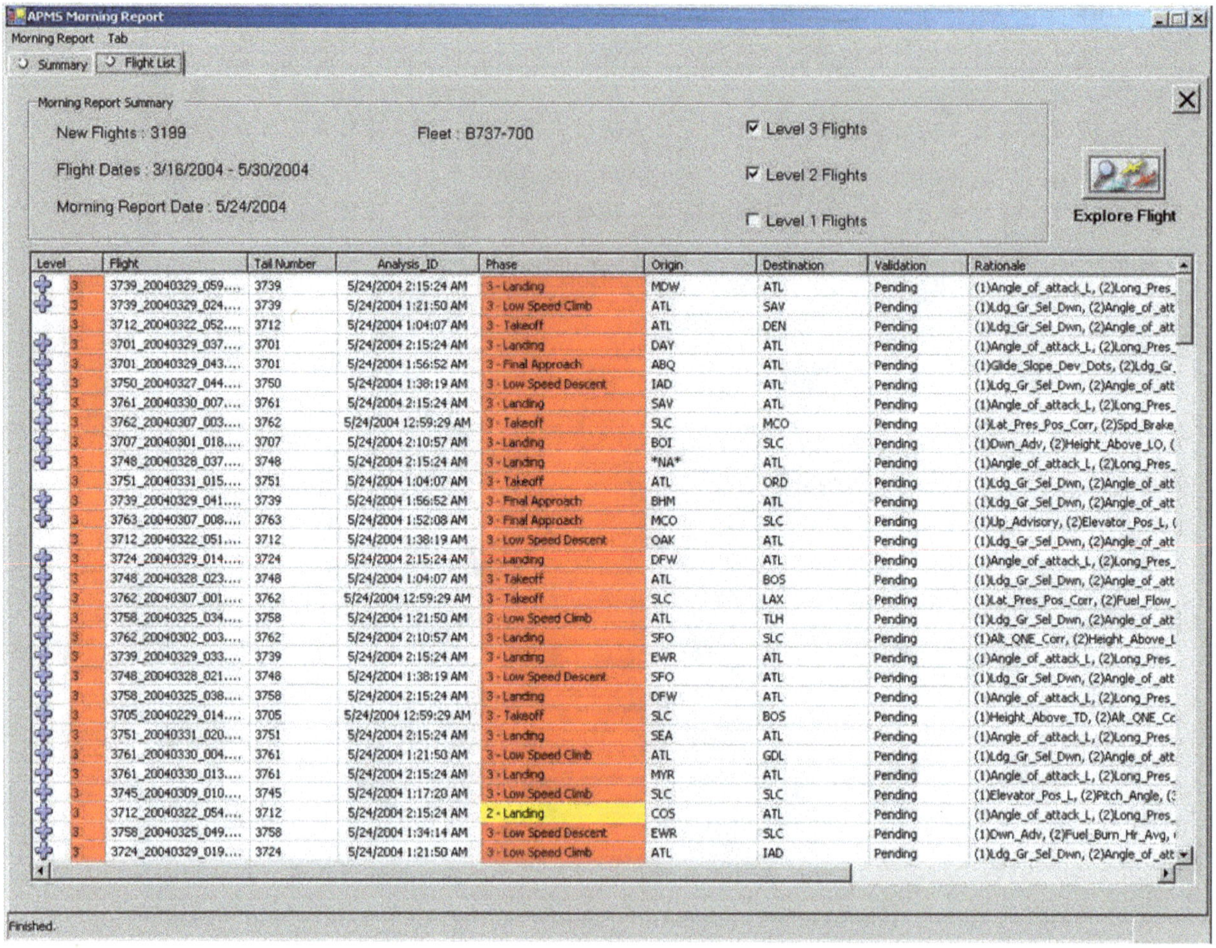

Figure 5.3.6. The Morning Report display of flights with level 2 or 3 atypicalities.

Of the 16,000 flights analyzed by The Morning Report, 3.1% fell within the criterion of having at least one phase in level 3, the most extreme 1% of atypicallity. A few of these level-3 atypical flights were truly unique (i.e., singletons). However, many of the flights that met this criterion could be grouped into similar patterns.

The frequent similar patterns fell into one of the following categories:

- High-energy state during the arrival phase

- Go-arounds

- Landing rollout anomalies (such as unusual braking or use of reverse thrust)

- Atypical climbs (such as leveling off at low altitude following departure)

- Turbulence and accommodation (the former appearing as noise in spatial and motion-related parameters, the latter in reduced airspeed for the flight phase)

- Unusual arrival paths (such as very short downwind, base, or final approach segments)

- Traffic Collision Alert System (TCAS) resolution advisories accompanied by escape maneuvers

72

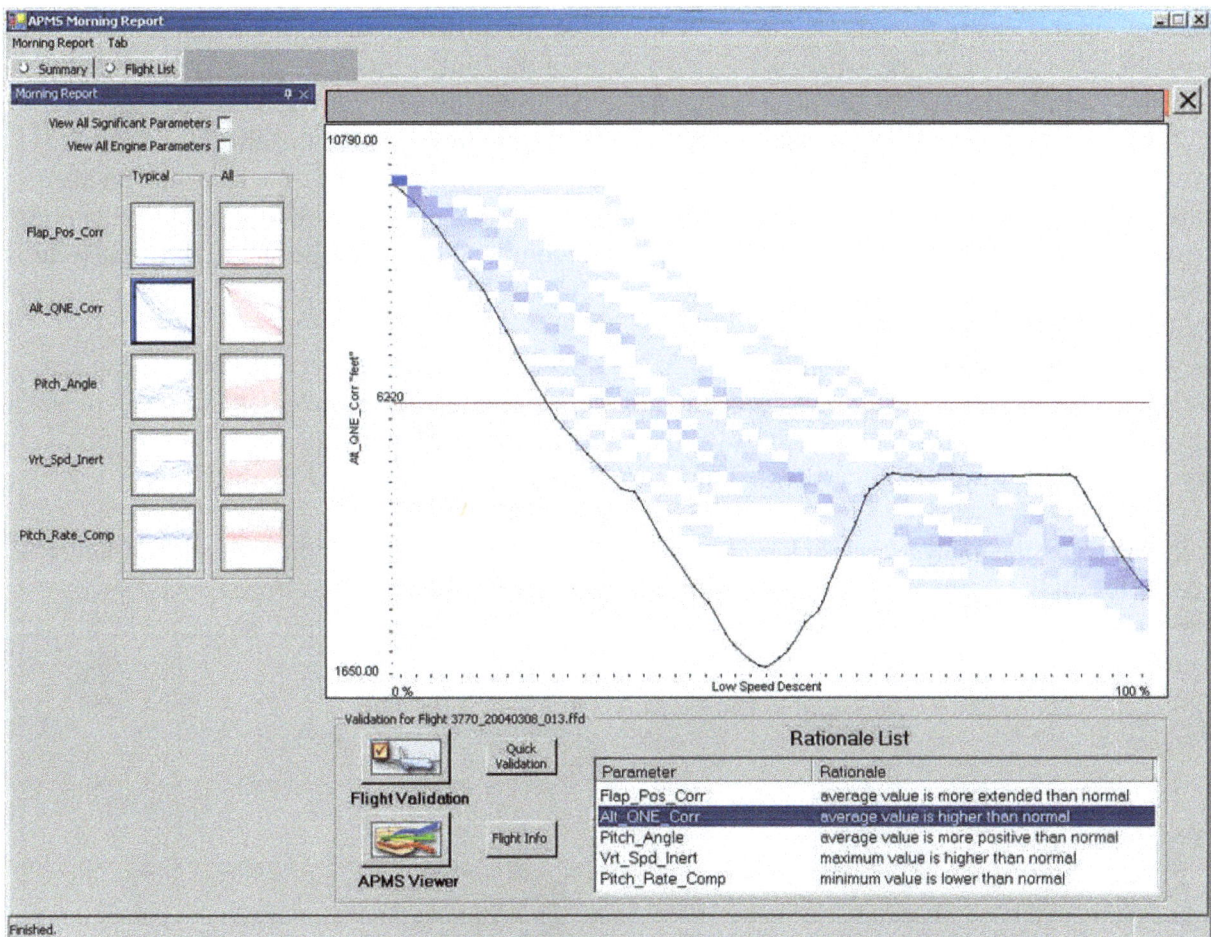

Figure 5.3.7. The Morning Report display of atypical vs. typical parameters.

Truly unique events included a speed-recovery maneuver observed in the arrival terminal area (slow speed in clean configuration followed by decreased pitch and increased thrust but no stall warning) and a firm landing. Neither the maneuver nor landing events was discovered by the current FOQA exceedance analysis. Importantly, none of these events that were empirically discovered by The Morning Report had been identified by any previous analyses.

Figure 5.3.7 shows a comparison of the parameters of a selected atypical flight to those of typical flights or to those of all flights during the same phase. On the left side is a listing of five specific parameters that caused the flight to be identified as atypical by The Morning Report. To the right of each of these are thumbnails of comparisons of each parameter to "Typical" and "All" performance envelopes. The thumbnails under the heading "Typical" are the comparisons to the same parameters in the 80% most typical flights in the same phase. Those under the heading "All" adds the remaining 20% atypical flights. The comparison with the "Typical" envelope focuses on how the selected flight is atypical. The comparison with the "All" envelope shows whether this flight is truly alone in its atypicality. In the lower right corner "Rationale List" displays the explanation for having decided that this flight was identified as atypical by The Morning Report.

The flight in this example is atypical because it went around after encountering a high-energy state during the arrival phase. The x-axis is the percentage of the time through the flight phase; in this case "low-speed descent." At the origin (0%) is the first time the aircraft descends through 10,000 feet above mean sea level. At the right end (100%) is the last time the aircraft descends through 2500 feet above field elevation (afe). Flights vary in the length of time to complete each phase, hence the use of percentage of time through the phase is used to normalize and enable comparisons. This flight took much longer than typical to complete the phase because the flight went around; it includes two approach descents.

The y-axis in the example shown in figure 5.3.7 is selected to be altitude in feet above mean sea level. It can be seen that the aircraft descended to a low altitude, then climbed and descended again to join the approach. This "shape" is always the signature of a go-around detected by this tool, validated by the view of this flight presented in figure 5.3.10.

From this page, the analyst can click on the "Flight Validation" button to validate, invalidate, or make notes on the event, or click on the "Flight Information" button to make use of ADIS for weather, or click on the "APMS Viewer" button to explore the flight in detail, as shown in figure 5.3.8.

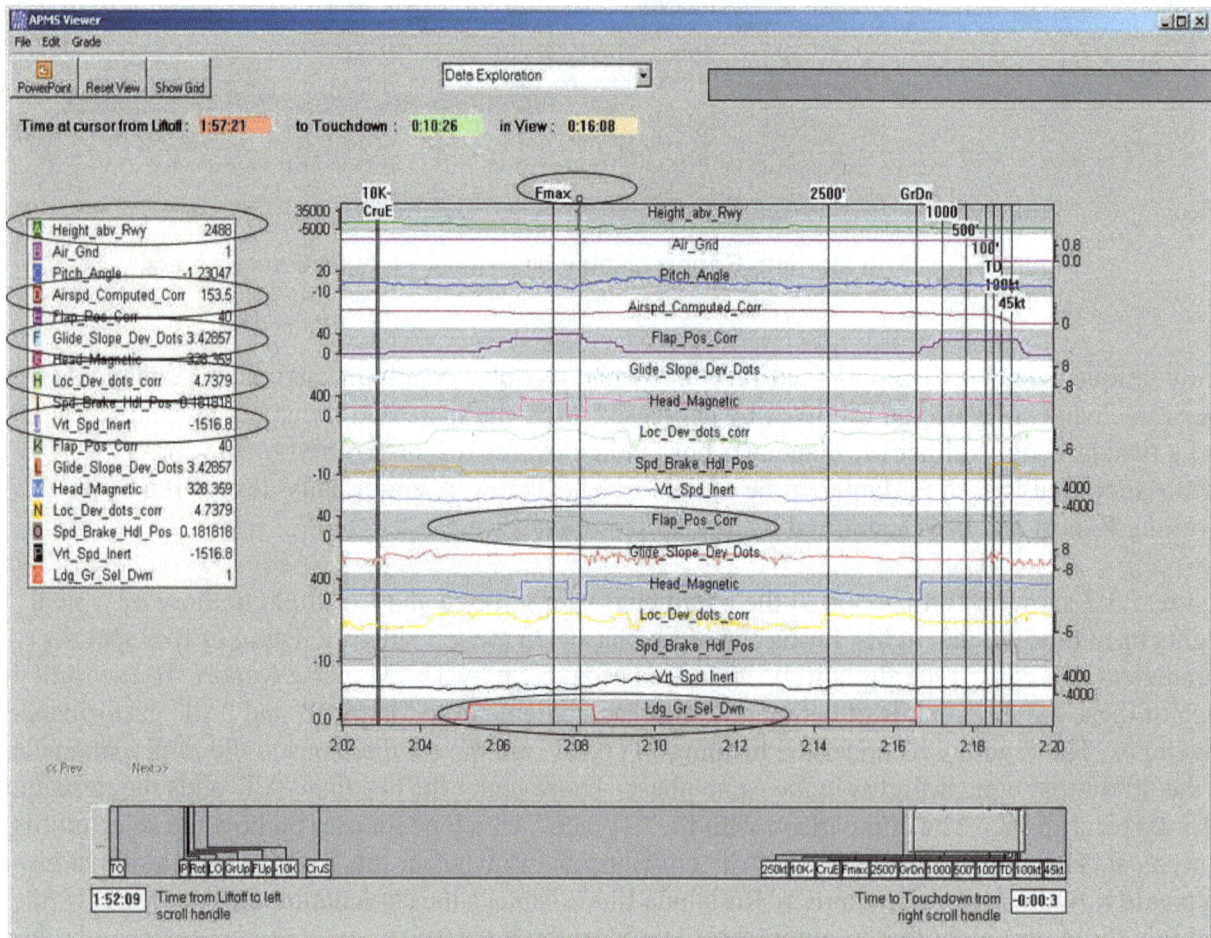

Figure 5.3.8. The APMS Viewer of flight parameters.

The APMS Viewer for this flight as shown in figure 5.3.8 clearly indicates that the flaps and gear were fully extended and then they were retracted, confirming that a go-around occurred.

In figure 5.3.8, the cursor (i.e., the vertical line at about 2.08 hours) has been placed just before the go-around, at which point in time the values of the approach parameters are automatically displayed in the data box at the left of the screen. The aircraft was at 2488 feet above the runway (Height_abv_ Rwy), at 153.5 knots (Airspeed_Computed_Corr), an appropriate airspeed for this point and configuration. However, it was more than 3 dots above the glide slope (Glide_Slope_Dev_Dots), not established on the localizer (Loc_Dev_dots_corr), and descending at 1516 feet per minute (Vrt_Spd_Inert). The aircraft has high potential energy (i.e., it is high above glide slope), and somewhat elevated kinetic energy (due to the descent rate) at this point, both of which would have to be dissipated before landing. With the aircraft in landing configuration (and therefore, maximum available drag) at a low altitude close to the runway, the pilots had no options available to dissipate that energy in the time and distance remaining, and wisely elected to go around.

On the Menu line of the screen shown in figure 5.3.8 is an option called Graphical Airspace Design Environment (GRADE). By selecting that option, this flight or just this phase of the flight can be exported to an airspace visualization and animation tool, as shown in figure 5.3.9.

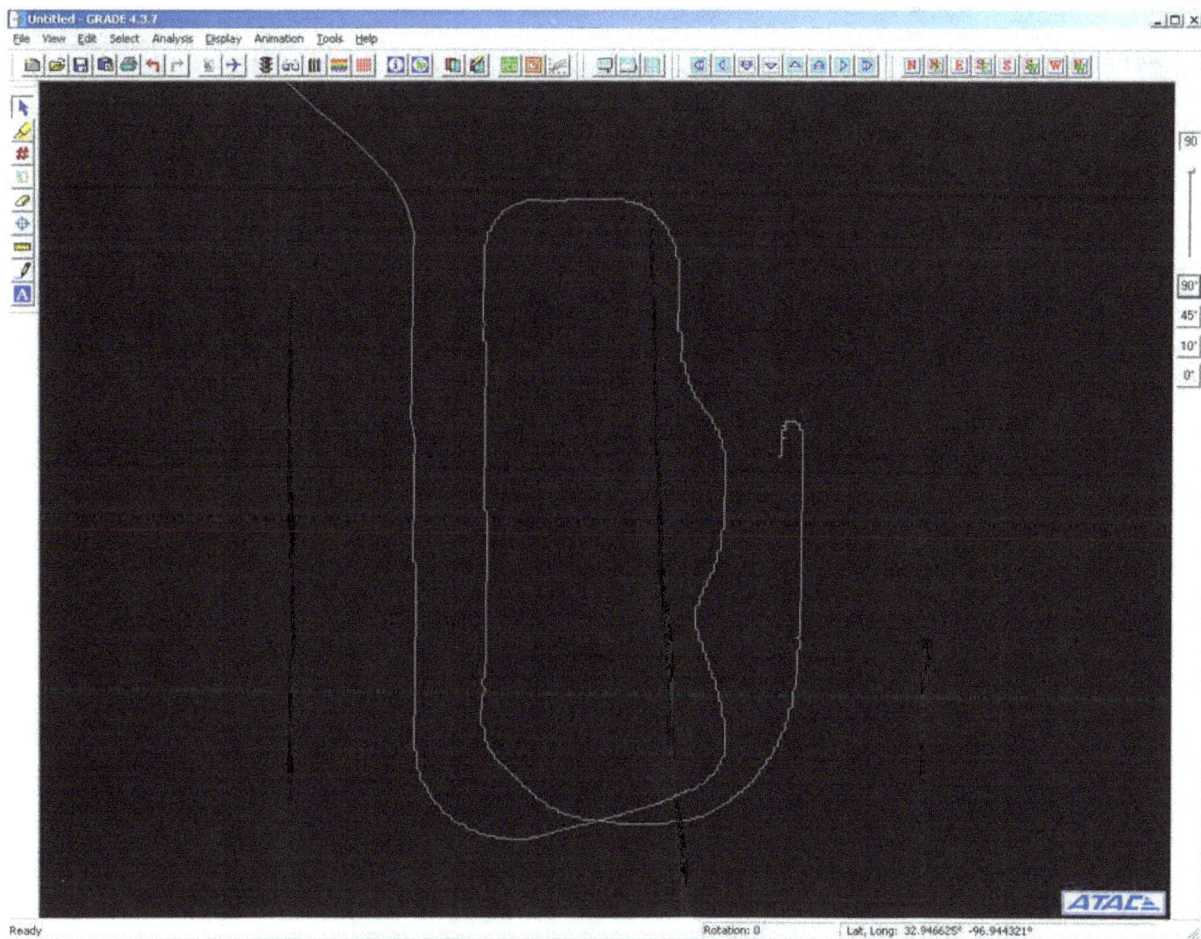

Figure 5.3.9. Horizontal view of the flight in figures 5.3.7 and 5.3.8 in GRADE.

Figure 5.3.9 is a horizontal view of the flight in GRADE (where the x- and y-axes are longitude and latitude) showing the aircraft arrived from the northwest and completed a downwind and dogleg base to final approach. The first final approach includes "s-turns" indicating the pilots' attempt to become established on the desired glide path. The aircraft then completes upwind, crosswind, downwind, base, and final legs to a parallel runway. This view further confirms the go-around.

Figure 5.3.10 shows a vertical view of the same flight. Notice first that the x-axis (horizontal distance from touchdown) is compressed relative to the y-axis (altitude) (in this case 6:1). This display is conventional among such visualization displays of air traffic because, although an aircraft may traverse the world laterally, commercial aircraft rarely exceed 8 miles in altitude.

With this view from the west, the path of the first, abandoned approach can be compared to the second approach, from which the landing was completed. This view confirms the aircraft was well above the desired path, in a high-energy state.

When the ADIS tool was accessed by clicking on "Flight Information" on the screen shown in figure 5.3.7, this flight was found to have been conducted in visual meteorological conditions as shown in figure 5.3.11.

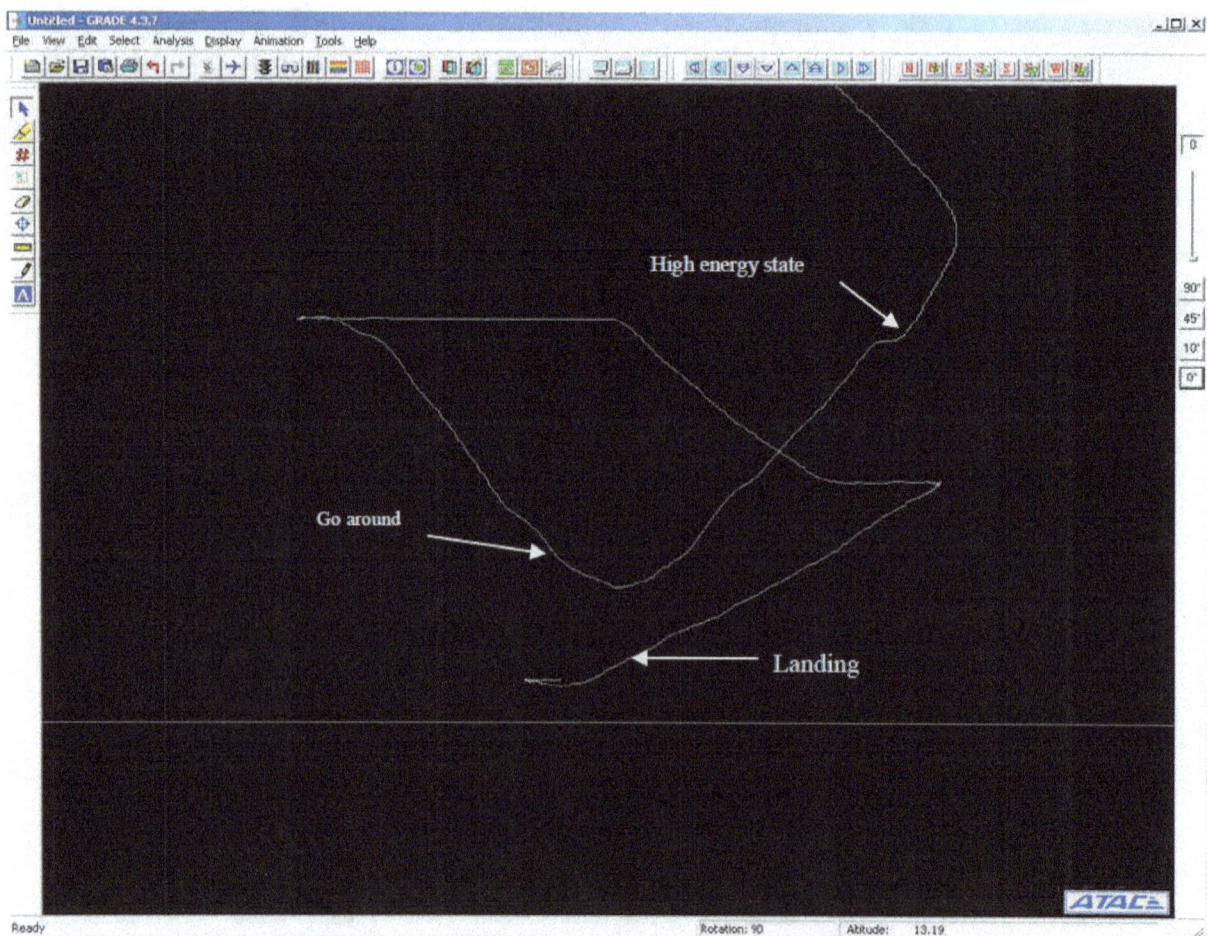

Figure 5.3.10. Vertical view of the flight in figures 5.3.7 and 5.3.8 in GRADE.

> Recorded at: -65

Visibility:10 (mi),Temperature:18 (c),Dewpoint:0(c),Wind Direction:290 (deg),Wind Speed:13 (kts),Altimeter:2996 (in).Sky Condition:CLR

ATIS Message:TWO NINER NINER SIX. VISUAL APCH TO, RWY nnX, RWY nnX, RWY nnX, RWY nnX. ACFT LNG RWY nnX CAN EXP TO HOLD SHORT OF TWY nnnn FT AVAIL, ACFT LDG RWY nnX CAN EXP TO HOLD SHORT OF TWY nn, 9050 FT AVAIL.. NOTAMS... KEEP XPNDR ON WHILE ON TWYS. Bird Activity VCNTY ARPT. READBACK RWY HOLD SHORT Instructions. ...ADVS you have INFO O.. .

Flight Category:VFR

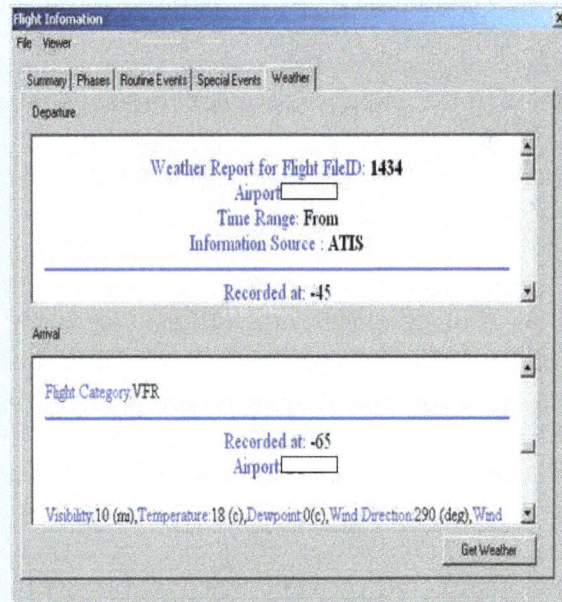

Figure 5.3.11. The ADIS display of weather information for the flight of figure 5.3.7.

As described previously, the most frequent reason for a flight to be atypical (within the sample data set from three air carriers) was a high-energy state between 10,000 and 2500 feet afe. This finding was of great interest to the three airlines participating in the research and to their FOQA vendor as well. The APMS team, the vendor, and the airlines perceived a need to develop a tool to allow routine examination of arrivals, approaches, and airports for high-energy states. NASA entered into collaboration with the vendor and the airlines to develop an energy-state analysis tool.

High-energy arrivals have several common characteristics, including the aircraft were at high speed (200 to 250 knots) as they approached 2500 feet afe, they were above the desired glide path during low-speed descent and final approach, and they were descending onto the glide slope rather than capturing it from below. Analyses of these flights (discussed further later in this section) revealed that descending onto the glide slope was a frequent precursor to unstable approaches. A typical search for exceedances alerts to an unstable approach, but it does not identify the descent onto the glide slope above 2500 feet if the aircraft is stabilized before 1000 feet afe.

Aircraft energy-state indexing and display tools were developed that provide a simple format for understanding the energy state of an aircraft arriving into a terminal area and airport. For any single flight, the index enables an analyst to assess quickly and accurately whether the aircraft was on a proper profile and at an appropriate speed throughout the arrival and approach, by reference to a single index value at any point in flight. Figures 5.3.12 (a) and (b) show examples of displays of the energy-state indexing tools for arrivals at Seattle-Tacoma Airport (SEA-TAC) and Los Angeles International Airport (LAX), respectively.

Three charts are useful to the analyst. The top charts in figures 5.3.12 (a) and (b) depict the statistical distribution of the energy-state Index for the user-selected group of flights operating into the selected airport (in this case SEA-TAC and LAX). An Index value of 100 (the gold line) is considered to be ideal and is called the Gold Standard. Shown on these top charts are the distributions of energy-state Index values (in shades of blue) for these flights from 30 miles to near the point of landing measured at 5-mile increments to 10 miles from landing, then at 2-mile intervals. The shading of the blue blocks indicates the density of flights; the darker the shade of blue, the higher the number of flights. The blocks containing 80% of the flights (what the APMS program considered to be the typical range) at these snapshots are connected with the green band. The dark green line is the median value of the energy-state Index. The superposition of a flight on this Gold Standard Energy Index chart allows the analyst to visualize the relative state of compliance with normal operations in regard to aircraft energy state during approach at the selected airport.

The two charts at the bottom of figures 5.3.12 (a) and (b) provide supporting information that aid the analyst in interpreting the Gold Standard Energy Index chart. Aerodynamic drag and engine power effect changes to the energy state of the aircraft. The Flap Position chart indicates the level of pro-gressive flap configuration (and thus level of drag) that occurs during the approach. The N1 Engine-1 chart indicates the level of engine power being used during the course of the approach. The information provided in these two charts helps analysts understand contributions from these two primary factors to variations that occur in the Gold Standard Energy Index chart. An experienced flight-data analyst understands the relative variations in these two charts and their associated effect on total aircraft energy state. The distributions (the green bands) for these two charts are displayed in operational values relative to their respective function. Engine power is displayed in percent of N1 RPM (100% = Takeoff power) and Flap Position is displayed in degrees of flap travel from zero (Zero = No flaps extended).

Such an aircraft energy-state indexing display tool also provides a simple format for understanding the energy state of groups of aircraft arriving into any terminal area or airport and for comparing operations among those facilities. Corresponding drag and power index values displayed in conjunc-tion with energy state allow analysts to explore how an out-of-desired energy state developed and progressed.

Energy-state displays can be used to provide graphical depiction of the distribution of energy, power, and drag indices along the descent path for any selected group of flights, along with underly-ing descriptive statistics, enabling airline-safety analysts to compare flights by airport or runway of arrival to any other airport or group of airports, thereby facilitating the identification of potentially problematic arrivals or approaches.

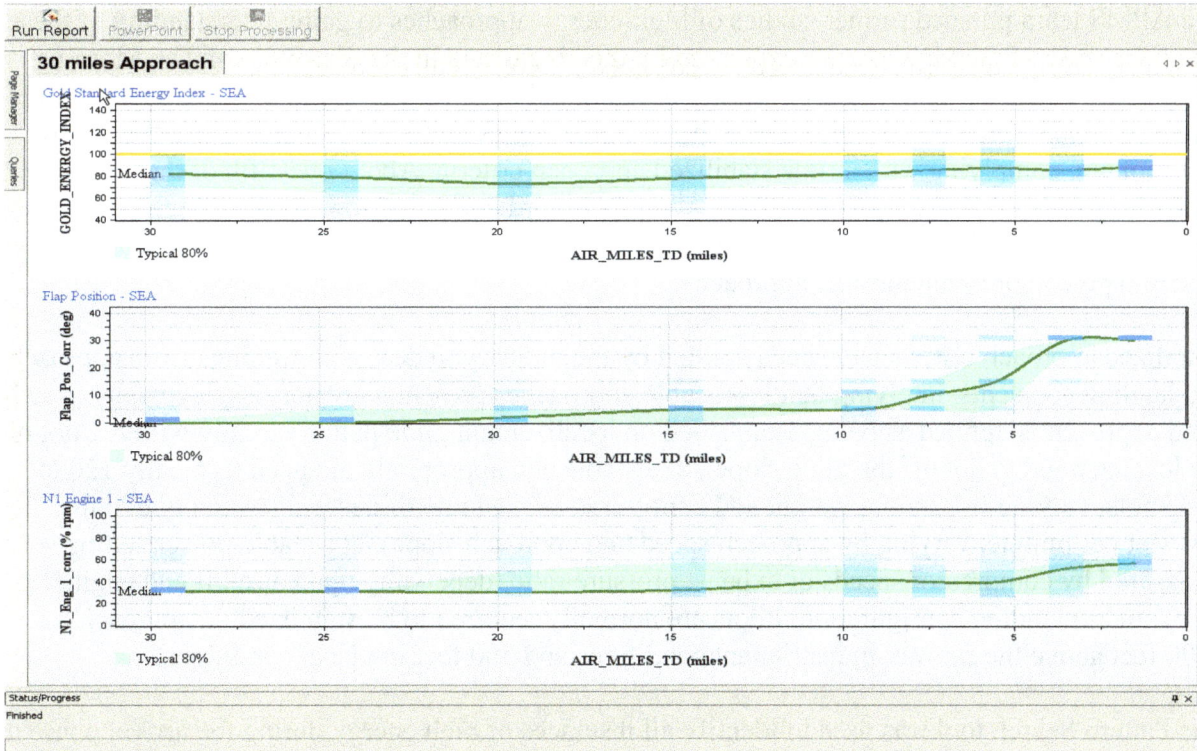

Figure 5.3.12 (a). Energy-state display and indexing tool (SEA-TAC arrival).

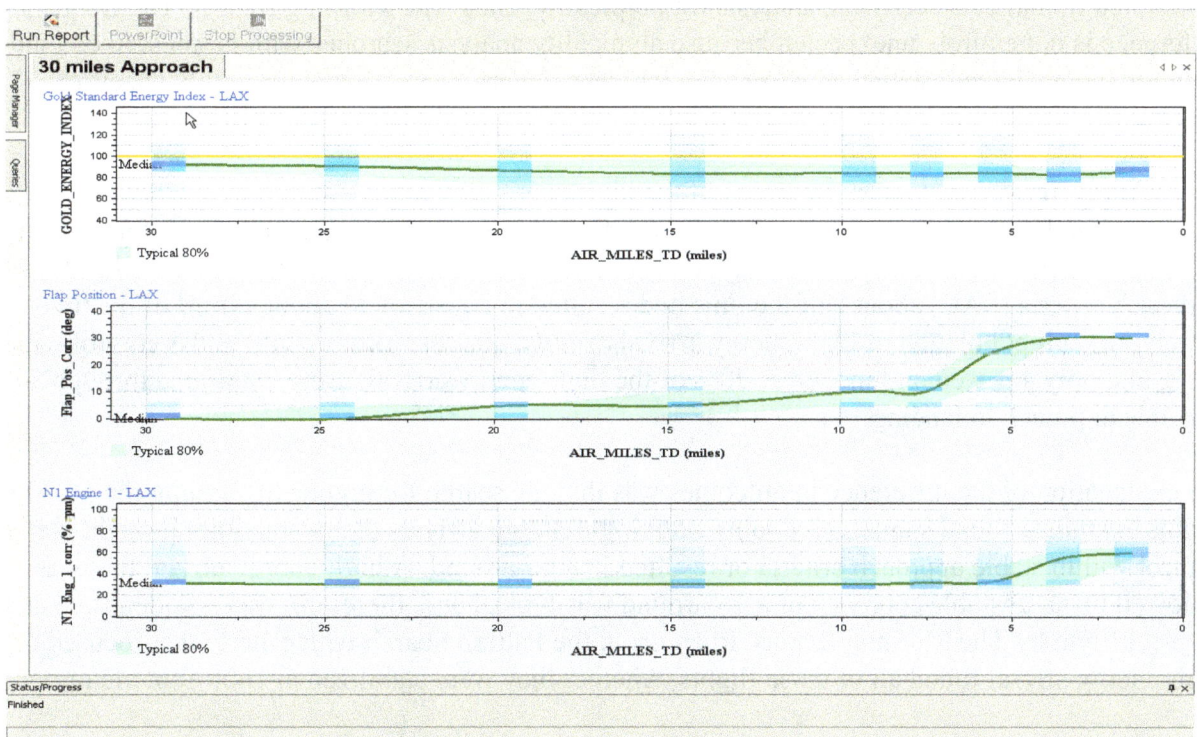

Figure 5.3.12 (b). Energy-state display and indexing tool (LAX arrival).

The APMS team pursued further studies of high-energy approaches to gain understanding of these events because of the high interest expressed by the air carriers in the discovery of The Morning Report. Less than one-half of 1% of all flights submitted to atypicality analysis were deemed atypical because of high-energy state. Among these flights, about half were brought under control by 500-foot afe, complying with airline stabilized approach criteria. About one-fifth of these high-energy arrivals ended in a go-around. But, about one-third continued to landing, representing an unstable approach. Based upon these observations, a high-energy arrival state was postulated to be a potential precursor of an unstable approach.

Stabilized approach criteria have been applied by the industry to deal with findings from approach and landing accidents, implying that less-stable approaches present a greater risk. Typically, a stabilized approach is defined as being established on localizer and glide path (typically within 1 dot off the localizer and ½ dot off the glide slope), at a stable and appropriate airspeed (typically 120 to 150 knots, varying by landing weight and configuration) and rate of descent (typically 600 to 900 feet per minute, varying by approach speed and approach slope), at a stable power setting (measured by compressor speeds or exhaust pressure ratio, depending upon aircraft and engine type), and in landing configuration. Pilots are normally required to be stabilized on approach by 1000 feet above the runway in instrument conditions and 500 feet in visual conditions.

The Pattern Search tool was used to identify all instances of high energy during the arrival phase to demonstrate to the airlines how the APMS tools could be used to assess whether high-energy arrivals are a precursor of unstable approaches. The Pattern Search on the flight dataset from one of the airlines revealed that high-energy arrivals were three to four times more frequent than had been discovered in that data set by the analysis for atypicality using The Morning Report. The frequency difference is not entirely unexpected because atypicality analysis searches with no prescribed guidance for flights that are statistically extreme, whereas Pattern Search looks for highly specific criteria.

Surprisingly, however, the consequences of the high-energy arrivals identified by the atypical analysis were found to be different from those of the full population of high-energy arrivals found with the Pattern Search. Those detected by atypicality were brought under control within stabilized approach criteria (SAC) about half the time and resulted in a go-around about a third of the time. Among the full population of high-energy arrivals, only a quarter were brought under control within SAC, and only 4% went around. Over 70% of the high-energy arrivals were continued through an unstable approach to landing.

An explanation of the difference in outcomes was that the search for atypicality among the 16,000 flights examined found something doubly rare: high-energy arrivals that were either brought under control within stable approach criteria or resulted in a go-around. A high-energy arrival that was followed by an unstable approach or a go-around was typical and, therefore, was not identified as an atypical flight by The Morning Report. In contrast, the Pattern Search with criteria that focused on high-energy arrival found all of those flights, whether they were stabilized at 1000-foot afe or not.

Importantly, however, these results should not be generalized because the findings were based on the analysis of the 16,000 flights that were available for the study, and these were a nonrandom sample of flights from a nonrandom sample of airlines. This analysis was a demonstration, showing how airlines could use atypicality analysis, and that airlines must apply all the APMS tools to get definitive statistics.

The demonstration was carried a bit further for a few flights to show how ADIS could be used to understand the context of high-energy arrivals. Four selected flights were examined at LAX, an airport with an average rate of high-energy arrivals in the dataset. Two of these flights were brought under control within SAC (and, therefore, had no exceedances identified by the FOQA program), and two were not brought under control within the SAC, continued to landing, and had multiple unstable-approach-related exceedances identified by FOQA.

The GRADE tool was used to depict their lateral and vertical paths in the terminal airspace, and ADIS was used to examine the weather and air-traffic conditions to which these four flights were exposed. The air-traffic data were obtained through the ASMM Project's relationship with the FAA Office of Performance Analysis in the collaboration on the PDARS. PDARS provided one-hour of data for each flight, when the participating airlines gave permission to disclose the date and hour in which the selected flights arrived.

There was nothing statistically or operationally significant about the following weather data shown for these four flights except that the two flights exhibiting exceedances occurred in moderate wind conditions. All four flights arrived in visual meteorological conditions.

- Nonexceedance flights
  - Exemplar 1 – Night VMC

    Visibility: 8 (mi) Temperature: 12 (c) Dew Point: 11 (c) Wind Direction: 160 (deg) Variable Wind Direction: (deg) Wind Speed: 03 (kts) Wind Gust: (kts) Altimeter: 3021 (in) Sky Condition: FEW013 BKN180 SUN Position: Night Darkness

  - Exemplar 2 – Day VMC

    Visibility: 10 (mi) Temperature: 21 (c) Dew Point: 17 (c) Wind Direction: VRB (deg) Variable Wind Direction: (deg) Wind Speed: 05 (kts) Wind Gust: (kts) Altimeter: 2996 (in) Sky Condition: FEW018 FEW034 SCT080 BKN120 BKN180 SUN Position: Day Light

- Exceedance flights
  - Exemplar 3 – Day VMC (moderate wind)

    Visibility: 10 (mi) Temperature: 16 (c) Dew Point: 08 (c) Wind Direction: 250 (deg) Variable Wind Direction: 240V350 (deg) Wind Speed: 16 (kts) Wind Gust: (kts) Altimeter: 2988 (in) Sky Condition: FEW100 SCT200 SUN Position: Day Light

  - Exemplar 4 – Day VMC(moderate wind)

    Visibility: 10 (mi) Temperature: 18 (c) Dew Point: 12 (c) Wind Direction: 230 (deg) Variable Wind Direction: (deg) Wind Speed: 12 (kts) Wind Gust: 19 (kts) Altimeter: 2998 (in) Sky Condition: FEW008 SUN Position: Day Light

The GRADE graphical displays in figure 5.3.13, with the perspective of the profile view, portray the vertical paths of the four exemplar flights in white and the vertical paths of all other flights landing in the same hour in green. The concentrated green band can be interpreted as aircraft on the glide slope, both for straight-in arrivals and flights turning from downwind to final (the case for the four exemplar flights). The two graphics at the top of figure 5.3.13 (high energy to touchdown and to 1 mile) are the flights that had exceedances. The two graphics at the bottom of figure 5.3.13 are the nonexceedance flights. (These graphics are screen dumps taken from the GRADE displays with no absolute scales. The relative scales of the flights are correct and are the only feature of significance to this visualization of the events.)

Examination of the air-traffic data revealed that three of the four flights were operating at a higher airspeed over the very-high-frequency (VHF) omnidirectional range (VOR) navigational-system fix point at Santa Monica than would be expected had the charted "expect" speed been applied. The VOR at Santa Monica is called SMO. The charted "expect" speed is the required speed that is published on the arrival chart for the segment crossing over SMO VOR. These aircraft crossed SMO at 250 knots indicated airspeed. The charted "expect" arrival speed is 230 knots, well before SMO.

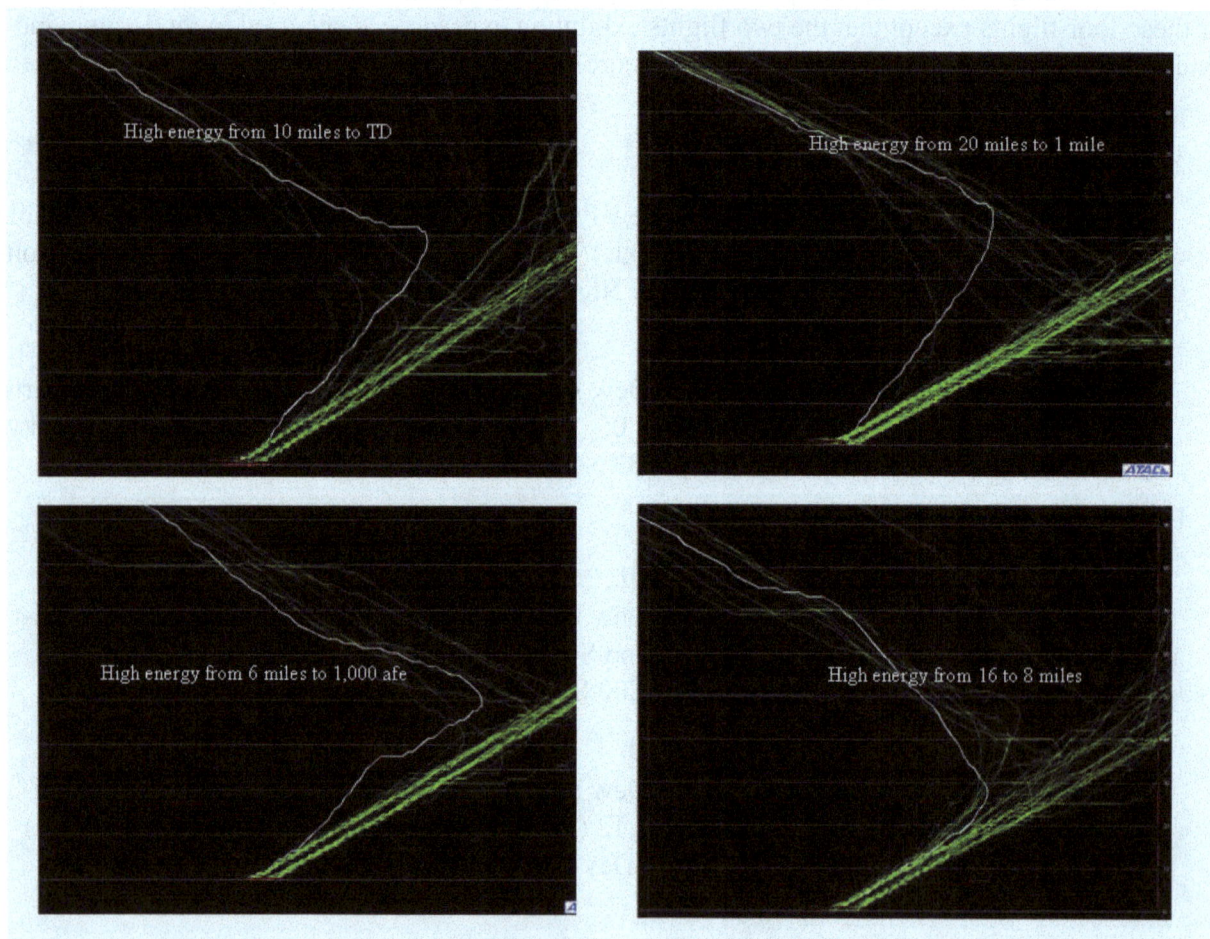

Figure 5.3.13. GRADE display of air traffic for four exemplar flights.

All four flights crossed SMO within 700 feet of the published altitude for either the associated instrument approach or charted visual approach. Altitude at this point was not a factor, but the graphics of figure 5.3.13 show that all four flights were higher than similarly situated flights at various points during the arrival and approach. All four flights joined the final approach well short of the charted visual approach path, indicating that they had either been vectored to the final approach course or had been cleared for a visual approach other than the charted visual. Of the two nonexceedance flights (at the bottom of figure 5.3.13), the night exemplar (on the right) was the only flight at the "expect" arrival speed of 230 knots as it crossed SMO, and the day exemplar (on the left) completed the longest of the four flight paths.

Only one of the flights had an observable air-traffic constraint, a B-747 five miles ahead of one exceedance flight. This information is informative because, in the absence of air-traffic information, the participating airline assumed that ATC had vectored its aircraft into high-energy states because of constraints of other departing or arriving aircraft. These exemplars suggest high energy may have resulted solely from difficulties in determining where a visual approach could be initiated from downwind to ensure an appropriate energy state. There is no guidance in real time to assist these decisions, and pilots must make them in varying conditions of wind, lighting, and visibility.

The APMS tools contributed importantly to two demonstrations of the value of integrating information from diverse sources to gain a full picture of an event. One was the study of ICAC (described in appendix B). The other was the study of metering of air traffic (described later in this report), in which APMS was used to assess the impact on flight operations.

The approach to developing the APMS tools was to respond to user needs, and then implement each as a prototype for testing at collaborating airlines. Maturing the tools to allow transfer to industry was accomplished by demonstrating their use, evolving them during evaluation tests with air carriers, disclosing tools as new inventions, patenting inventions where appropriate, and licensing patented tools to commercial FOQA software vendors. Table 5.3.1 is a list of the APMS tools that have been disclosed and their disposition at the time of the completion of the ASMM Project.

APMS generated a set of capabilities as a standalone system. Using data in any of the vendors' formats as inputs, APMS provides a suite of tools for conducting the following variety of basic and advanced functions:

- Exceedance analysis, management, and reporting,

- Identification, analysis, and management of statistically atypical flights,

- Operational studies using Pattern Search and Routine Event statistics,

- Depiction of flights of interest in airspace visualizations using GRADE under license from ATAC, and

- Examination of weather data for flights of interest.

TABLE 5.3.1. DISCLOSING, PATENTING, AND LICENSING INVENTIONS

Invention	Disposition
ARC -14362 Aviation Performance Measuring System (APMS) - 1999	Industry adopted APMS concepts of full-flight data storage and processing
ARC -14508 Automated Statistical Analyses of Flight-recorded Parameters at Routine Events - 2000	Industry adopted APMS concept of "snapshots" at routine events for statistical analyses
ARC-14981 Pattern Search Tool for Analysis of Aircraft Flight Data- 2002	Industry implemented capability to search for patterns of parameters
ARC-15036 Aviation Data Integration Sytem (ADIS): Secure Information of Aviation Data with De-Identified Flight Data - 2002	Patent Pending. ADIS archive transferred to the FAA for access by all FOQA carriers. ADIS encrypted de-identification schema licensed to SAGEM Avionics.
ARC-1504 Morning Report Atypicality Tool for Analysis of Aircraft Flight Data - 2003	Patent Pending. Non-exclusive license to SAGEM Avionics.
ARC-15086 Phase of Flight Performance Report Tool for Analysis of Aircraft Flight Data - 2003	Industry is using some of the concepts. Process and statistics included in DNFA technology.
ARC-15356-1 Indexing and Display of Aircraft Energy State for FOQA Programs - 2004	Patent Pending. Innovators: NASA, Delta, JetBlue, & SAGEM; SAGEM implementing.
ARC-15356-2 Display of Aircraft Energy State for FOQA Programs - 2004	Patent Pending. Innovators: NASA, Delta, JetBlue, & SAGEM; SAGEM implementing
ARC-15356-3 Real-time Analysis and Display of Aircraft Approach Maneuvers - 2004	Patent Pending. Offers potential aircraft display making use of energy indexing.
ARC-15546-1 Flight Data Translation Utilities for the Distributed National FOQA Archive (DNFA) – 2005	Interface technology designed for DNFA; allows automated conversion from proprietary to NASA data format.

These functions will be converted for use in the Distributed National FOQA Archive (DNFA) under the Information Sharing Initiative (ISI) in FY05 and FY06 (the ISI project is discussed in section 7.0, "Toward the Future Vision"). The ISI opens the door for expanded use of the capabilities developed under both the APMS and the PDARS activities of *Intramural Monitoring*.

In the near term (perhaps with enhancements of the National Archives started under ISI), APMS can provide to an air carrier capabilities to provide:

- Daily alert to atypicalities and exceedances,

- Automated access to weather and air-traffic conditions without reidentifying flights,

- Ability to search for prior and monitor for subsequent similar events, documenting frequency and location, and

- Basis for directing its resources toward well-documented and researched problems.

In the midterm, as the interest grows for information integration across distributed diverse data sources, use of the APMS tools is envisioned as a prime capability in an operational concept in which:

- Automated issue-level data integration frees safety analysts from specific data sources,

- Review issues across diverse types of data (such as FOQA and ASAP) are available to the airlines,
- Integrated data allow understanding of systemic issues identified in any given event or data source.

In the long term, this concept would evolve to a national operation in which:

- Information is integrated across organizations,
- Airlines compare their operations to de-identified industry aggregates to understand context of operations,
- Each airline becomes a node on a system that can establish industry norms, and
- Each airline can readily determine whether identified issues are unique to their operation or common across airlines, allowing airlines and industry to act at the appropriate level.

When the ASMM Project ended, the following barriers to realizing the long-term objective remained unresolved:

- Gaining access to other sources of data for routine analysis
  - Air-traffic radar data belongs to the FAA. Permission for limited access was gained for demonstrations, but no permission was granted, nor was a system developed for accessing air-traffic data routinely.
  - Similar problems exist even within organizations. For example, air carriers have no permission to link their ASAP reports with FOQA data.
  - There is a broad future challenge to being granted access routinely to extract and integrate information from diverse, distributed data sources for systemic analysis.
- Each new data source is likely to pose a new technological challenge to extracting and using its information.
  - Conceptually, many APMS tools generalize to other digital sources, but, except for a very limited experiment with radar data described later in this section, this generalization has yet to be demonstrated.
  - Automated analyses of textual data are a special concern because of the large number and great variety of such sources of valuable safety information.

Implementation of APMS to overcome some of these barriers proceeded on two paths. The path to *intramural* implementation was through commercial vendors of software. As shown in table 5.3.1, APMS tools have been transferred through concept adoption or licensing to vendors.

The path to *extramural* implementation of the capabilities developed as APMS tools that will overcome some of the barriers identified previously is the ISI, which is discussed later in this report. ISI was initiated in the final year of the ASMM Project. ISI will focus on designing hardware, software, and networking to enable information sharing among air carriers and the FAA by transferring an operational system, instead of commercializing individual tools. Although the initial demonstration will have limited analysis capabilities, including Pattern Search, Routine Events, and Energy

Analysis, eventually The Morning Report will follow in subsequent years and be retailored to focus on atypical groups of flights rather than individual flights.

5.3.2 Performance Data Analysis and Reporting System

The other primary product developed under the *Intramural Monitoring* element of the ASMM Project was the PDARS. From its outset, PDARS was a close collaboration among NASA, FAA ATO-PA, and the FAA's contractor, the ATAC Corporation. It was the amalgamation of the FAA's interest in measuring system performance and safety and NASA's interest in developing technologies for monitoring and enhancing system safety. The NASA Ames Research Center and the FAA Office of Performance Analysis developed PDARS in response to two motivational factors—the accident-reduction goals of NASA's Aviation Safety Program and the Government Performance and Results Act (GPRA), which required government agencies to measure performance and demonstrate proof of improvement. A good marriage of those motives, PDARS is the counterpart to the APMS that enables the FAA to measure the capacity, performance, and safety of its air-traffic services.

FAA ATC facilities record large volumes of daily data, but, prior to PDARS, there was little ability to generate facility-specific metrics for improving safety and efficiency. PDARS provides ATC decision-makers at the facility level with a comprehensive set of tools and methods for monitoring the performance and safety of the day-to-day operations of their facility. In addition, the PDARS data and toolset can be used on a more macroscopic level by regional or national managers to monitor the performance and safety of the NATS as a whole (den Braven and Schade (2003)).

An activity of PDARS was to test and evaluate advanced concepts, methodologies, and software tools for analyses of ATC radar-track data. The objective is to provide technologies to identify, analyze, and manage conditions and events that could compromise safety, while meeting the projected requirements of increasing air traffic. Tools such as PDARS can be used to facilitate a proactive reaction within the ATC facilities to mitigate the circumstances surrounding such events. It is necessary to be able to merge information produced by PDARS (for the perspective of air-traffic control) with information developed by APMS/FOQA (for the perspective of the aircraft) to understand all the factors that may have been involved in an incident or unwanted trend that had been identified by either one.

The FAA already uses many systems with which to monitor particular operational aspects of its air-traffic-management system. PDARS is a unified capability for monitoring many aspects of a facility's performance. It provides the capability to:

- Collect, extract, and process ATC operational data,

- Compute quantitative operational performance measures on a regular basis relating to safety, delay, flexibility, predictability, and user accessibility,

- Conduct causal analyses and operational problem identification and analyses,

- Access design and simulation tools for "what-if" analyses and for identification and emulation of system improvement options, and

- Archive performance statistics and basic operational data for use in research development and planning studies.

PDARS incorporates innovative technology for rapidly processing large volumes of complex air-traffic data, for extracting and visualizing information from these data, and for a secure network that enables facilities across the continental United States to share and discuss this information. Data are accessed daily from each of the sites participating in the test and evaluation of PDARS, the data are processed overnight, and customized reports are routinely delivered to each of the participating facilities each morning. Figure 5.3.14 shows how PDARS automatically and continuously collects, processes, and analyzes ATC information. PDARS provides to each ATC facility the capabilities to:

- Routinely collect, process, and analyze air-traffic-management operational data,

- Generate daily reports,

- Share data and reports among facilities,

- Provide software tools to support exploratory and causal analysis studies, and

- Archive basic operational data and measurement.

The core processing modules are integrated directly with a data-management system that allows for easy retrieval and reporting on the collected information. Data are stored locally at each FAA ATC facility as well as at a central location for archiving and additional reporting. A secure wide-area network (WAN) allows for data transfer between the central and remote sites. Reports are available in the form of Microsoft Excel workbooks or as interactive Web pages. The ability to monitor continuously, convert the collected data into reliable information, and share that information for collaborative decision-making is the basis for a proactive approach to identifying and alleviating life-threatening aviation conditions and events.

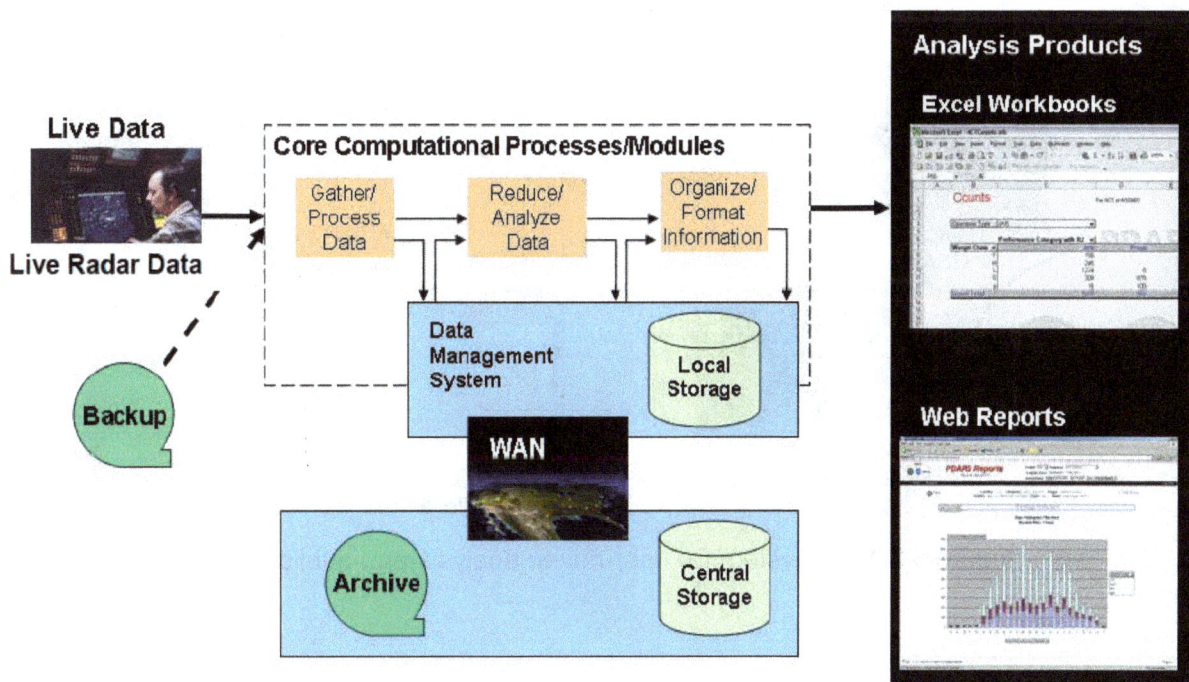

Figure 5.3.14. PDARS: A system for automatic data collection, analysis, and reporting.

The front-end user tools of PDARS allow for easy access to information with a focus on short turn-around times and professional style publishing as depicted in figure 5.3.15. Here the reporting system serves as a conduit through which access to additional analysis capabilities can be obtained. These additional capabilities include other office tools, commercially available statistics packages, and the custom GRADE visualization software.

As indicated in figure 5.3.16, the typical setup of PDARS equipment at each FAA ATC facility includes one data-acquisition PC that connects to the data source through a gateway. In addition, at least one PC designated for data analysis allows users at the facility to access the processed information using the PDARS software tools.

PDARS provided an important contribution to the ICAC study described in appendix B that demonstrated the value of integrating information from various sources to give a complete picture of an event. In the case of ICAC, that air-traffic information was particularly significant. Table 5.3.2 shows the results of the PDARS analysis for San Francisco (SFO) and Los Angeles (LAX). One of the interesting results was that no conclusive evidence was found to show that go-arounds were more likely to result during an approach with an ICAC.

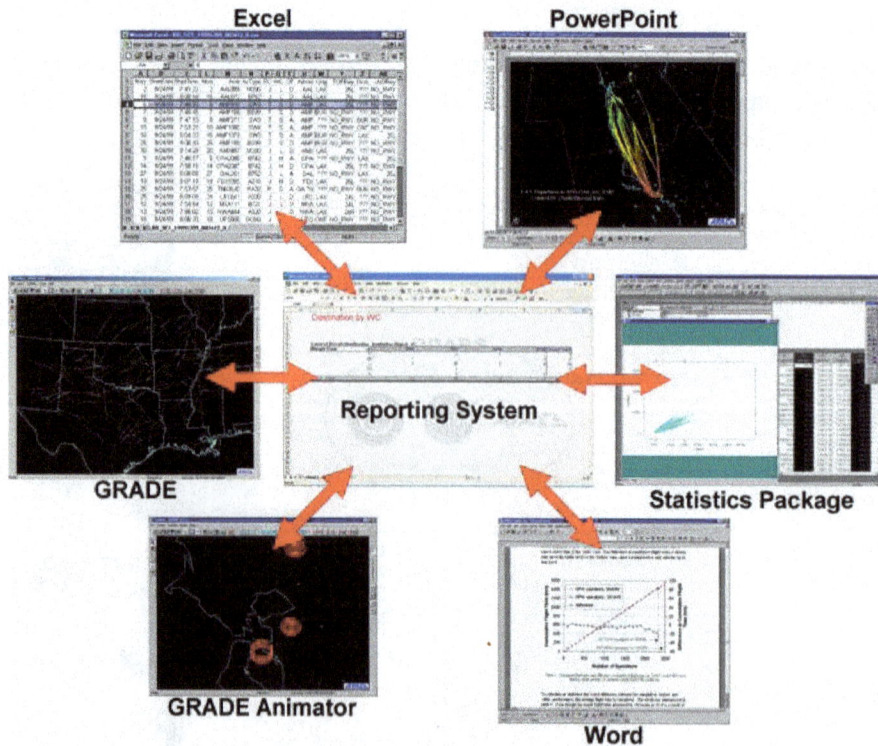

Figure 5.3.15. A system for user-driven analysis and publishing.

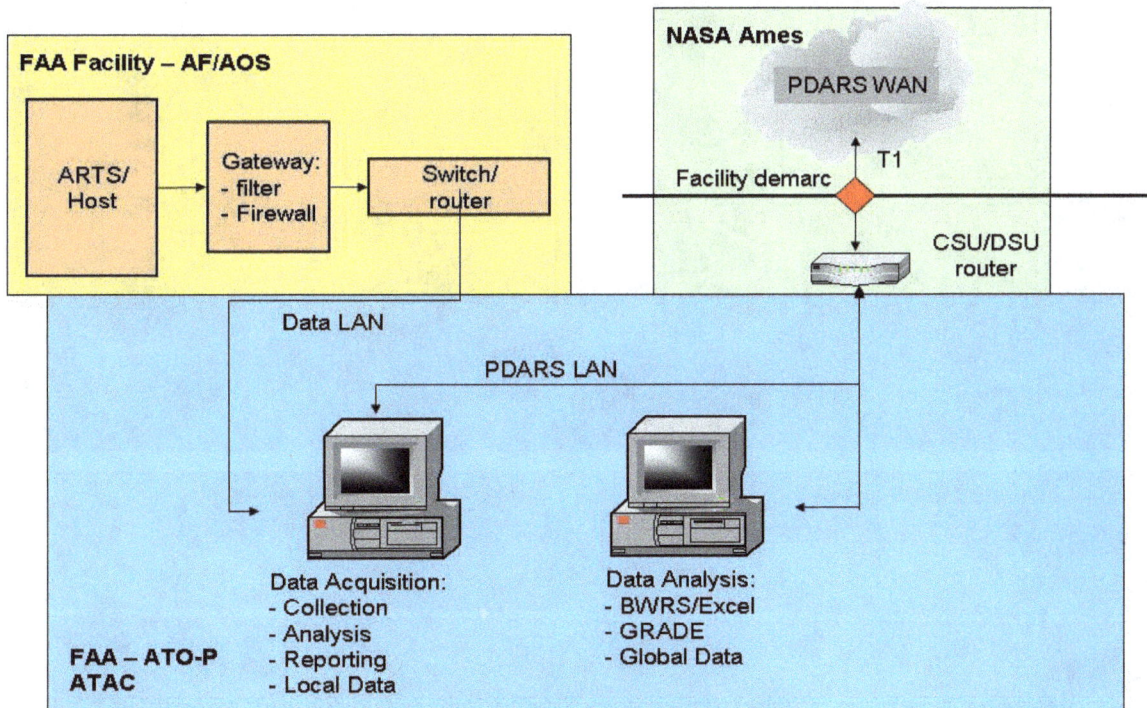

Figure 5.3.16. The PDARS network.

TABLE 5.3.2. COMPARISON OF ICAC ANALYSIS FOR SFO AND LAX

	LAX	SFO
Average number of ICACs per day	21.9	11.3
Std. dev. number of ICACs per day	7.6	12.2
ICAC rate per 1000 arrivals	35.6	26.5
Average ICAC time to end of track	74 sec	76 sec
Average threshold distance at ICAC	2.7 nm	3.1 nm
Number of go-arounds	34	34
Average go-arounds per day	1.1	1.1
Std. dev. number of go-arounds per day	0.94	1.27

The clustering and visualization technology from The Morning Report of Atypical Flights (described under APMS previously) was adapted to radar-track data as shown in figure 5.3.17. This adaptation required a system integration of PDARS data management, APMS-style clustering component, The Morning Report Display, and GRADE visualization. The top half of the slide shows a top-down view of aircraft arriving at LAX. The bottom half of the slide shows the same aircraft in a profile view. Each cluster of similar tracks is given a separate color.

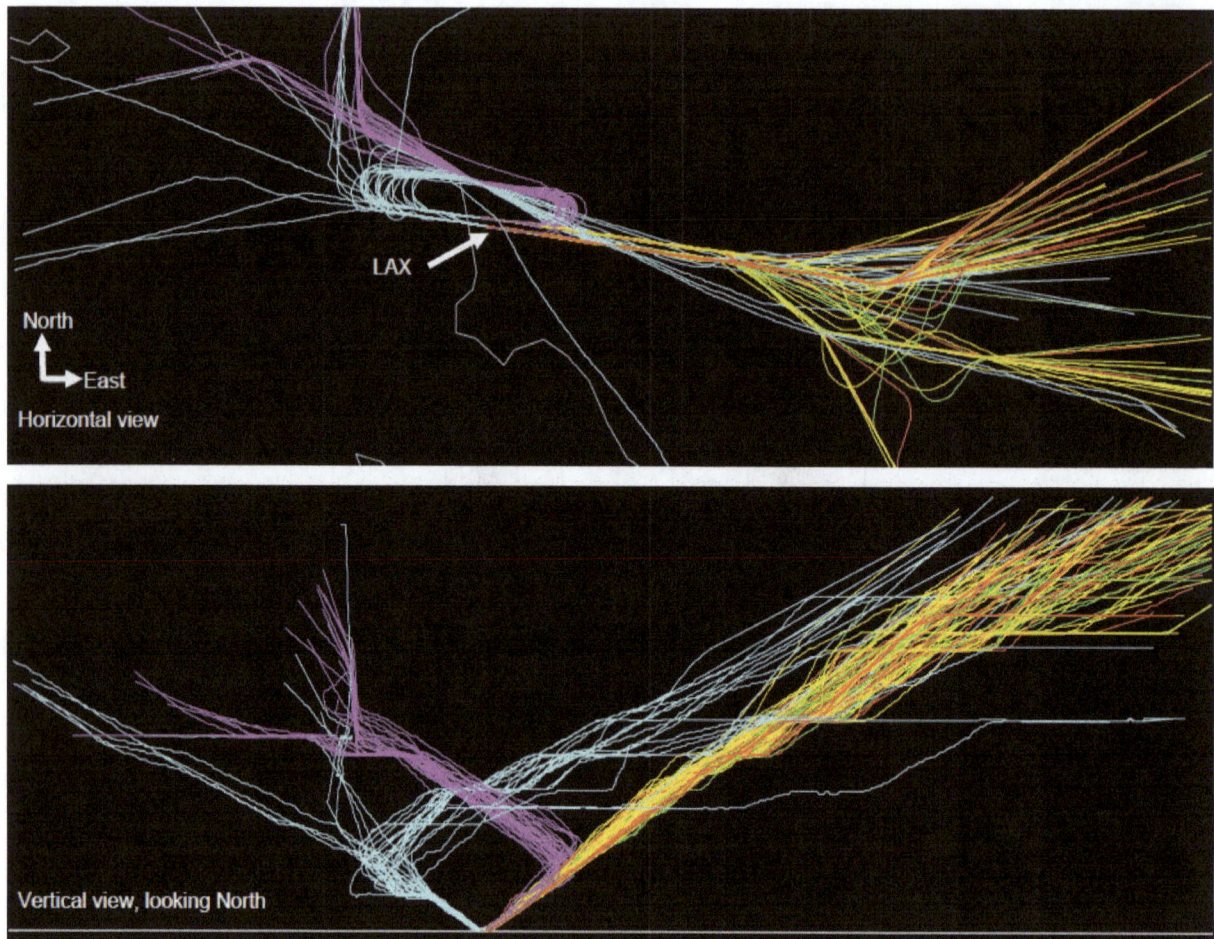

Figure 5.3.17. Cluster visualization with GRADE.

The Morning Report displays for the PDARS version are an Executive Summary screen and a List of Atypical Flights that are very similar to the displays for the APMS version that were shown in figures 5.3.5 and 5.3.6. Just as with APMS, upon entering the flight-list page of The Morning Report for PDARS, the user/analyst can examine atypical flights. Each flight on the flight list can be compared to its contemporaries using numerous density plots called performance envelopes. The individual flight can be plotted against its "cluster mates," versus the 80% most typical flights, or against the entire population for numerous parameters. The example of figure 5.3.18 is a plot of heading in degrees relative to the direction of the runway versus the percentage of time to touchdown.

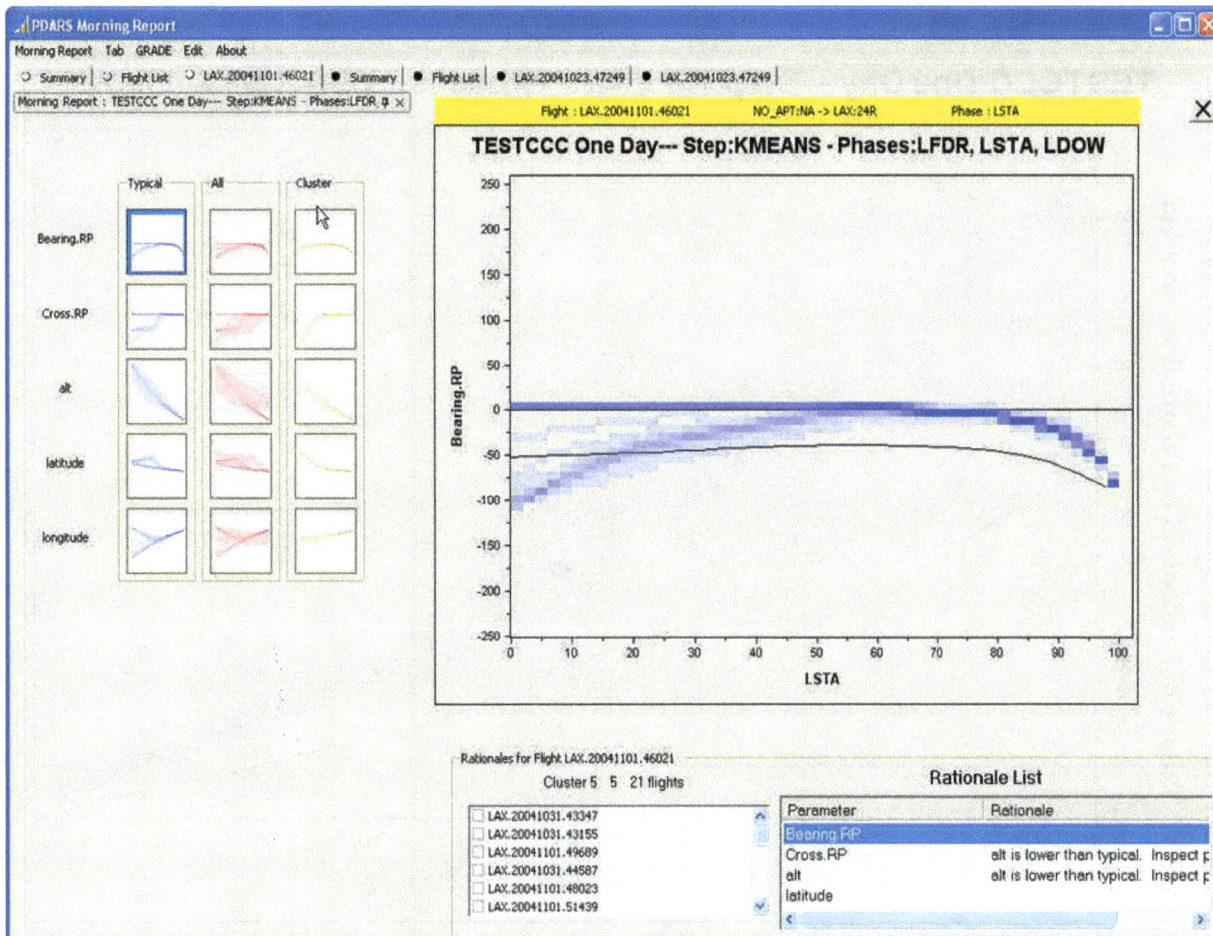

Figure 5.3.18. The Morning Report display from PDARS data.

The particular performance envelope example of figure 5.3.19 shows altitude versus percentage of time to touchdown of an atypical flight (black line) compared to the distribution of "typical" flights (blue shading). This flight was probably atypical because of low altitude during this phase.

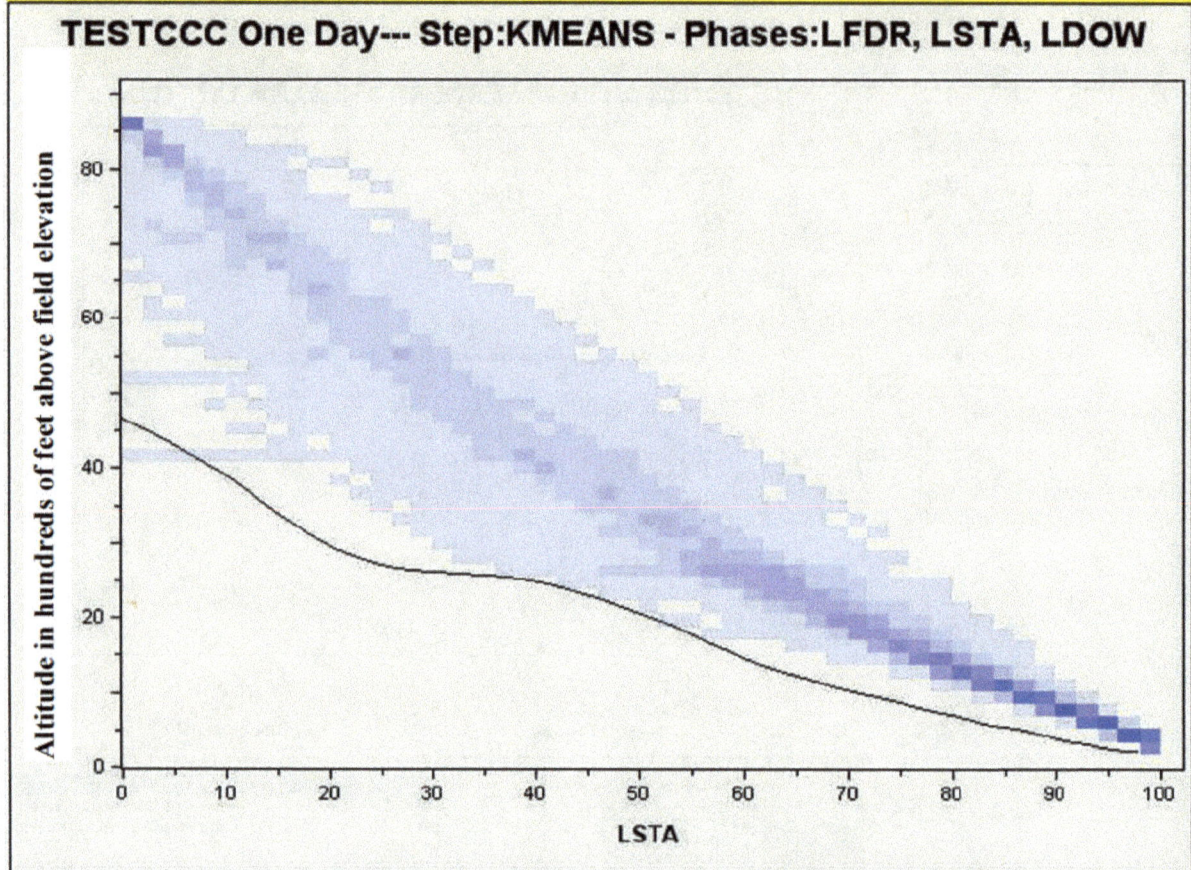

Figure 5.3.19. Performance envelopes.

Continuing on with the analysis, The Morning Report for PDARS allows for flights of interest to be displayed in GRADE and shown colorized by different criteria, as shown in figure 5.3.20. In this example, the flight of interest is shown along with the other flights that are in the same cluster. However, only their associated parameters during the "colored" section of their tracks were used for clustering analysis. In this example, the grey portions of the flight tracks are considered to be outside the clustering "phase" and not considered during the clustering computations.

92

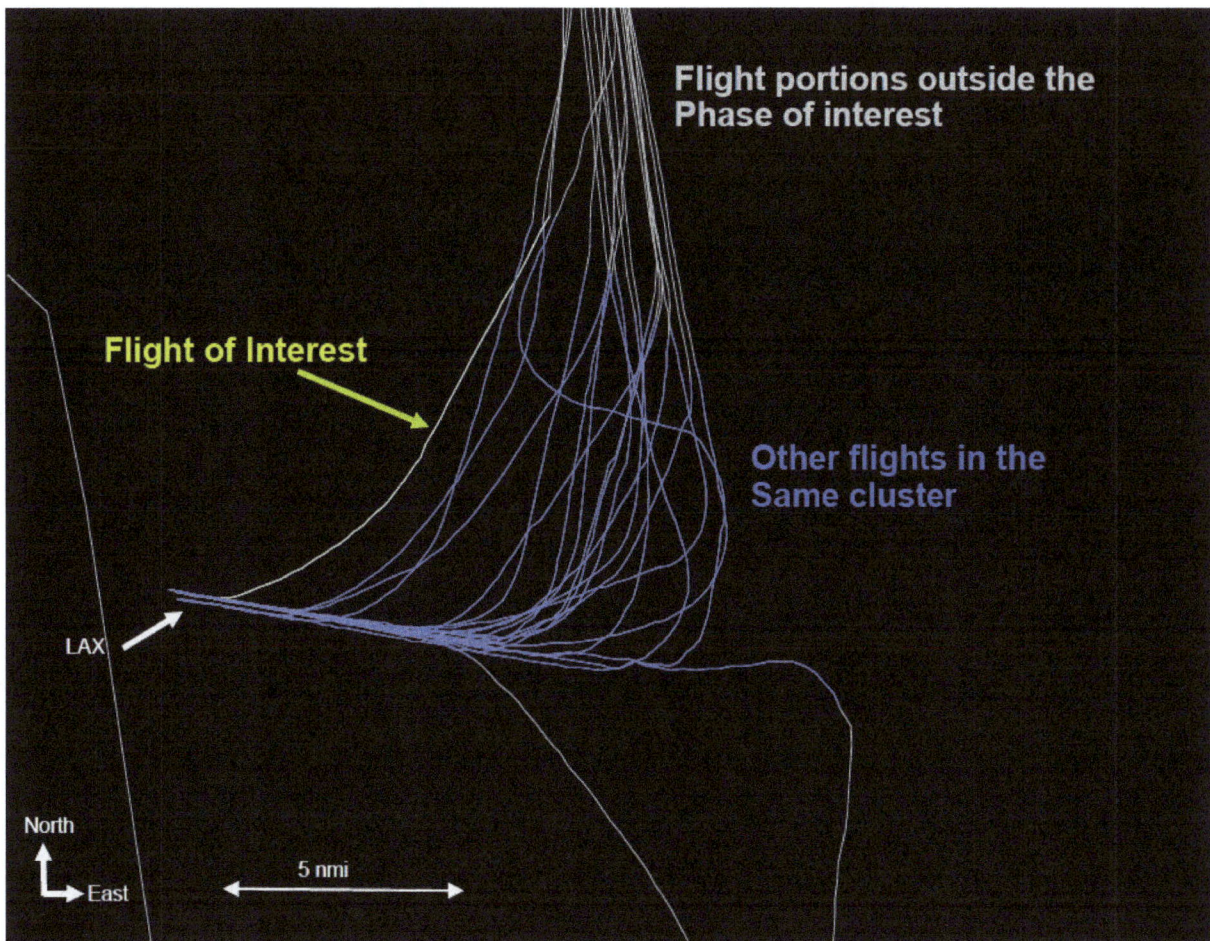

Figure 5.3.20. Clustering of flights shown in GRADE.

PDARS has been used by the FAA ATC facilities to:

- Measure, explain, and improve facility operation,
- Communicate with the airlines, military, public, and other stakeholders,
- Analyze staffing requirements,
- Create and validate training scenarios,
- Support airspace redesign, and
- Find lost aircraft.

PDARS has also demonstrated value to

- Airlines:
  - Provide quick, clear, and decisive response to questions and complaints.
  - Measure, explain, and improve airline operation.
  - Provide detailed analysis that has led to schedule improvements.

- Military/Security:
  - Provide quick, clear, and decisive response to questions.
  - Support setup of Air Defense Identification Zone (ADIZ), Temporary Flight Restriction (TFR), and Special Use Airspace (SUA) volumes.
- Public:
  - Provide clear information about traffic operations.
  - Provide better response to noise complaints.

PDARS merges many data sources to achieve complete tracks for terminal-to-terminal coverage and provides increased accuracy over traditional flight-track-data sources. PDARS enables safety and performance measurements on the inter-facility, regional, and national levels. In the following paragraphs and figures are some examples of how PDARS can and has been used.

PDARS enables trending analysis, as exemplified by figure 5.3.21, which shows the trend over a period of two years of the flight time spent in Oakland center airspace for SFO to LAX flights.

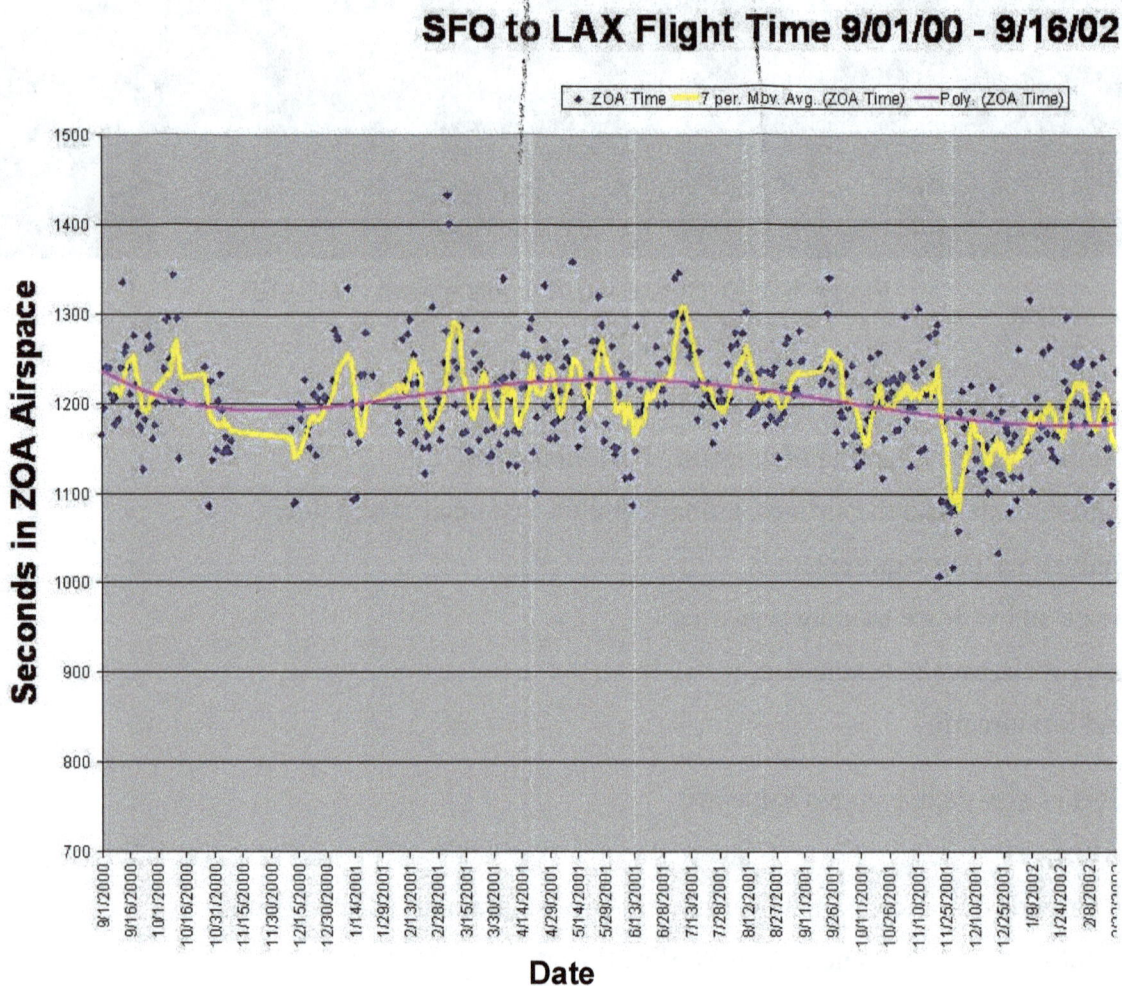

Figure 5.3.21. Trending analysis with PDARS.

PDARS tools allow users to quickly go from summary tables to query tables to charts to three-dimensional (3-D) visualization to animation playback, as indicated in figure 5.3.22. All these capabilities support exploratory and causal analysis using the underlying dataset.

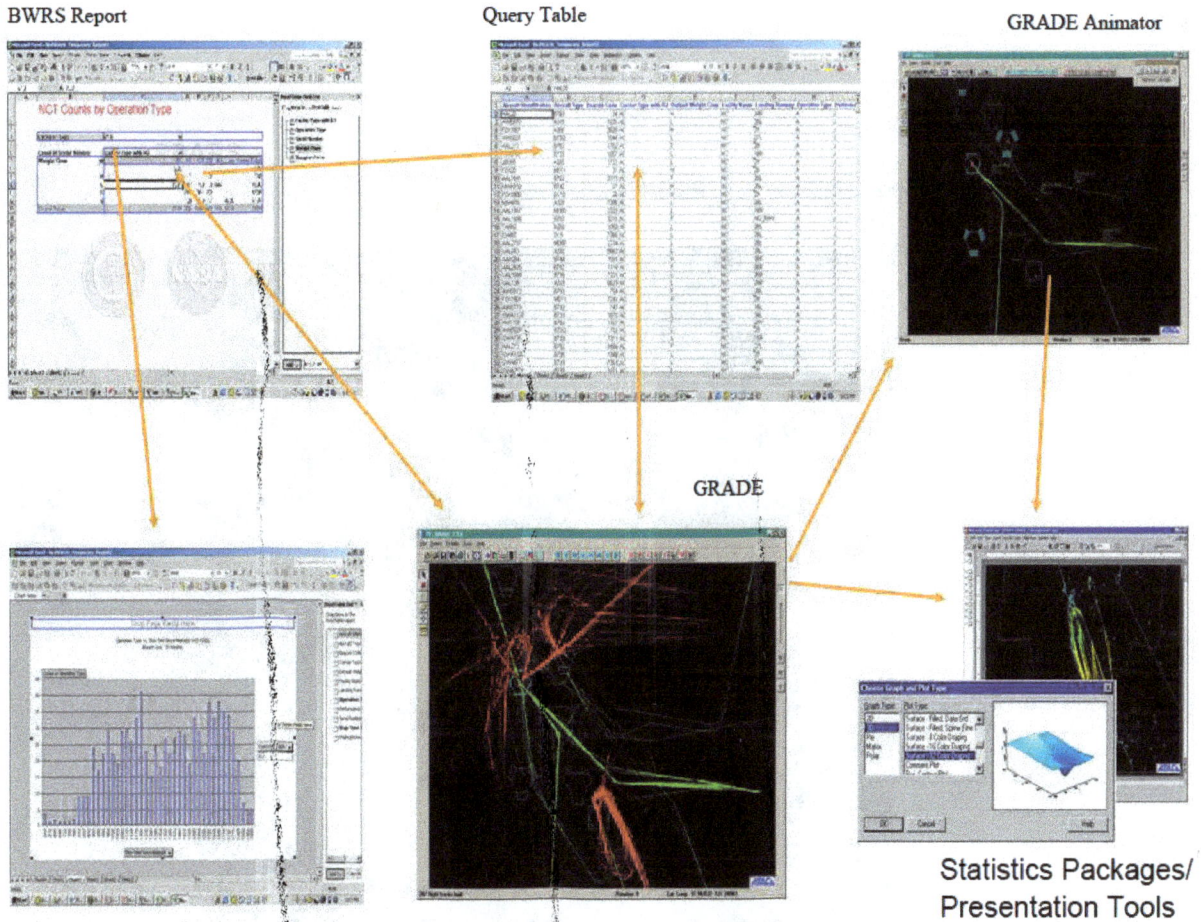

Figure 5.3.22. PDARS enables exploratory analysis.

Figure 5.3.23 highlights some of the visualization capabilities of GRADE. In this figure, GRADE shows flights arriving and departing SFO. Green lines represent the horizontal trajectories of arriving flights, and red lines show departures. The 3-line data blocks show the positions of arriving and departing traffic (in green and red, respectively), at a particular point in time. Grade capabilities include 3-D visualization, custom colorization, and animation playback. Users have found these visualization features highly effective in communicating complex ATC concepts to other aviation stakeholders.

Visualization

Colorization

Animation

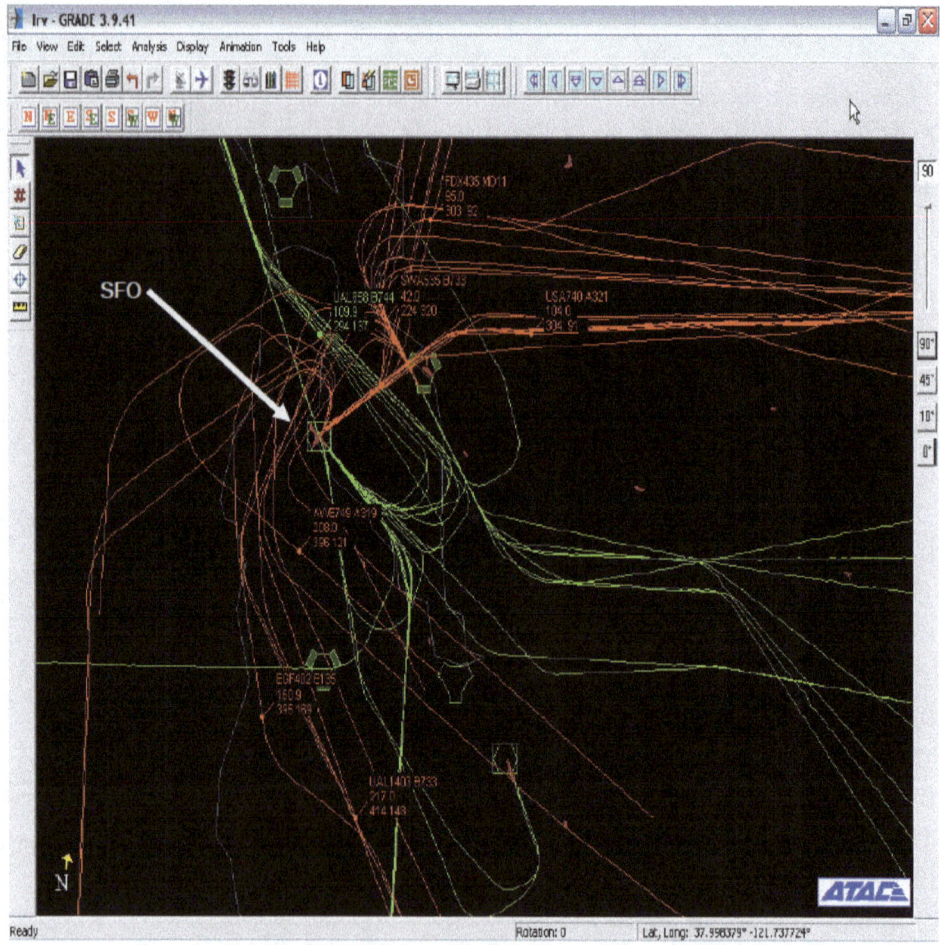

Figure 5.3.23. Visualization of PDARS information using GRADE.

PDARS enables analysis of daily operation of the National Airspace System on local and inter-facility levels in response to a great variety of queries. PDARS allows facilities to identify situations that can be changed or improved and to quantify the consequences of changes from safety and efficiency perspectives. For example, figure 5.3.24 shows an arrival stream to LAX runway 25R. PDARS has computed "aircraft state" events at predetermined intervals (2000-, 1000-, and 300-feet afe for use in analysis of unstable, high-energy, or high-rate-of-descent approaches. The "snapshots" taken of each aircraft at the specified altitudes (called the "EV_STATE" in figure 5.3.24) is the point at which aircraft conditions (such as airspeed, heading, and glide slope) can be measured for subsequent analyses.

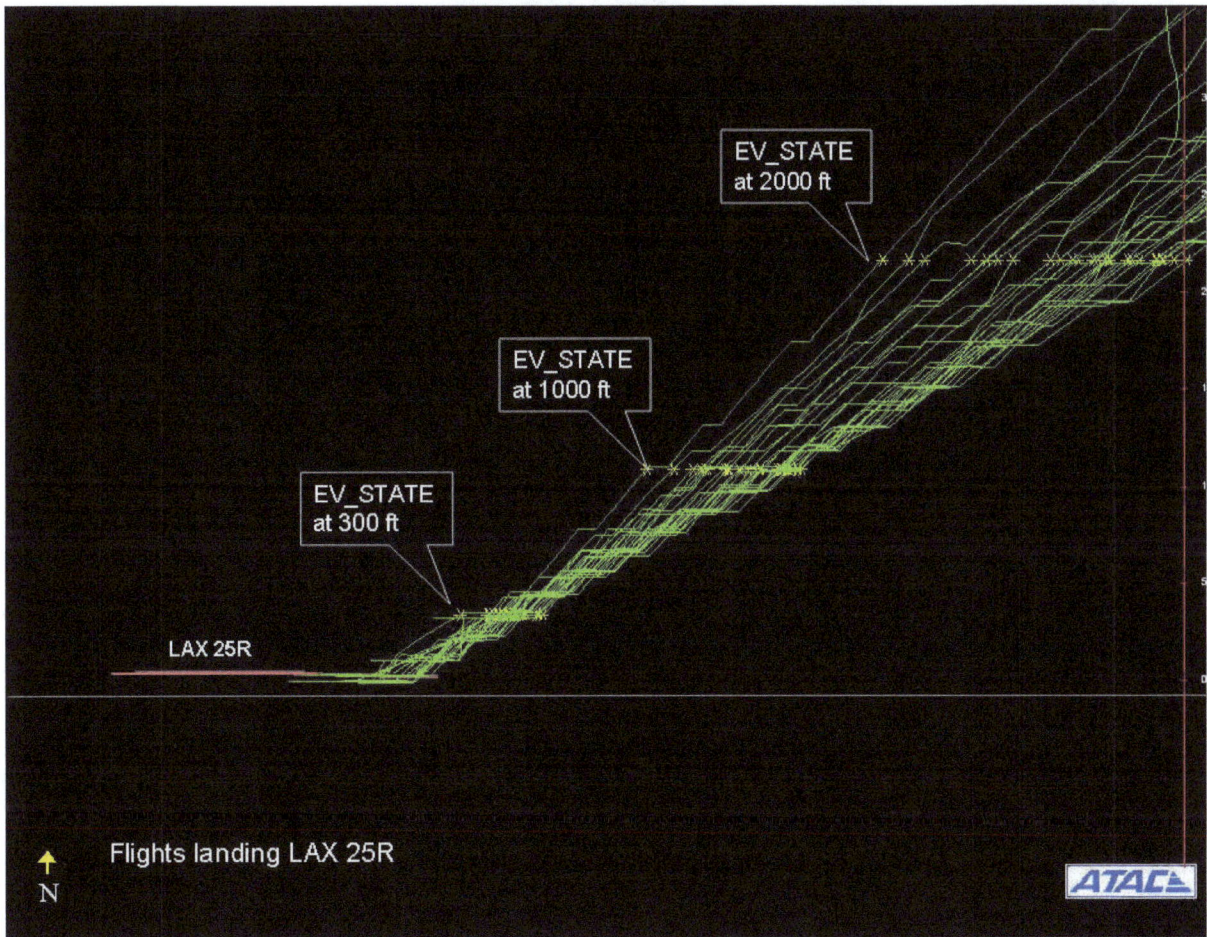

Figure 5.3.24. PDARS supports above ground level (AGL) analysis.

PDARS enables analysis of sector density both numerically and visually, as indicated in figure 5.3.25, which is a horizontal view of a snapshot of the distribution of aircraft relative to sector boundaries.

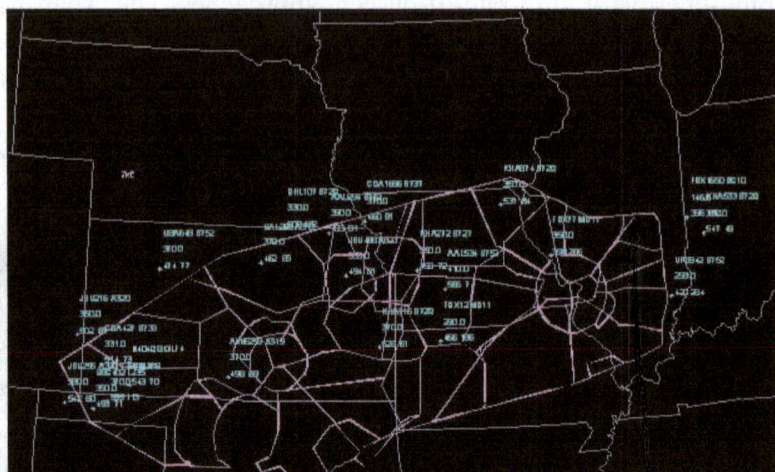

	Count					
	Acid	Altitude	Ground Speed	Latitude	Longitude	Total
8	AAL534	410	506.7	38.910232	-93.387776	
9	AWE259	370	468.6	37.381854	-98.580975	
10	COA1686	370	226.4	40.107822	-95.33336	
11	COA427	331	511.0	37.520675	-101.511883	
12	FDX12	290	470.4	38.006022	-93.526941	
13	FDX77	350	372.3	39.398939	-90.822552	
14	FFT387	370	748.0	36.867598	-101.368332	
15	JBU216	350	501.6	37.986761	-102.18171	
16	JBU255	330	779.0	36.806298	-102.352737	
17	KHA272	350	469.3	39.25	-94.044908	
18	KHA533	330	630.1	39.583012	-87.096715	
19	KHA715	370	845.8	37.928566	-94.98431	
20	N404QS	410	538.0	37.170324	-100.886295	
21	UPS942	294	416.4	38.547642	-88.000642	
22	USA648	370	224.5	39.194353	-100.278192	
23	Grand Total					

Figure 5.3.25. PDARS supports sector-density analysis.

Figure 5.3.26 shows a graph from a PDARS report that depicts instantaneous traffic count through-out the day for one en-route sector of an ARTCC. Also plotted is the Monitor and Alert Parameter (MAP) value for that sector, which is used loosely as a measure of sector capacity.

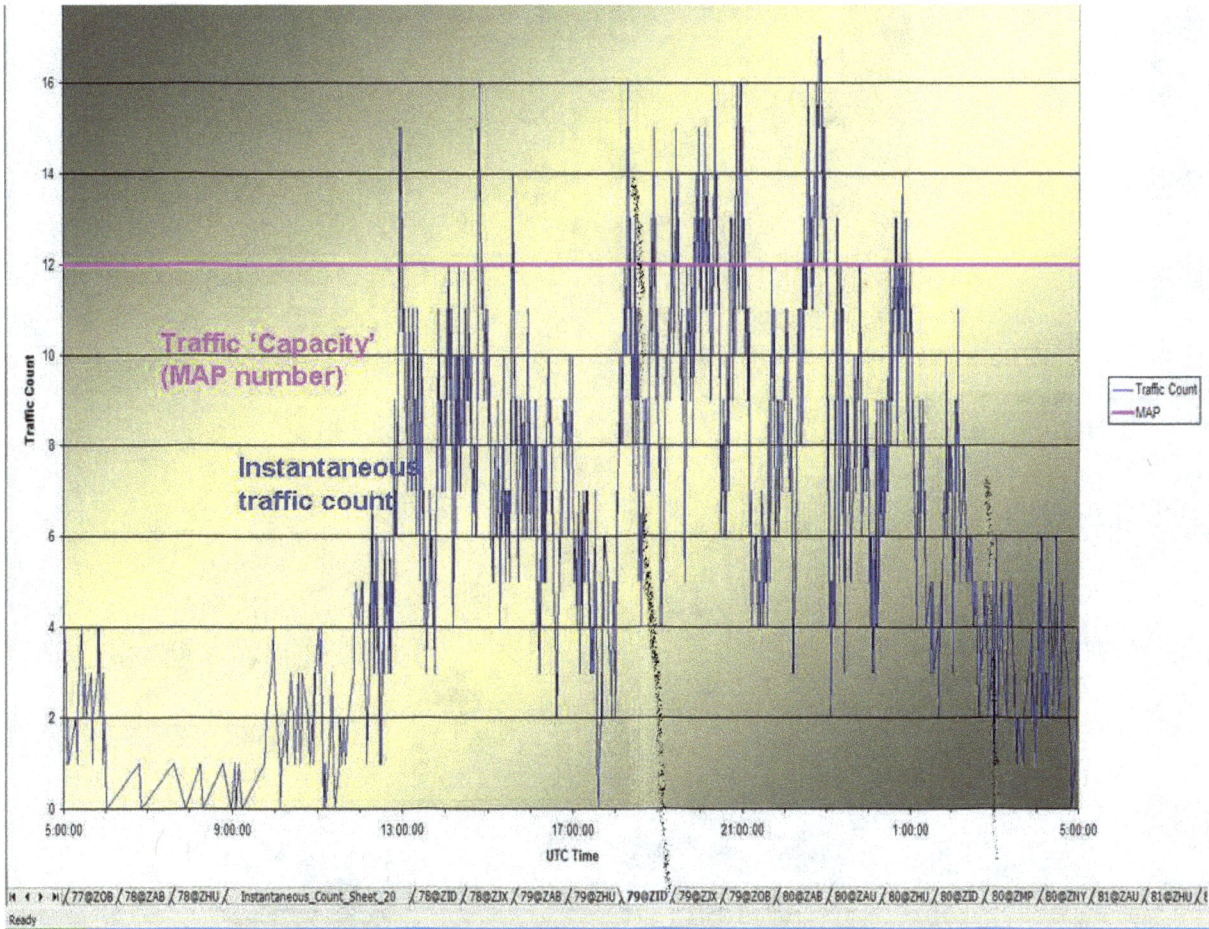

Figure 5.3.26. PDARS supports sector capacity analysis.

Figure 5.3.27 demonstrates analysis of a track compared with its corresponding flight plan route using PDARS. The white line shows the actual track of a flight; the blue dashed line shows its flight-plan route. PDARS maintains information about all the filed, amended, and updated flight plans of en-route transiting aircraft that can be used to compute a variety of track vs. flight-plan metrics.

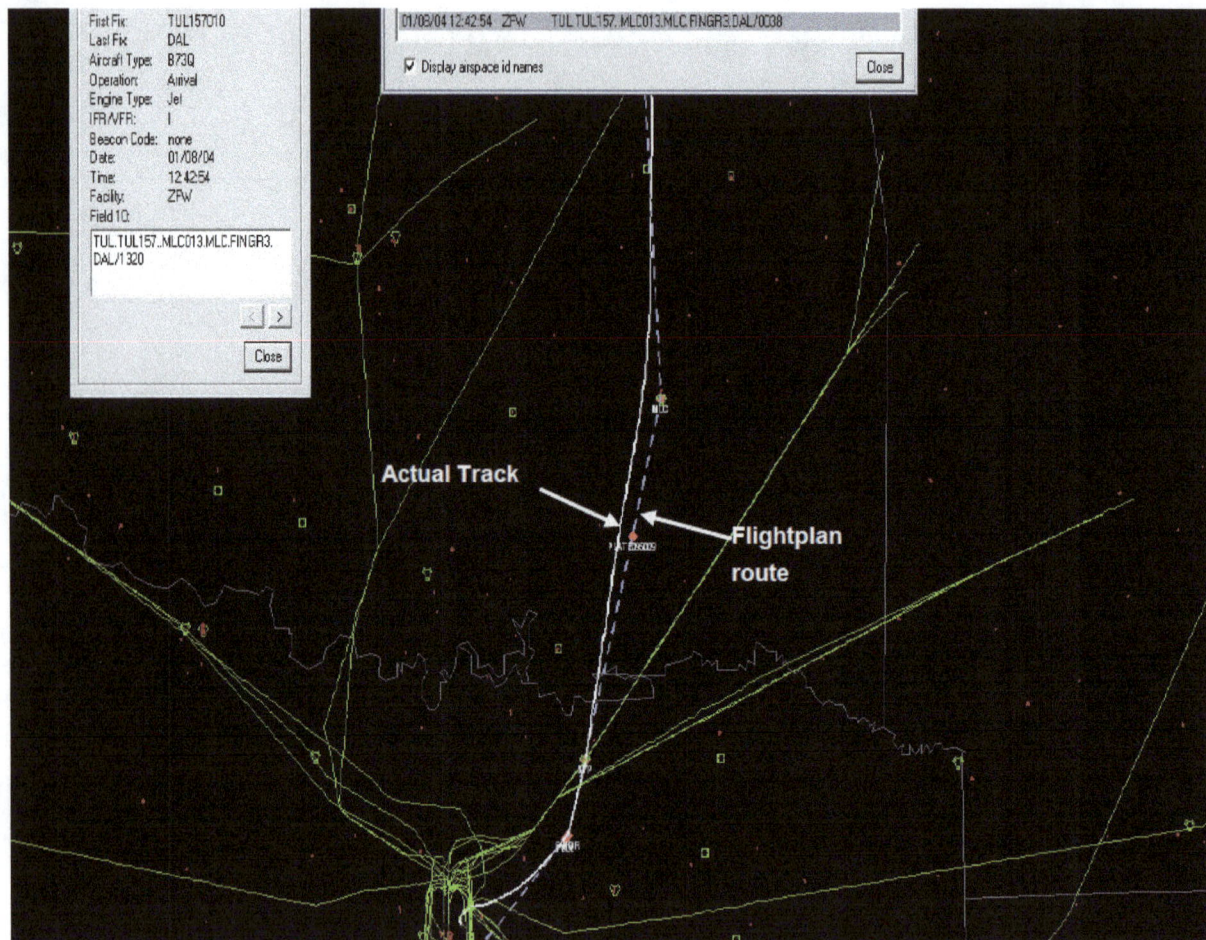

Figure 5.3.27. PDARS supports track vs. flight-plan analysis.

PDARS supports proximity analysis and can numerically and visually show horizontal and vertical distance between aircraft, as shown in figure 5.3.28. The horizontal separation distance (hsep) between the two aircraft is 6.99 nautical miles and the vertical separation distance (vsep) is 6.3 FL. (An FL, or Flight Level, is 100-feet pressure altitude in standard atmosphere.)

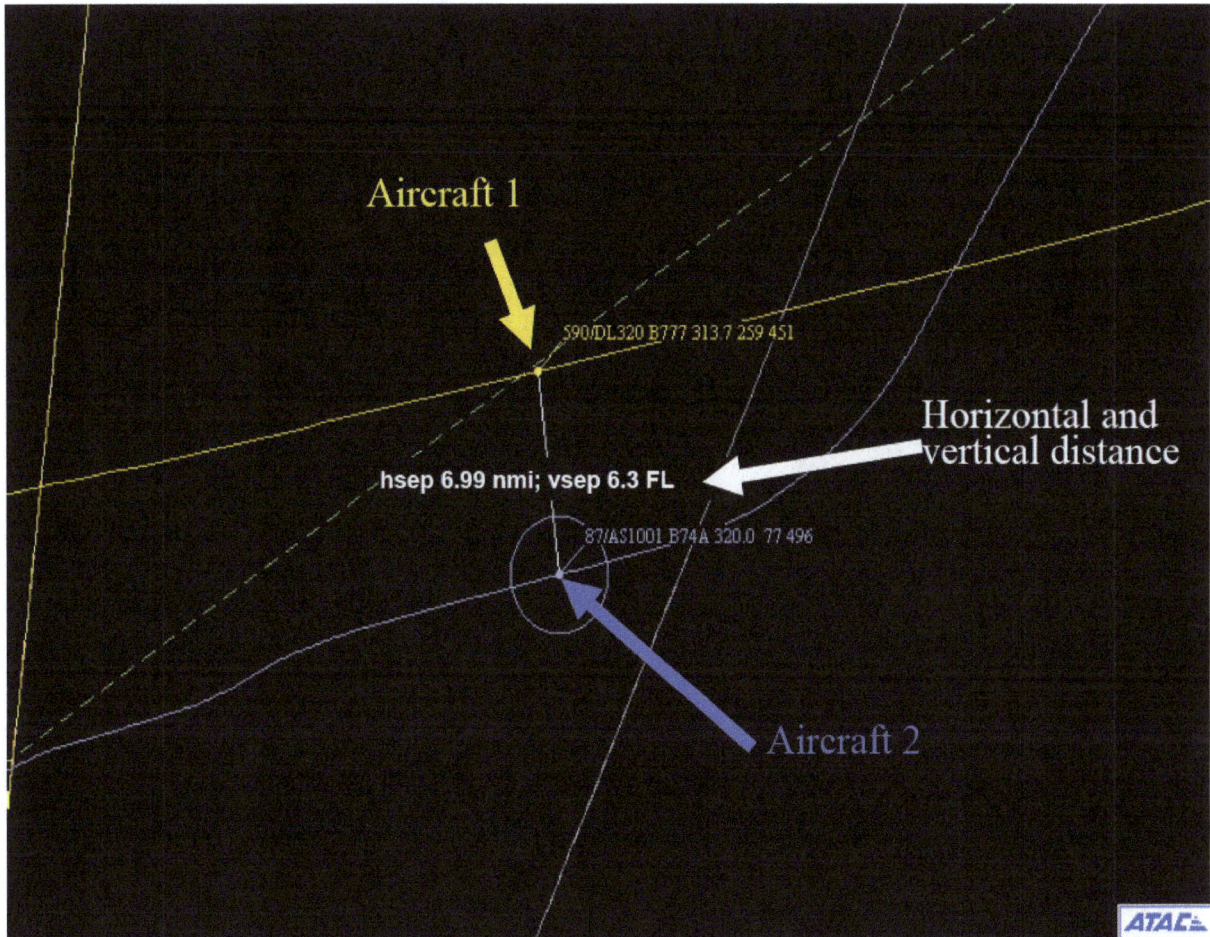

Figure 5.3.28. PDARS supports proximity analysis.

A proximity analysis applied to a large number of flights, and arrivals at SFO over a 24-hour time period, is shown in figure 5.3.29. The tracks are color-coded by distance to the closest flight. A light blue color signifies more that 5 nautical miles 3-dimensional separation distance, whereas red indicates less than 3 nautical miles. The pattern illustrates the typical reduction in separation of flights arriving to closely spaced parallel runways.

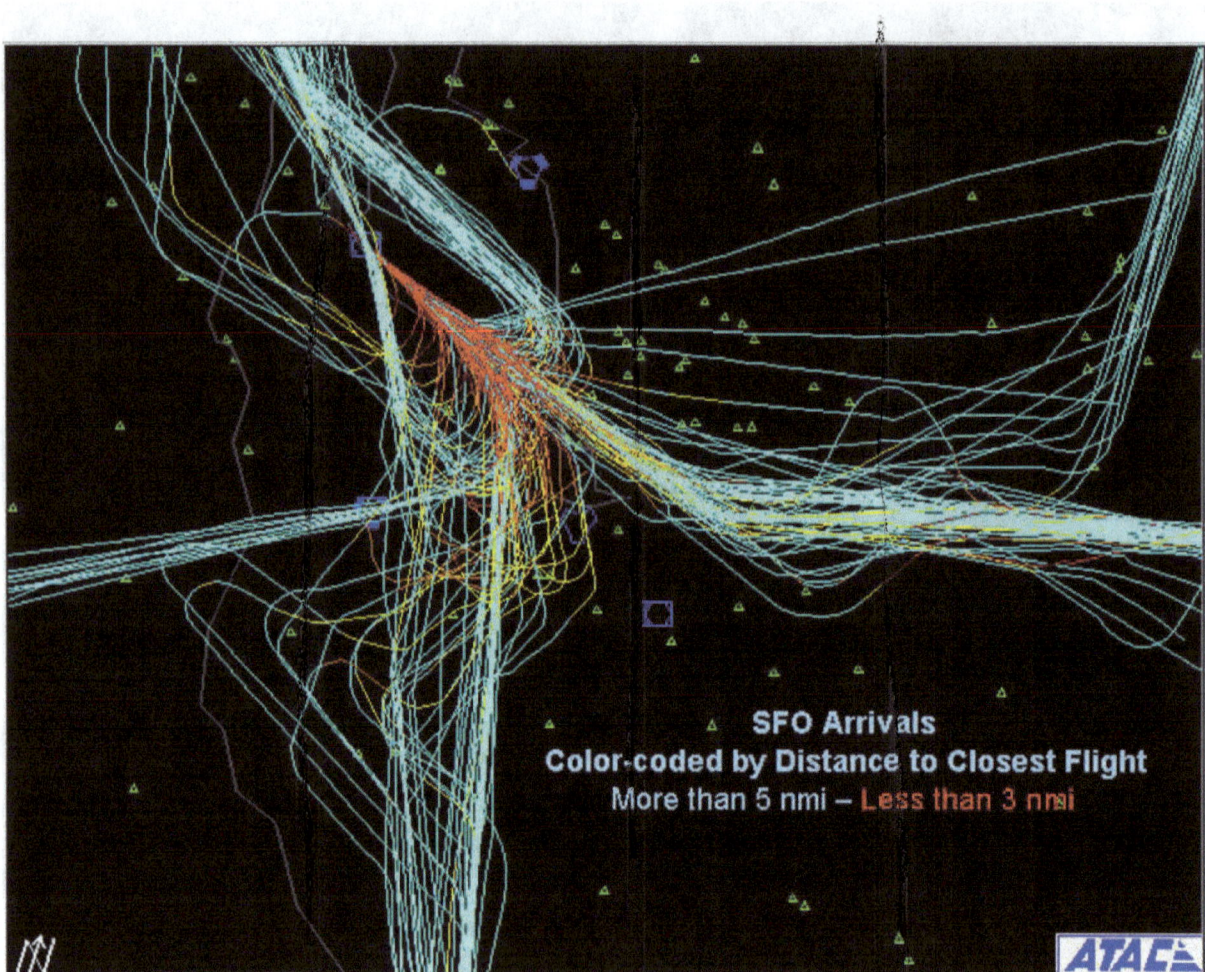

Figure 5.3.29. An example of proximity analysis with PDARS.

Figure 5.3.30 displays how PDARS data can be used to illustrate and analyze locations with a higher than usual number of TCAS Resolution Advisories (RAs), as reported by TCAS-equipped flights. Such RAs tell pilots to take evasive action to avoid collision with other aircraft. The figure shows the area where the RAs occur (the white rectangle) with the trajectories of the aircraft involved (shown as colored lines) and the boundaries of the local Class B airspace (in yellow). This profile view shows that the Class B airspace, which is supposed to separate arriving and departing commercial aircraft from general aviation traffic, does not extend far enough to the east to protect this arrival traffic flow.

When PDARS was turned over fully to the FAA at the end of the ASMM Project:

- It had been in operation for six years.

- Nineteen users' meetings had been convened.

- It had 33 FAA data feeds (20 ARTCCs, and 13 terminal radar approach control facilities (TRACONs)).

- There were 39 FAA facilities connected (20 ARTCCs, 13 TRACONs, 5 Regional Offices, and the Air Traffic Control System Control Center (ATCSCC)).

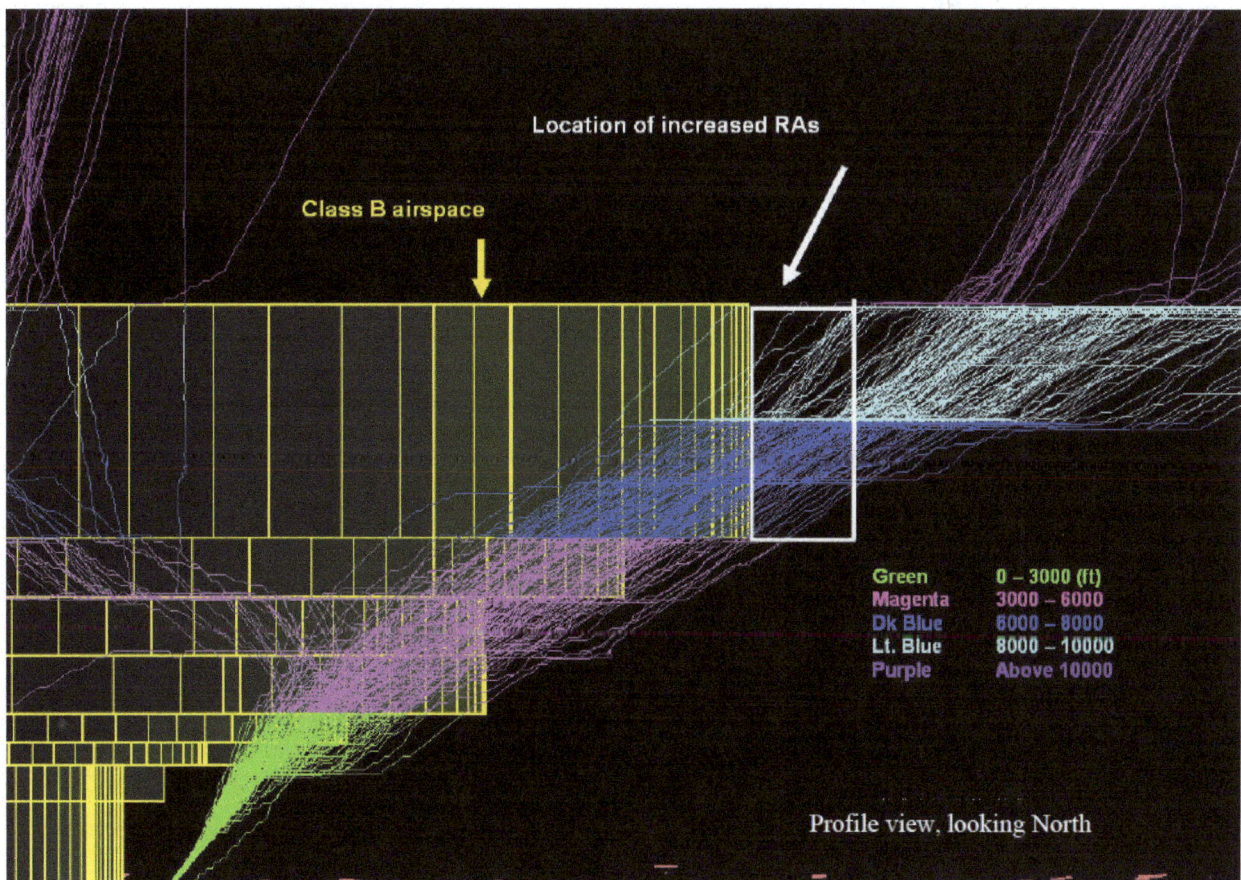

Figure 5.3.30. PDARS supports TCAS RA analysis.

- There were over 200 FAA users.

- More than 400 reports were being generated daily (192 on site; 239 centrally).

- More than 23,000 facility days of data had been archived.

The approach taken in the development and usage of PDARS was a reason for its success:

- Engaged the "user" from the outset of the concept and continuously in its evolution,

- Designed the applications to respond to the users' perceived needs first,

- Provided support, training, and "hand-holding" continuously,

- Responded quickly to users' concerns and recommendations,

- Identified a "champion" able to deal with the "political" issues in the user community, and

- Exploited NASA's reputation, objectivity, and expertise.

The PDARS team had to overcome many political and technical challenges, including:

- Ensuring an acceptable, secure WAN,

- Reconstructing a flight from diverse data sources of segments (analogous to "unshredding" paper),

- Automatic processing for next-day, high-quality, reliable reporting,

- Quickly accessing reports and processed data,

- Using low-cost infrastructure (i.e., standard PC technology),

- Advancing visualization and exploration capabilities to produce meaningful displays,

- Building an evolving system to grow and adapt as technologies advance, as informational needs change, and as users learn more about their needs and the capabilities available in PDARS.

In addition to expanding to all the major TRACONs in the continental United States (from the 13 at the time of transfer), future plans for PDARS at the time that NASA transferred full responsibility to the FAA, included:

- Expanded use of The Morning Report for PDARS,

- Further data integration,
    - Geographic information system (GIS), weather,
    - Facility logs,
    - Scheduling and demand information,
    - Voice communication, and
    - Airline data (if made available),

- Enhanced data mining tools,
  - More efficient algorithms and automation for specific-use cases ("issues"), such as ICAC, and
- Baselining and trend detection.

Moreover, when FAA took the "hand off" of total responsibility for PDARS at the end of FY05, their plans included:

- Continuation of the expansion of PDARS applications,
  - Further expansion of PDARS installations,
    - Plans to cover the 35 destinations included in the Operations Evaluation (OEP),
    - Plans to add oceanic coverage,
  - Further development of metrics,
    - Safety, Capacity, Efficiency, and Cost,
    - Runway, Airport, TRACON, Center, and National Airspace System (NAS),
    - Traffic, terrain, weather, airspace, and obstacles, and
    - Trend Analysis.

At the time of the transfer, the FAA expected that PDARS would continue as a key application for performance and safety measurement and enhancement. In the near term, PDARS was expected to provide the capability for interconnected ATC facilities to collect, extract, process, and merge ATC data routinely; compute quantitative performance measures; disseminate performance measurement reports; access system design and simulation tools for what-if studies; and archive basic operational data and performance statistics. In the longer term, PDARS was expected to be nationalized as an operational capability in the FAA. Radar data from all ATC facilities would then become sources for identifying operational issues and would provide a key element for understanding issues that are identified in other data sources or by human analysts. PDARS would also provide the archive of data on which to base airspace redesign studies and will be the source of the measures of performance to meet the requirements of the GPRA.

At the termination of the ASMM Project, NASA and the FAA were in negotiations exploring ways in which NASA could be permitted to access PDARS data for NASA's future activities. The vision of the future is that PDARS would potentially become a node of a source of information that could be integrated with information from other data sources (such as FOQA/APMS). The ASIST Concept Paper provided in appendix D is a clear statement of the need for such information, not only by the FAA for measuring their ATC performance as stated previously, but also for sharing and integrating with other data sources to identify and gain insight into systemic issues.

In summary, the *Intramural Monitoring* element of the ASMM Project tested, evaluated, and brought to operational capabilities advanced concepts for processing numerical data in the suite of APMS tools for flight-recorded data and PDARS for ATC radar-track data.

5.4   Modeling and Simulations

5.4.1 Fast-Time Simulations

The *Modeling and Simulations* element of the ASMM Project had as its objective the reliable and validated identification of operational hazards, their causal factors, and their risks by simulating the interactions of the multiple human and nonhuman agents operating the aviation system. Successful computational modeling approaches were combined to provide a suite of interactive investigative, evaluation, and assessment tools for system-risk assessment. These developments were motivated by the requirements of the ASMM objectives for a human-centered system model of the interactions and mutual expectations among multiple human and nonhuman agents to enable risk-precursor identification based on human-system capabilities and limitations. This approach provided an essential capability for proactive management of safety risk using a predictive analysis of the systemwide impacts of interventions or local changes in the system. During the Identification step of proactive risk management, safety incidents are identified by APMS, FOQA, PDARS, NAOMS, ASRS, ASAP, or other sources. Fast-time simulations provide guidance in conducting additional investigation of each identified incident and a preliminary risk assessment based on approximate likelihood and probable consequence. During the Evaluation step, fast-time simulations assist in acquiring insight into the operational significance of an identified event or trend, explore for the important contextual factors of that event conducive to failure and human error, and assess the probability of its risk. During the Formulation step, these tools assist in predicting the systemwide impacts of proposed interventions to support the selection of the most appropriate intervention.

One of the most interesting methods of predicting the future behavior of a system is simulation. Simulation has become an important tool to engineers during conceptual and preliminary design for investigating and predicting behavior of complex systems. Its value has been demonstrated when, for whatever reason, it is impractical to exercise the actual system in its true environment, and when absolute fidelity is not essential to studying the system behavior. Simulations entail analogues, replicas, mathematical representations, or combinations of these factors that have been validated and verified at, perhaps, an individual component level.

Investigations of various fatal aviation accidents over many recent years have frequently resulted in their attribution to "human error." Two of the questions that must be answered in order to formulate the appropriate intervention are "Why do highly trained professionals make mistakes?" and "What features of the system operations are conducive to human error?" Identifying when equipment and procedures do not fully support the operational needs of the human operators of the system is key to a proactive approach to reducing error and improving flight safety. Consequently, analyses to identify causal factors and to understand why the accident occurred must account for human performance. Serious human errors that result in aviation accidents are rare events, and the low probability of their occurrence makes the study of these errors difficult to investigate in the field and in human-in-the-loop tests. Human errors characteristically result from a complex interaction of unusual circumstances, subtle "latent" flaws in system design, procedures, training, organizational cultures, and limitations and biases in human performance. Such systemic problems can lead to the fielding of equipment or the implementation of procedures that put the system safety at risk, particularly when they are used in a manner or under circumstances that were not envisioned or tested. Currently, analysis of the causal factors of systemic issues is based on third-party observations, experiential reporting, and, less frequently, human-in-the-loop tests. These tests are not only difficult, labor intensive, and expensive, but they seldom shed satisfactory insight into why the human performed as (s)he did, particularly when it entails some low-probability confluence of conditions and contexts.

Computational models of human performance in large-scale and complex systems have served engineers in predicting system performance for over 20 years, and have been used to predict requirements for aiding systems to augment human performance and ensure safe system operation. Human-performance models have also served the Human-Factors and Cognitive-Sciences communities by establishing a platform for the embodiment of architectural and functional representations of key human operator characteristics. In order to define safe and effective processes and procedures for an evolving system concept, the human operators' performances must be appropriately and consistently included in the design of the new operation and of any automated aiding that is proposed to help the distributed human operators in their activities. In this model paradigm, the response of the humans in the system is a function of the demands of environmental stimuli mediated through the current state of the humans.

The models of human performance in systems then look to two basic interactions to answer that question. First, there are models of the basic information-processing elements, functions, and resources that the human operator uses during an activity (Card, Moran, and Newell (1986) and Wickens' Multiple Resource Theory, Wickens (1984)). In these models, limitations (perceptual, memorial, cognitive, or motor) are provided in order to anticipate how humans might be overloaded by task demands. Humans do not have infinite bandwidth and, in fact, have some unique attributes that determine what types of tasks they can and cannot do simultaneously. Second, there are models describing how the functions of a human operator change in response to environmental stressors (either internal or external to the operator). These models have played a large role in human-reliability analyses as "performance-shaping functions" (Swain and Guttmann (1983)).

The human operator's function in the distributed air/ground air-traffic management (ATM) system includes visual monitoring, perception, spatial reasoning, planning, decision-making, communication, procedure selection, and execution. The level of detail to which each of these functions needs to be modeled depends upon the purpose of the prediction of the simulation model. Human operators sharing control and information with automated aiding systems and with other operators in the control of complex dynamic systems will, by the nature of those systems, need to perform several tasks within the same time frame or within closely spaced time frames. Multiple operators performing multiple tasks are a challenge to the state-of-the-art in human-performance representation. What makes it absolutely essential to model multi-tasking and task management stems from the fact that there are limits to human bandwidth, information storage/retrieval capacity, and the ability to focus attentional resources that clearly affect performance. This effect can lead directly to safety-critical performance by the operators. In other words, when humans are asked to do too many things at once, they change what they do and how they do it. To accurately predict human performance, it is important to understand when the multiple task demands exceed human bandwidth and how humans respond under such circumstances. The number of activities, the complexity of the tasks, and the diversity in human capability all contribute to the complexity of the analysis and are the foundation for research into the understanding and modeling of human multitasking behavior.

Fast-time simulations incorporating models of human-system performance are useful in understanding and predicting the sensitivities of human performance to changes in design, procedures, and training. These simulations examine the effects of parameter variation on human-system performance by small perturbations of each parameter from a nominal value. It is now possible to perform on the order of a million simulations using a modern computer and rapidly generate extensive probability distributions describing the human-system response. Statistical analysis indicates that the uncertainty in the probability distribution decreases as more simulations are performed with sufficient numbers of iterations of the situation being simulated and its variations based in human-system interactions to allow for the distribution of performance to be developed. From these distributions, the probability of the highly infrequent "erroneous" performance can be estimated, and the causal structure of those rare errors can be examined in detail. Potential interventions can also be explored by examining the attributes of the causal structure, making preventative adjustments in selected parameters, and then performing additional simulations to assess the human-system response.

Fast-time simulations have been shown to provide reliable data-driven assistance with prediction to support formulation of the most appropriate interventions to prevent incidents. (See, for example, Aldrich et al. (1989), Blom et al. (1998), Cacciabue (1998), and Blom et al. (2003).) Fast-time simulation is a powerful technique to uncover a design or procedural or training flaw that may induce human error under some low-probability confluence of conditions and contextual factors and, thereby, compromise the safety of operations of the national aviation system. When combined with human-in-the-loop testing in nominal and off-nominal scenarios, fast-time simulations provide a complementary technique to develop systems and procedures that are tailored to the humans' tasks, capabilities, and limitations. Fast-time simulations enable the more expensive and less efficient human-in-the-loop simulations to focus on those situations most likely to be of high safety risk. At the same time, the human-in-the-loop simulations provide bases for validating and/or modifying the models used in the fast-time simulations.

The *Modeling and Simulations* element of the ASMM Project focused on simulations to reveal the systemic causal factors behind human performance when controller, pilot, and automated systems interact in specific flight scenarios. While several effective simulations exist for exploring questions related to capacity of the aviation system, this activity was uniquely designed to explore its vulnerability to events or trends that could compromise its safety. Human-performance models were integrated with existing models of other elements of the aviation system to simulate the interactions among the multiple human and nonhuman agents that operate the NATS and to examine the performance of the system and reveal potential hazards. Specifically, this activity addressed identification of the intended and the unintended consequences of changing the information control balance and demands among human and automated operators in the system.

The *Modeling and Simulations* activity used as the starting point for its objectives the significant investment within NASA and in other agencies in complex human-system-performance representation. Specifically, an activity of another AvSP project called Systemwide Accident Prevention (SWAP) developed the models of human performance used in the Fast-Time Simulations. (A description of these models is provided in section 5.4.3 and appendix E.)

The incorporation of human-performance models is necessary when examining the performance limitations and boundaries of a complex system involving many human-machine interactions. Human performance must be considered to assess potential incidents, determine if a hazard exists, and forecast the potential of a failure in the interest of proactive management of safety risk.

Proactive management of safety risk has the following needs, some of which can be met only with simulation:

- Determine human and machine requirements for successful task performance (perceptual, cognitive, motor, and informational),

- Describe possible mechanisms for that performance,

  - Define causal relationships between the human operators and the context in which they operate,

- Predict performance of human-system model involving multiple interacting humans and machines under nominal operating environments,

- Identify hazards and their causal factors in new operating environments for which there are no data or experience, and

- Determine risks of identified and unexpected hazards.

There are several sources of "safety risk" that can be assessed using a computational model, including emergent behavior from interaction between the human operators and the requirements of the system within which they operate. Analyses of these sources entail imposing a stress-induced risk by running the human-system model in simulations at, or slightly beyond, the performance boundaries that would normally be considered safe. A different sort of emergent behavior occurs in the tails of the probability density functions of performance corresponding to rare events in operations. These events are typically the outlier incidents that entail the confluence of numerous unusual system states and human responses to those states. There is also the emergent behavior of the "novel" response to a hazardous situation that occurs as a result of unique operator behavior (or unique operator-system interaction). All of these sources of risk were modeled in the analyses that were conducted as case studies under the *Modeling and Simulations* element of the ASMM Project.

Several different perspectives and approaches have evolved in developing simulation models of human performance because human performance is complex. The diagram shown in figure 5.4.1 illustrates this diversity, ranging from general theoretical models at the top of the figure to objective-oriented pragmatic models at the bottom, and from biological physical models at the left to process task models on the right.

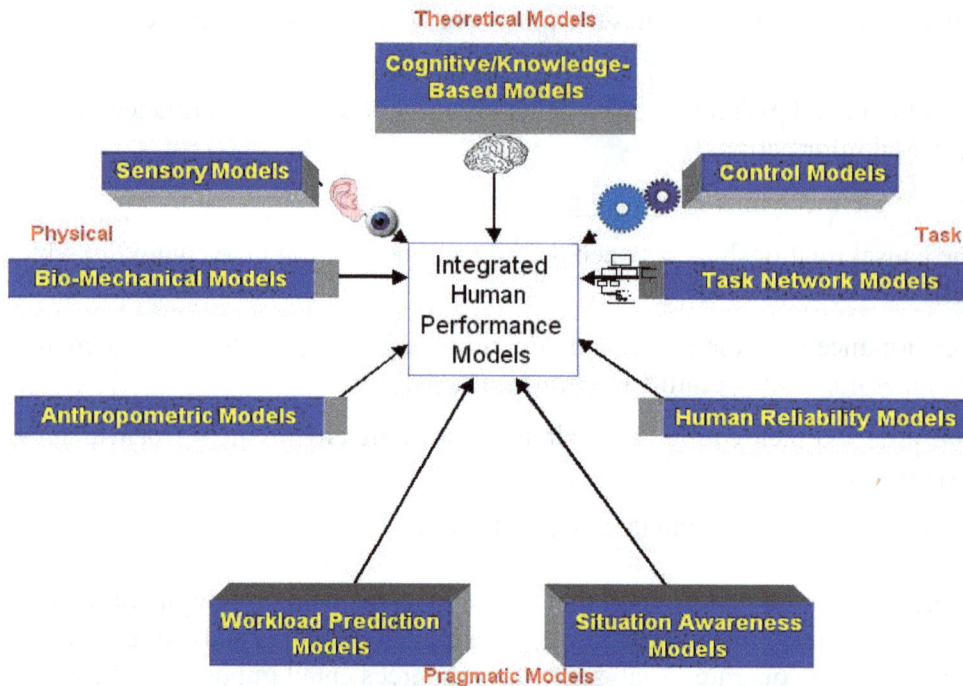

Figure 5.4.1. Perspectives and approaches to modeling human activity.

The state-of-the-art of human performance modeling is described in section 5.4.3, in appendix E, and in Foyle et al. (2003, 2005). (See also the bibliography in appendix E.) The human-performance model selected for the experiments conducted under the *Modeling and Simulations* element of the ASMM Project was the Air MIDAS (Air Man-Machine Integration Design and Analysis System) model developed at San Jose State University, San Jose, California. Of the five predictive models described in appendix E, each with its own unique characteristics, Air MIDAS was found to fit best to the scenarios of these experiments and to the integration with the simulation of the rest of the system required in each scenario. Air MIDAS is described in appendix E as an integrative multi-component cognitive model designed to examine workload, memory interference, and misperception and is most closely related to the "Workload Prediction Models" of figure 5.4.1. Air MIDAS is driven by a set of user inputs specifying operator goals, procedures for achieving those goals, and declarative knowledge appropriate to a given simulation. These asserted knowledge structures inter-act with, and are moderated by, embedded models of cognition for managing resources, memory, and action.

The diagram shown in figure 5.4.2 represents the primary model components in an integrated, closed-loop simulation of the aviation transportation system and human operator. On the left side of figure 5.4.2 are the "domain" models that represent the portion of the aviation transportation system, the equipment, and the characteristics of the portion of the real world with which the operator inter-acts. The use of the term "World" in the figure and in the subsequent discussion is meant to imply this domain, which is the full extent of the world within which the human model operates in a given simulation. It is the representation of everything external to the human operator(s) essential to the scenario of the simulation.

110

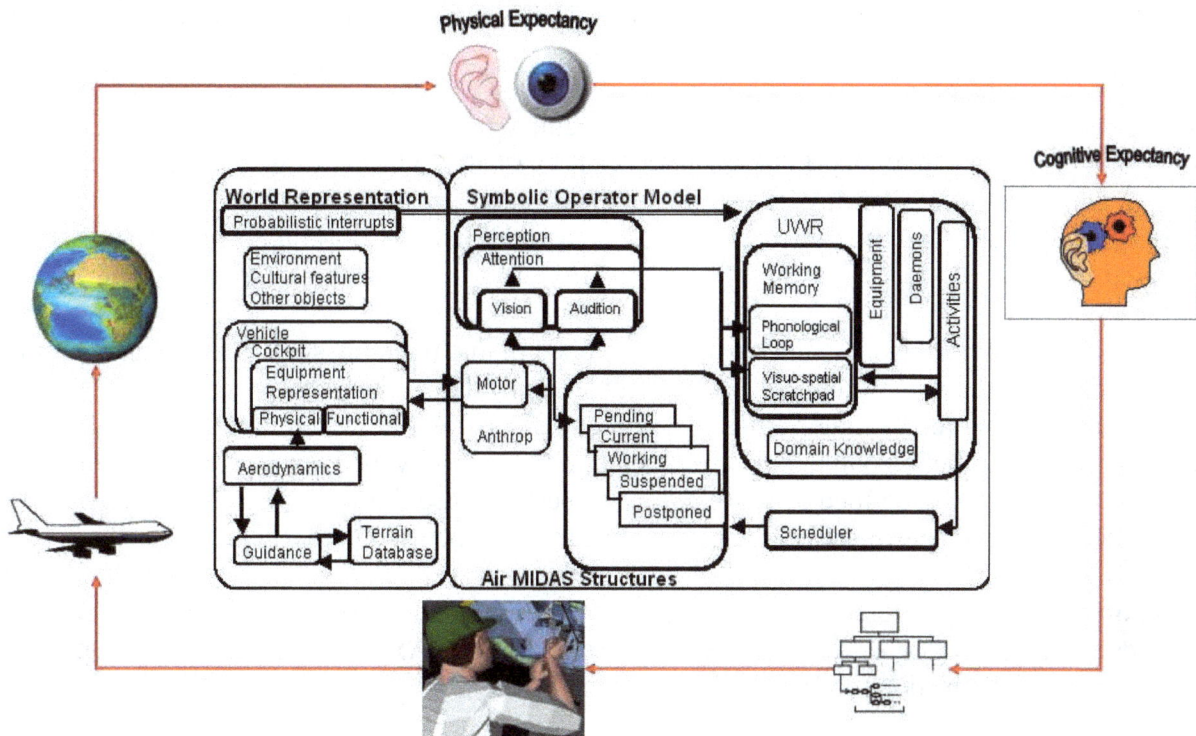

Figure 5.4.2. Representation of the models simulating the aviation transportation system and human operator (Air MIDAS).

On the right side of figure 5.4.2 are models describing components of human-operator performance. The human-operator model functions as a closed-loop control model with inputs coming from the world of the model, i.e., from the representation of the aviation transportation system, and with outputs representing action taken in this world.

The Air MIDAS software is performance-prediction software that uses models of human performance within an integrated computational framework to generate workload and activity timelines in response to operational environments. As shown in figure 5.4.2, the main components of the model are the simulated operator's world representation and the symbolic operator model (SOM) representing perceptual and cognitive activities of an agent. Within the SOM, the Updateable World Representation (UWR in figure 5.4.2) contains information about the environment, cockpit, vehicle, physical constraints, and terrain. Updates of the states of these elements are provided through the perceptual and attention processes of the SOM.[9] The world representation serves to trigger activities in the SOM to serve mission goals or respond to anomalies. The UWR also contains the Working Memory (WM) of the simulated operator, the domain knowledge, and a goal-based procedural activity structure. In Air MIDAS, the internal UWR provides a structure whereby simulated operators access their own tailored or personalized information about the world in which this model operates. The structure and use of the UWR is akin to human long-term memory and is one of the aspects of Air MIDAS unique among most human-system modeling tools. In order to capture the central

---

[9] Like a human's, the model's representation of its world may not be consistent with reality.

role of schema and internal representation, the Air MIDAS model has an elaborate representation of both declarative and procedural information.

Activities to be performed are managed through a queuing process and scheduled according to priority and resource availability. Four resource pools (Visual, Auditory, Cognitive, and Psycho-motor) are checked for availability in response to the demands by the required tasks. Depending on the resources available and the task requirements, an activity may be working, pending, or postponed.

Tasks or activities available to an operator are contained in that operator's UWR and generate most of the simulation behavior. Within Air MIDAS, a hierarchical goal representation is used. Each activity contains slots for attribute values describing, e.g., preconditions, temporal or logical execution constraints, satisfaction conditions, estimated duration, priority, and resource requirements. The durations of activities are defined by distributions of performance times and, at any given performance, a stochastic mechanism selects the performance time to be used in that schedule cycle.

To support adaptation and response to unanticipated stimuli in the mission, a continuum of contingent or decision-making behavior is also represented in Air MIDAS, following the skill, rule, knowledge-based distinction reported by Rasmussen (1983).

Consequently, within the Air MIDAS model are stochastic representations of human limitations in capacity resources (visual, auditory, cognitive, and motor), in storage (working memory and procedural queue), in look-ahead capability (decision horizon), and in interruption and resumption management. Air MIDAS also models the inherent stochastic variability in activity performance and in context mode variability in activity selection. Further, it represents the stochastic nature of the adaptation in response as a function of context and of the adaptation of workload and priority levels based on the operator's state. For a detailed discussion of the implementation of the contextual model of Air MIDAS, see Verma and Corker, 2001.

The Air MIDAS model provides psychological plausibility in the cognitive constructs of long-term, working memories (with articulation into spatial and verbal components of these models), with sensory/perceptual and attentional components that focus, identify, and filter information from the simulated world for the operator's action and control. The cognitive function is provided by the interaction of context and action. The human and task environments are considered to be interactive and inter-responsive rather than linear components in an information-processing system. The human operator modifies the process of task performance as active adaptations to his/her perceptions of his/her own capabilities and the task environment demands. These adaptations include an understanding of context and a requirement to include context and its potential impact explicitly in the system-performance model. The explicit inclusion of context-responsive behavior in human-performance models of the Air MIDAS has been demonstrated by Corker and his colleagues (Verma and Corker (2001), Corker et al. (2000)). The Air MIDAS model takes the construct of context to guide the selection of activities to be performed to reach the operator's goals in simulation. In order to have these activities play out in the simulation world, the Air MIDAS model has a scheduler that accounts for the resource constraints of the human operator; i.e., the human-performance model is limited in the cognitive, perceptual, and motor tasks that can be performed simultaneously. Activity in this model is assumed to be concurrent unless constrained by resource or information.

The recorded output of the Air MIDAS version of the Operator model generally takes two forms: the output associated with measurement of the performance of the model in the task domain, and that associated with measurement of the modeled (internal) human-performance characteristics, i.e., the psychological functions that are responsible for the activities of the model. Psychological performance is recorded by "biographers" that keep track of the perceptual demands, operator's attention demands, cognitive loading, context-control switching, and memory representations.

System performance is also recorded as the modeled operator performs functions to meet mission goals in the domain of the simulation. These external measurements of task-related performance are scheduling, degradation, and shedding of tasks; time to complete a task; and timeline information on tasks performed. This information includes such things as the flight tracks of aircraft, the aircraft pilot's and the air-traffic controller's commands, the number of incidents, etc.

The Air MIDAS model of human performance was used in three missions, or case studies, to identify hazards and assess risks. Each of these different studies was also used to develop and explore the capabilities of the fast-time simulations. Each study had unique characteristics to demonstrate the applicability of the Air MIDAS model and to evolve its capabilities. The three studies to be discussed here addressed:

- Clear Air Turbulence Predictions (CAT),

- Miles-in-Trail and Time-Based Metering,

- Surface Movement Operations.

In the first two studies, Air MIDAS was linked to the Reconfigurable Flight Simulator (RFS) developed by the Georgia Institute of Technology (Georgia Tech) to identify the hazards and their causal factors in each scenario. (See Shah et al. (2005) for a description of RFS). The RFS was used to represent the left side of figure 5.4.2, which is called the "World Representation" in the figure. As stated previously, this term is simply meant to imply everything external to the human operator(s) with which the operator models are required to interact in the scenario of the simulation. For the last study, entailing a risk-assessment study of surface-movement operations, Air MIDAS simulations provided inputs to the Traffic Organization Perturbation Analyzer (TOPAZ) produced by the Dutch National Aerospace Laboratory (NLR) (Blom et al. (1998) and Blom et al. (2003)).

The CAT study analyzed the impact on the aviation transportation system of a modified commercial aircraft cockpit that had a display of CAT sensors. Three scenarios were studied: one in which there was no advance warning of CAT (the current state), one in which the CAT system provided an alert 1 minute in advance of entering turbulence, and one in which the CAT system alerted 5 minutes in advance of an encounter. The simulation examined the impact of these sensors on ATC in a sector with 12 to 16 aircraft. It was assumed that all the sensors worked reliably (i.e., no misses and no false alarms) and that all the aircraft were equipped with these sensors. New procedures describing the flightcrew response to CAT alerts were provided in the simulations as required.

In the CAT study, models of the controller and the pilots (all of whom were represented by Air MIDAS models) were linked to Georgia Tech's RFS (Pritchett (2002)). The RFS was used to represent the portion of the external world with which the human-operator models had to interact, as shown in figure 5.4.2. RFS was linked to Air MIDAS via a run-time interface through the DOD-developed High Level Architecture (HLA) to establish real-time integration between the models, and synchronize their operations through coordination of their independent internal clocks. The human-operator model representing ATC "looks" at the airspace provided by RFS, makes decisions on what actions need to be taken to manage the traffic in that airspace, and then communicates those decisions to the human-operator models who pilot the aircraft according to the ATC commands.

Figure 5.4.3 shows the number of "monitoring" processes that the air-traffic controller model was able to complete as a function of the prediction time of the CAT alert. The monitoring process is the method used by the controller model to gain information about the state of his/her world. The model is programmed to move its point of visual regard in a pattern based on the scan pattern used by controllers to ensure that information is not missed as they scan their radar screen Display System Replacement (DSR). The model simulates the movement by interrogating the equipment model (radar) for information at an x,y point on the surface of the radar display. (In three-dimensional equipment, as, for instance, cockpit instrumentation, the model also has a z-plane dimension that refers to whether the information displayed is in the focal plane.) At the point of inquiry, the equipment model provides information about its state (radar signature) at that time. If the model dwells on the position for a sufficient amount of time (which varies as a function of the type of equipment being scanned, see Landy (2002), then full information about that point on the equipment display is provided. If the model does not achieve a full dwell time, then a reduced set of information is passed to the UWR of the model. A controller's scan pattern of the radar display starts with the periphery of his/her sector boundary, then to any crossing points for reroutes at co-altitude, and then to each quadrant of his/her sector in a clockwise process. It is important that this process be as continuous as possible.

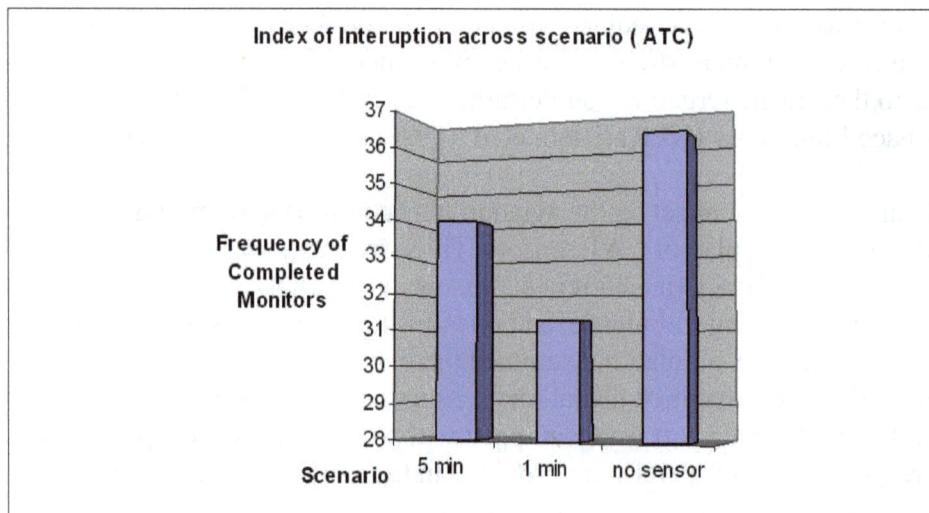

Figure 5.4.3. ATC frequency of interruptions.

Figure 5.4.3 shows that the 1-minute look-ahead sensor resulted in a reduction in the number of monitors completed. The 1-minute look-ahead was the most disruptive of the controller's monitoring process because of the link of the model to flight-crew behavior. If there was no alerting system, the flight crew penetrated the turbulence in sequence of their route of flight. They would call to the ATC to request a new altitude, but only after some (modeled) work to call to the cabin, reduce speed, and assure continued stable flight. The performances of these tasks are distributed in time because of variability in both the performance time and flightcrew scheduling of each task and, furthermore, distributed as a function of the flightpath of each aircraft relative to the area of turbulence. Similarly, in the case of the 5-minute look-ahead, the flightcrew model would schedule a call to ATC to change altitude as a lower-priority task than any other activity that was being performed. Consequently, in both the no-alert and the 5-minute-alert cases, although each aircraft would still ask for an altitude change, the requests would occur in a temporally distributed manner. However, when the CAT alert is received on the flight deck at 1 minute to penetration, each flightcrew model scheduled a call to ATC as a priority task as soon as they received the alert. Since many aircraft approached the turbulence field along different routes at about the same time, many near-simultaneous calls to the controller resulted, all of which had high priority and, therefore, they produced a period of interruption for the ATC. The introduction of these simultaneous high-priority tasks caused the reduction in the number of monitors completed.

The graph in figure 5.4.4 illustrates the time spent monitoring other traffic as a function of the sensor look-ahead characteristics. Again the 1-minute look-ahead provides the least time available to monitor because of the nearly simultaneous multiple priority calls for altitude changes described previously, whereas the controller's performance with 5-minute CAT alerts was about the same as with no alerts.

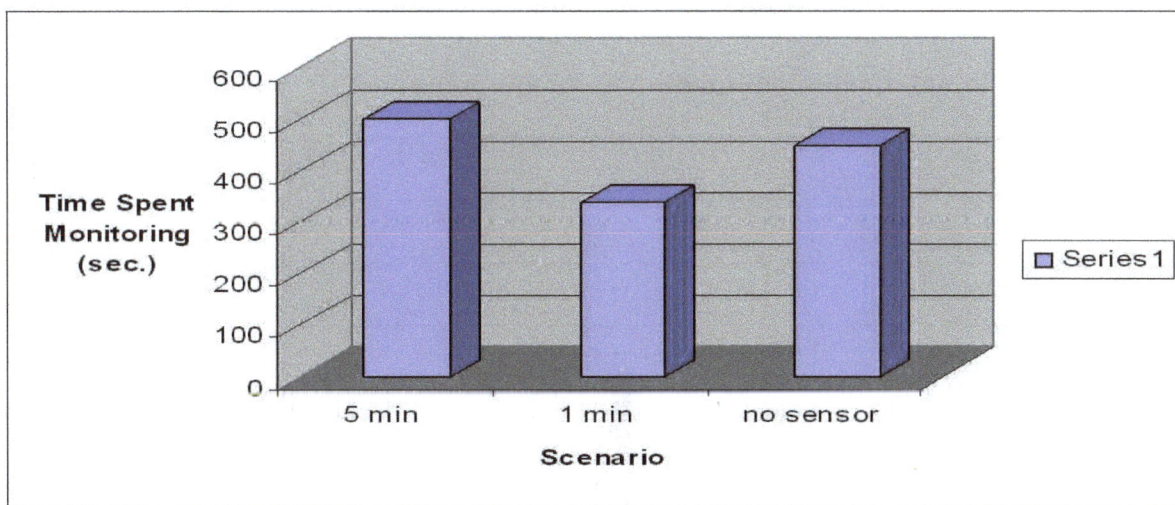

Figure 5.4.4. ATC monitoring time.

The graph in figure 5.4.5 illustrates the cumulative load on the air-traffic controller model as a function of the look-ahead time of the sensor. The condition of no sensor has the least impact on the controller's cumulative load. The 1- and 5-minute look-aheads have about the same impact on the workload of the controller model. The cumulative workloads presented here do not reflect the interruptions due to the massed calls at 1 minute to penetration nor the impact on the available monitoring time discussed previously. The workload calculation process of the Air MIDAS model does not have a function that accounts for the rate at which tasks are introduced (although, perhaps, this function should be developed). As each task is performed, the load associated with it is added to the cumulative workload. If there are too many tasks to be performed at given time, the lower-priority tasks are interrupted (as is the case of the controllers interrupting their scans in the 1-minute-alert scenario). However, if the deferred tasks are performed eventually, then their loads are added to the cumulative load in the model. It is the dynamics of when the calls for changes come that affects the interruptive behavior. A time line of second-to-second model load would show the spike associated with the management of the calls associated with 1-minute alerts.

The experiment indicated that onboard CAT sensing affected the simulated controller. This effect was observable because the Air MIDAS model integrated with the RFS enabled a full closed-loop-system representation that included representations of both air-traffic and flight-deck operations. The procedures that were implemented in the simulations were those anticipated by the designers of the CAT equipment. As described previously, the sequence of response on the flight deck differed as a function of when the alert was received relative to the penetration of the turbulence. The combination of those procedures and the look-ahead time in the system resulted in an interruption pattern that had its highest impact on controller performance at 1-minute look-ahead time. The model imposes limits on tasks to be performed based on load limits for scheduling of tasks by the operator. In the

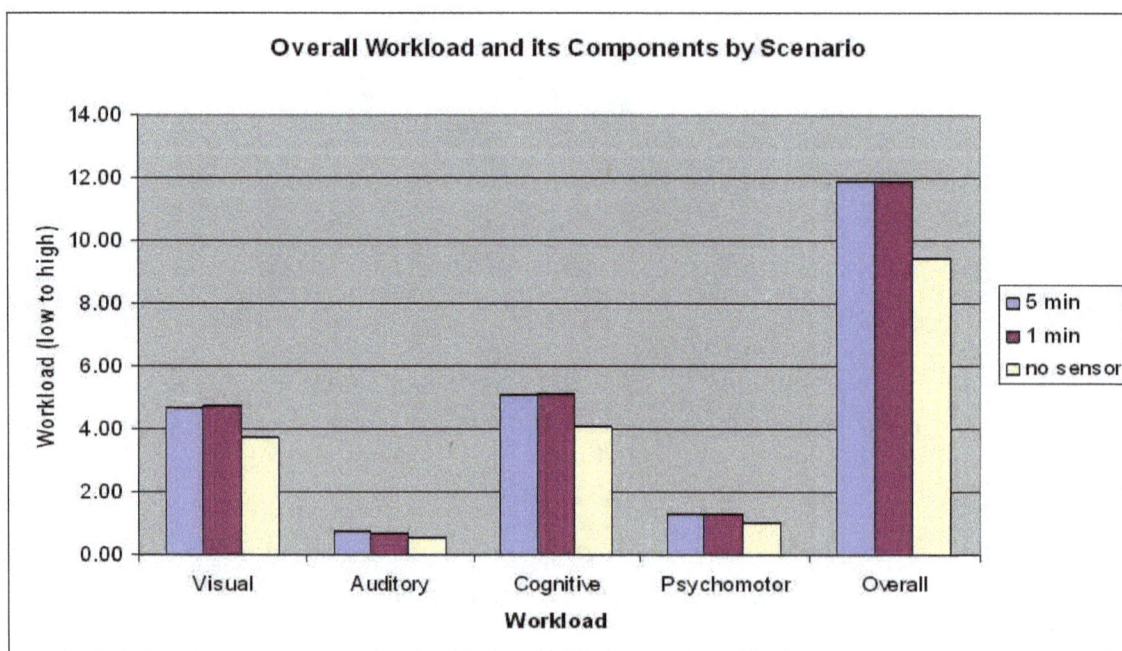

Figure 5.4.5. Workload across Scenarios.

experiment performed, the monitoring task of the controller for the 1-minute look-ahead design had to be interrupted more frequently than in the other scenarios to maintain the acceptable load level for the controller model. It is worth noting that 1-minute look-ahead was the design specification for the CAT system that was most likely to be implemented.

A second study was conducted to further develop and explore the capabilities of the Air MIDAS-RFS linkage to analyze statistically the systemic operational impact of a procedural change. This study was performed to explore the differences in workload resulting from time-based-metering (TBM) and miles-in-trail (MIT) metering of high-density traffic on approach to LAX through the Southern California TRACON and the Los Angeles ARTCC (ZLA).

In agent-based simulation, models of the agents relevant to the operational concept are created and executed at the same time in one large collective simulation environment, as diagrammed in figure 5.4.6. In the specific study of metering traffic flow, this representation included the environment of the arrival streams, and the agents of interest included the controllers of each of the center and TRACON sectors; communication channels for controllers emulating their VHF voice communication frequencies; surveillance technologies mimicking their radar systems; and the pilots and aircraft. The controllers were critical contributors to the overall system behavior in the scenarios relevant to this study, and thus were modeled using Air MIDAS. Multiple copies are made of each model type within the simulation run, each representing a single actor within the system. For example, if the arrival stream currently contains 50 aircraft, then the simulation has 50 aircraft models and 50 pilot models.

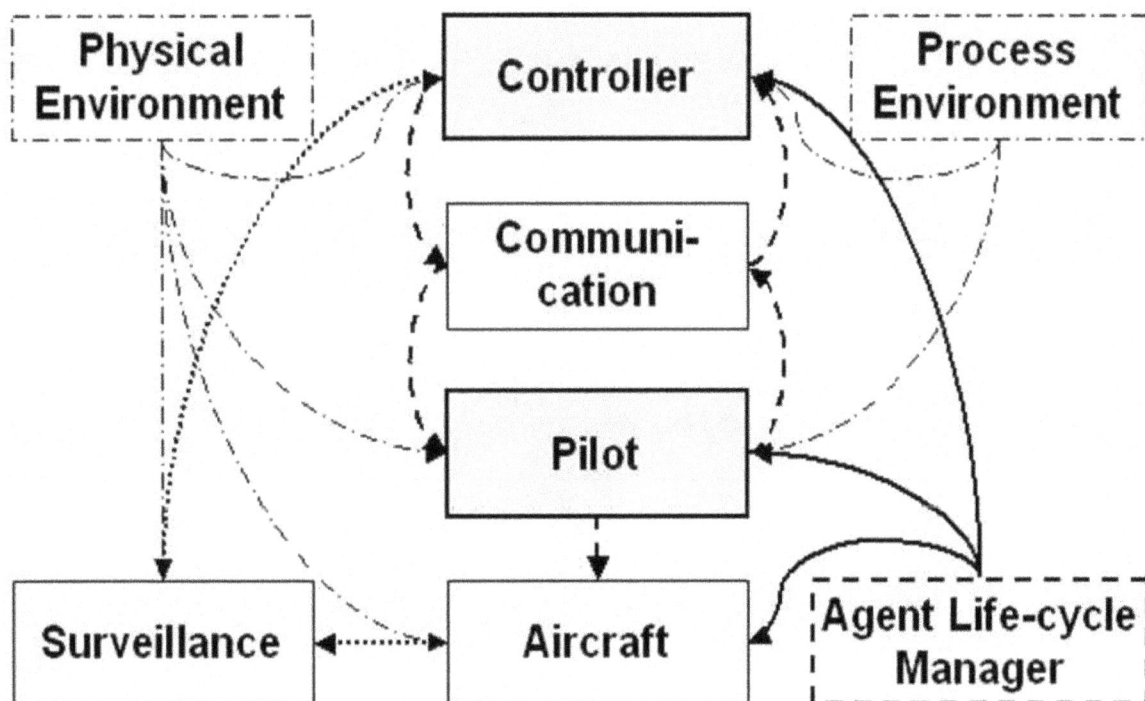

Figure 5.4.6. Agent-based simulations of managing arrival streams.

For MIT metering, the TRACON establishes distances to be maintained between aircraft during approach for landing. In TBM, Center establishes the times at which each aircraft is to arrive over a fix.[10] Procedures for both types of approach control were developed for two Center sectors and one TRACON sector. The air traffic recorded by PDARS for four days under each traffic control condition served as the validation traffic base. The simulation of controller performance was run using one hour of selected traffic with 54 aircraft in both approach and over-flight regimes. A nominal flightpath generated by using averaged flight parameters (i.e., latitude, longitude, altitude, and speed) at fixes and entry/exit points of the sectors was used as a reference.

The PDARS data in figure 5.4.7 illustrate the over-flight and arrival traffic patterns into LAX. The sector boundaries are sketched in purple. The over-flight traffic is in yellow, blue, and pink, and the arrival traffic is shown in green. The simulation included Air MIDAS models of the controllers of Center sectors 19, 20, and the lowered feeder sector (LFDR) and the air traffic flow was simulated in RFS.

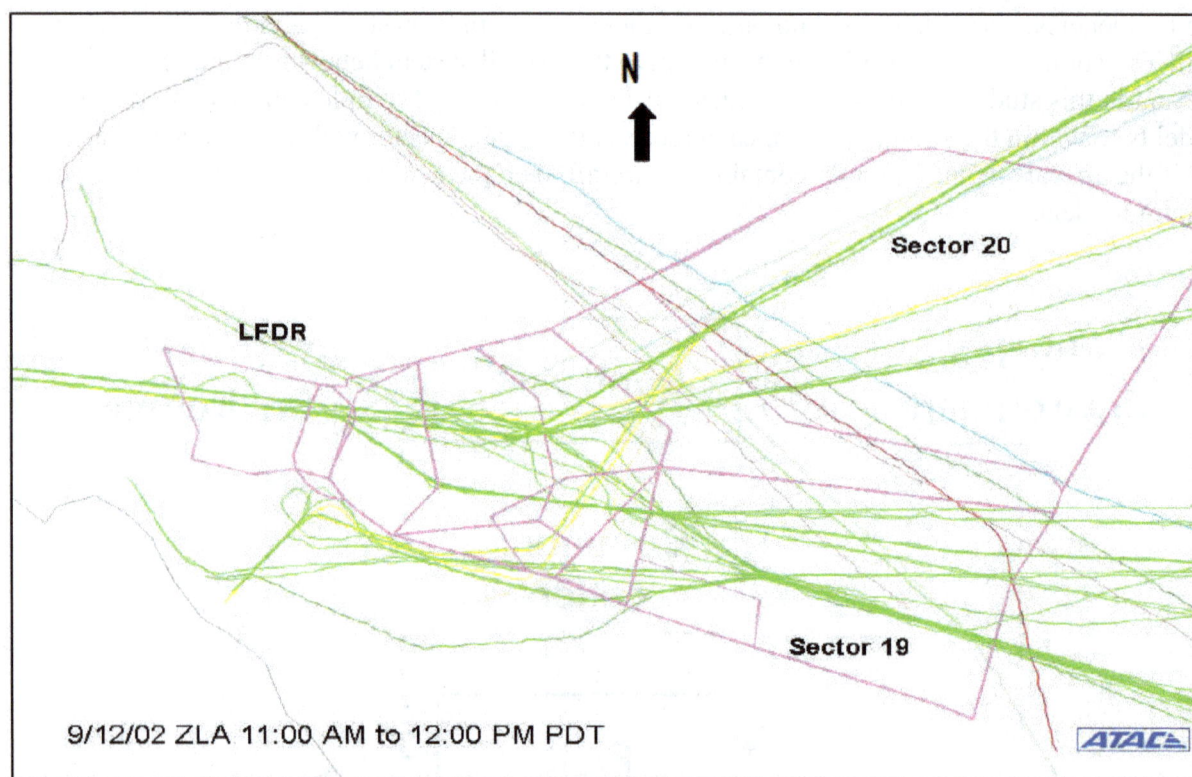

Figure 5.4.7. Example of PDARS data on flights through sectors 19, 20, and lower feeder sector (LFDR) into LAX.

---

[10] A fix is an intersection of two navigational (VOR) radials or the intersection of a navigational radial and distance measuring equipment (DME) used by ATC.

Figure 5.4.8 is an example of the correspondence of the modeled controller performance (white) to the human controller (green—from PDARS) in managing descent into LAX. (The y-axis represents the vertical flightpath of the descent in thousands of feet. The x-axis is the longitudinal distance on the approach.)

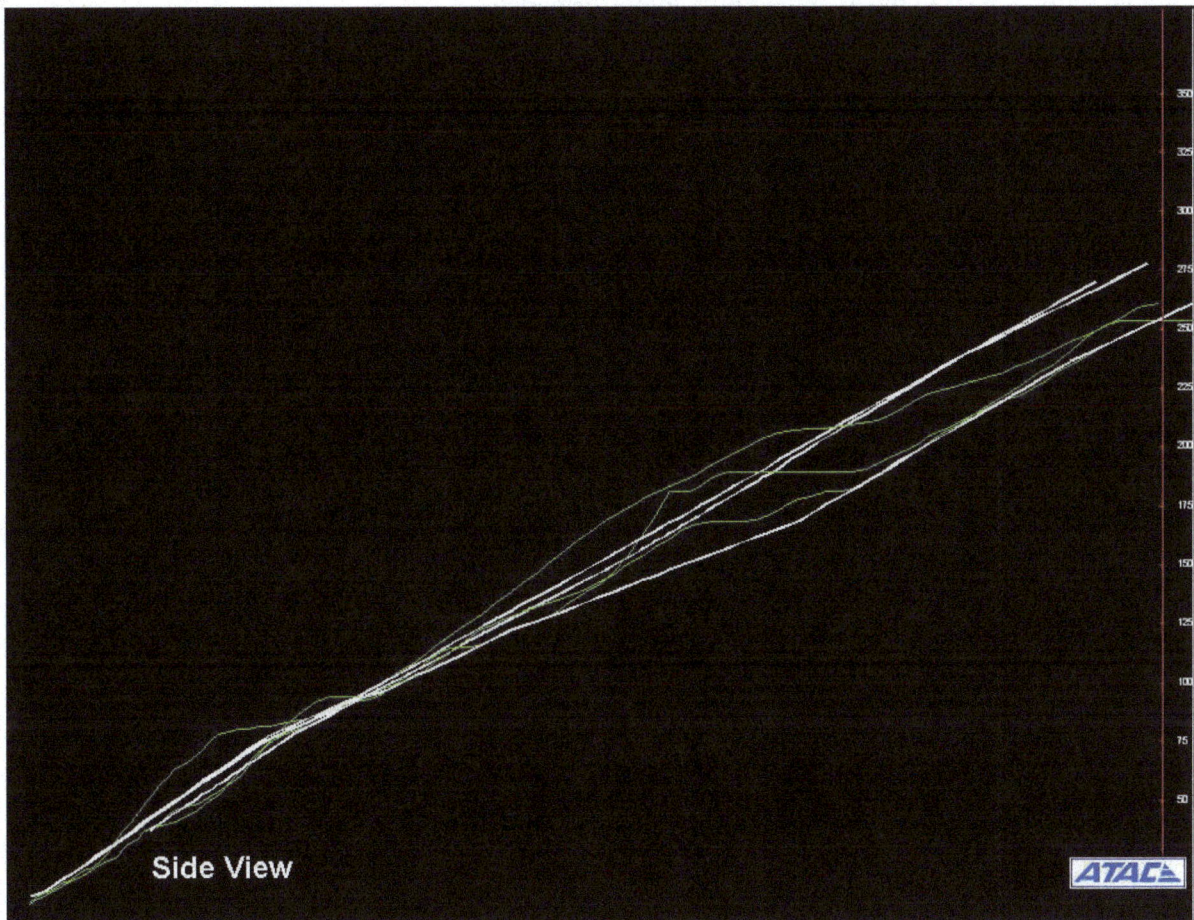

Figure 5.4.8. TBM Scenario: Actual compared with simulated tracks.

Figure 5.4.9 shows a plan view of one simulation in the TBM scenario. Note that the human controller (green track) takes an aircraft on a vector to satisfy time restrictions at the fix point (a merge waypoint on the ZLA approach indicated by the circle in figure 5.4.9.) The modeled controller (red) takes the aircraft to an empty slot on the other approach route to manage the same time restriction. The blue line is a nominal flightpath that was generated by averaging the aircraft tracks for 1000 flights. It is used to show the variation of an individual controller's performance as well as that of the model with respect to the averaged human performance.

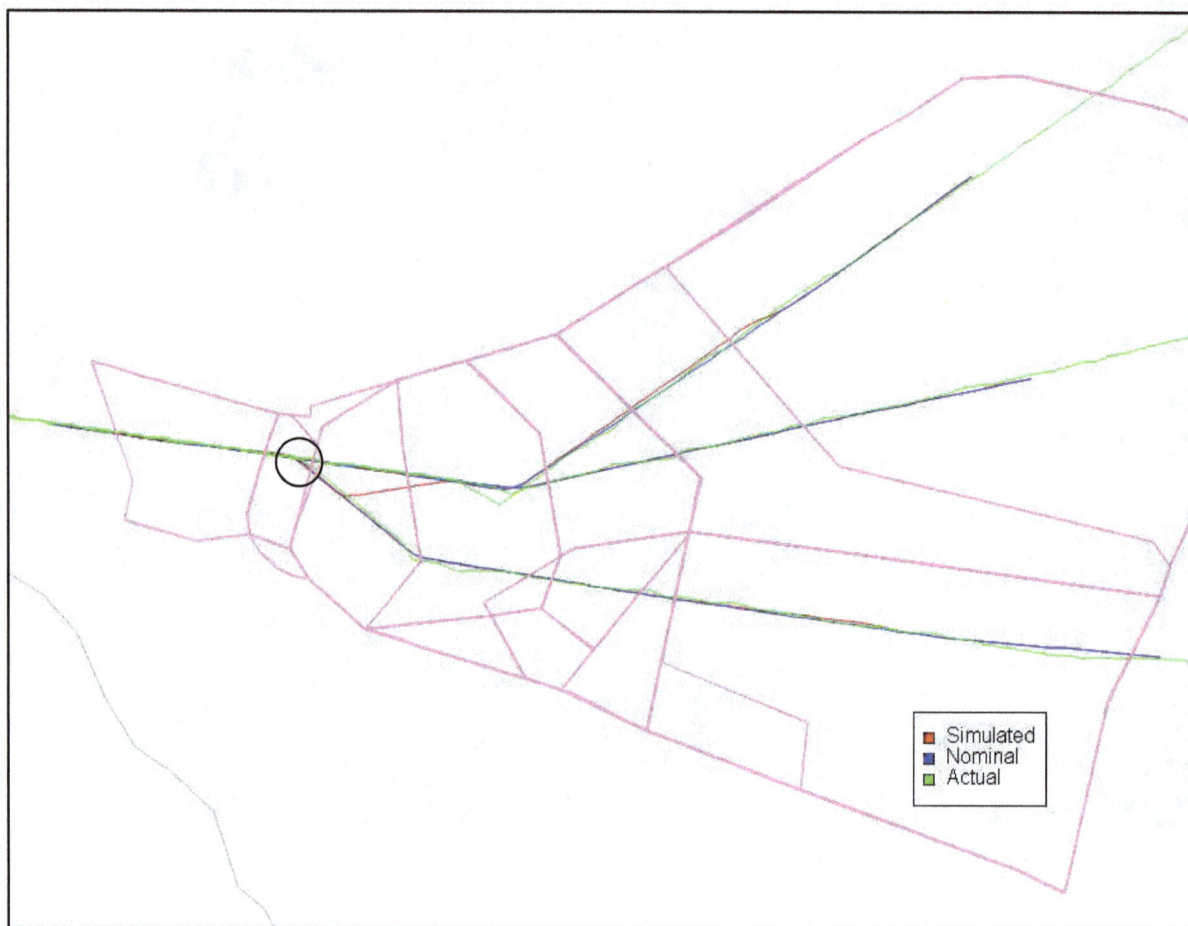

Figure 5.4.9. TBM Scenario: Nominal vs. actual vs. simulated tracks in plan view.

Figure 5.4.10 indicates the accuracy with which the modeled controller in a single simulation run managed to meet the required time at the fix compared with the actual controller. For individual aircraft (shown along the x-axis) the abscissa shows the simulated time over fix relative to the actual time over fix (shown as the zero axis). The ordinate is the simulated time either early (up) or late (down) relative to the actual time.

As shown in figures 5.4.11(a), (b), and (c), there was no statistically significant difference in workload of the modeled controllers operating TBM or MIT metering, even though more aircraft maneuvers were undertaken in the Center sectors under TBM control than under MIT metering control. Although the overall workload was the same for both TBM and MIT metering, it will be shown that the source and timing of the workload was redistributed as a consequence of different actions performed.

Means of Time Difference between Simulated and Actual tracks at Sector Exit Points - TBM

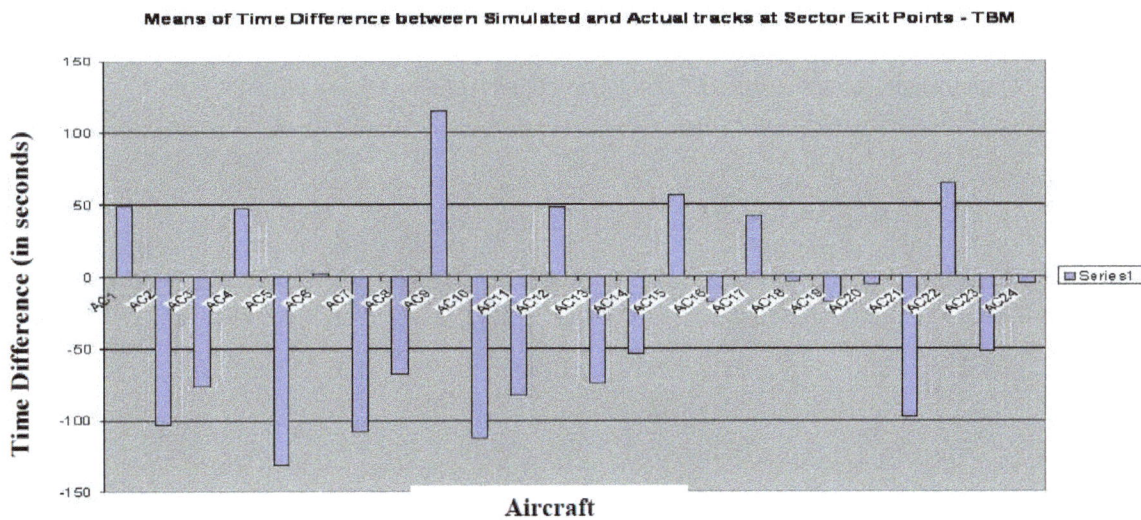

Figure 5.4.10. TBM: Mean values of time differences between simulated and actual tracks at sector exit points.

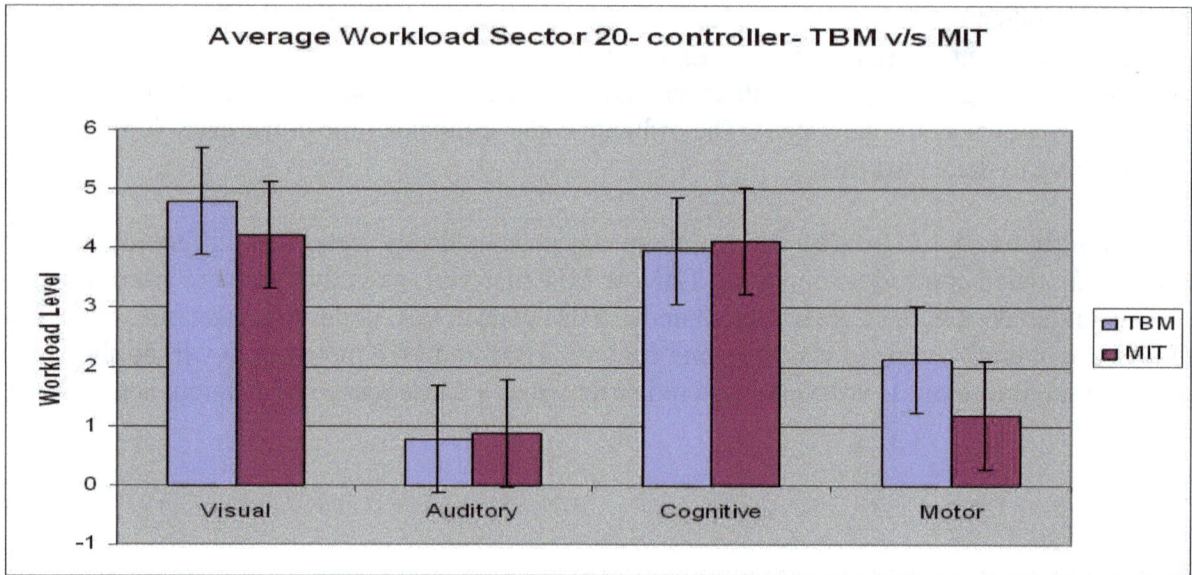

(a). Sector 20 controller.

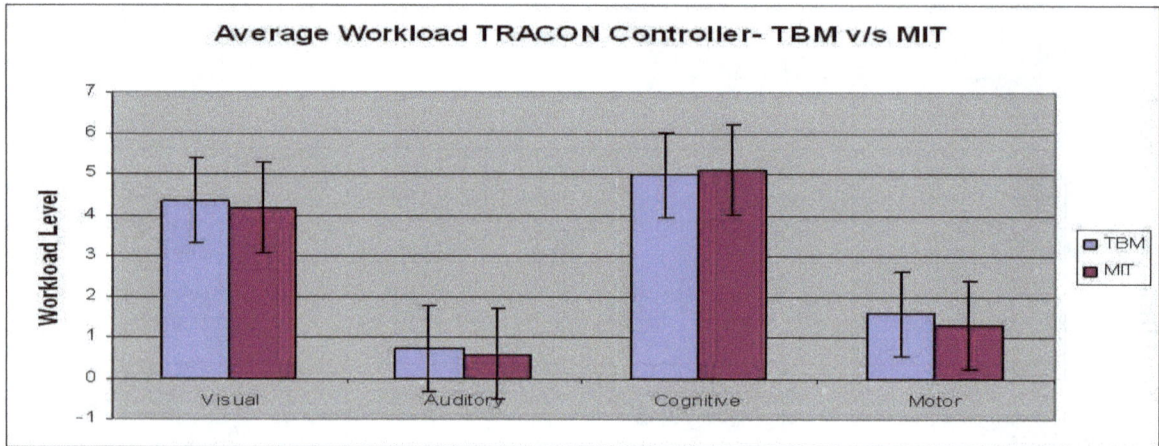

(b). TRACON controller.

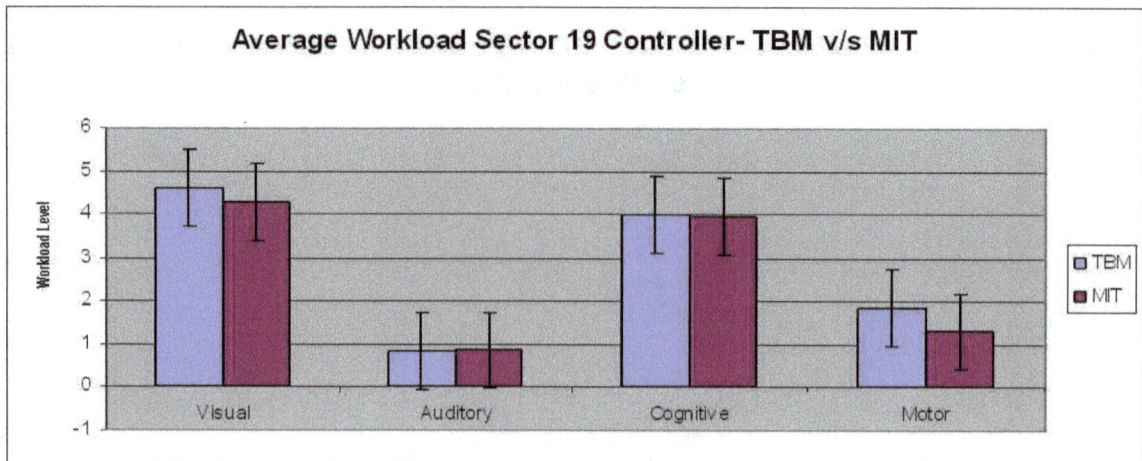

(c). Sector 19 controller

Figure 5.4.11. Controller workload – TBM compared with MIT.

The procedural sequences for both TBM and MIT metering were developed through interviews with controllers from the Los Angeles Center. Figure 5.4.12 illustrates the action sequences that produced the cumulative workload graphs in figure 5.4.11. Each of the boxes in figure 5.4.12 represents an activity in the controller procedure. Each of these actions has a load associated with its performance. The actions on the left of the figure are the TBM sequence and the actions on the right are the MIT sequence. The action sequences are programmed into the model to accomplish the tasks needed to achieve the goals for TBM or MIT metering.

The TBM sequence shows a high workload in the controller's initial interaction with the delay list and in planning the subsequent maneuvers. After that, a relatively low workload is experienced because of the conflict-free trajectories associated with the TBM. In the MIT sequence on the right, the controller has a moderate spike in workload at the hand-off, but also has a continuous load associated with monitoring separation conformance. In the TBM process, the controller is presented with a time to manage for an aircraft to cross a specific point in the airspace. The controller develops a plan to have that point reached at a specific time and provides that clearance to the flight deck. The flightcrew then enters that plan into its Flight Management System (FMS). From that point the controller monitors the conformance of the aircraft with the plan. Assuming relatively accurate conformance, the controller does not have that much to do in the rest of the trajectory to the merge point. In the MIT case, the controller has to monitor separation and issue clearances to the aircraft in the descent to ensure the required separation at the merge point, providing a set of tasks, each of which has an associated load. As in the case of the CAT experiment, the average loads across these tasks are roughly equal but differently distributed in time.

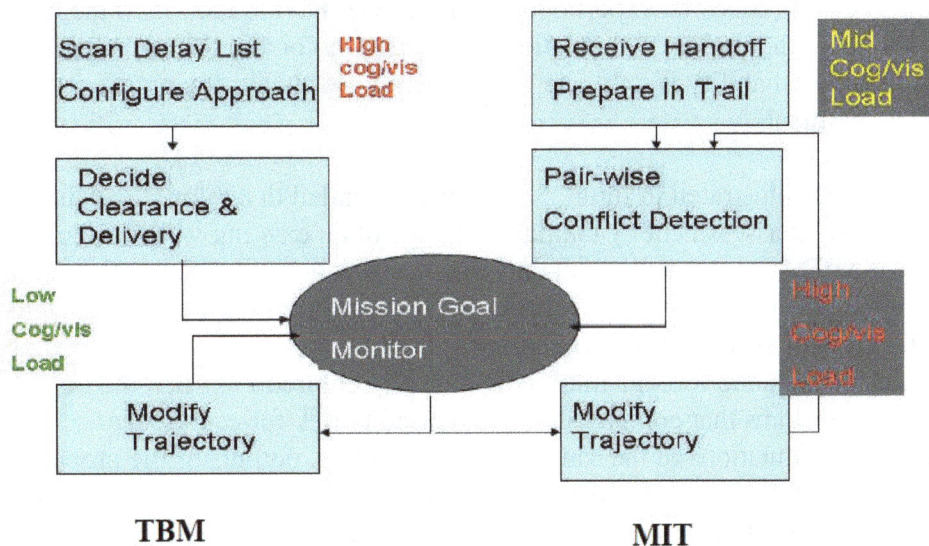

Figure 5.4.12. Action sequences for TBM and MIT metering for workload source analysis.

An analysis was performed on the full datasets to support a comparison between the actual traffic (sampled over four days), a "nominal" set of traffic, and the traffic produced by the Air MIDAS-RFS simulation. In summary, there were no statistically significant differences between the model-generated flight tracks and the actual flight tracks in terms of distance flown and time-in-sector for any scenario. For example, the F-test on the difference among the sources of the data tracks in time-in-sector and distance flown as dependent variables provided an F = 0.113 $_{(2,42)}$ (i.e., no significant difference).[11] Further analysis of the differences between MIT and TBM on those same variables showed a two-tailed F = 1.648 $_{(1,42)}$ (no significant difference). A Pearson Product Moment statistic in the analysis of all model factors (both source and type) found R^2 = 0.8.[12] An R^2 = 0.960 was found in the analysis of the sources of the data (i.e., actual vs. simulated). An R^2 = 0.004 was found in the analysis of the differences of variances among the actual and simulated flight tracks. Examination of the other recorded performance metrics showed similar close correspondence between model and actual human performance. For instance, with respect to the metric of miles flown in each sector, the simulated and actual controller performances were 99.6% in agreement. This study provided one of the unique opportunities to validate models of air-traffic-management performance against actual operational data.

In the first two studies described previously, the linkage of the Air MIDAS model of the human operator(s) with the RFS was used primarily to explore for possible hazardous situations and for their causal factors, although both studies also enabled further development and exploration of the Air MIDAS-RFS simulation capabilities. It was only in a later portion of the study of the TBM-MIT metering scenarios (described in section 5.4.2) that the fast-time simulation with Air MIDAS and the RFS linked dynamically was used to assess the safety risks of the identified hazards. Even then, the assessment was based on the predicted frequency of occurrence of the hazard, whereas the severity of its consequences was based on expert opinion. However, having identified the hazard, employing the many hundreds of thousands of simulation runs needed to obtain the probability of such rare events with adequate reliability is only one possible (and not always the most economical) approach to safety-risk assessment.

The third experiment (illustrated in figure 5.4.13) was intended to explore the value of another approach to safety-risk assessment by taking advantage of an existing well-developed tool. The capability of Air MIDAS to model human performance was used to enhance the human representation in the TOPAZ (Blom et al. (1998) and Blom et al. (2003)) developed by the Dutch NLR to assess the safety risk of specified aviation hazards. TOPAZ includes a simulation of a specified hazardous scenario using the formalism of a dynamically colored Petri net, the outputs of which are the variables of algorithms that compute the relative safety risk for each scenario. Independently, Air MIDAS was run in simulations of the same scenarios, and the outputs of the probability distributions of the model-generated human behavior were used to supply input parameters for performance of human operators in the simulations of TOPAZ. Although the representation of the human operator used in TOPAZ was improved by using the outputs of the simulations with Air MIDAS, this repre-

---

[11] The F-test is a statistical test to evaluate the equivalence of the two datasets. The validity of the F-test requires the underlying data to follow a normal (i.e., Gaussian) distribution.

[12] The statistic R^2, often referred to as the "coefficient of determination," measures the portion of the variability in a dataset that can be accounted for by a statistical model such as a regression equation. An R^2 value of 1.0 indicates that the fitted model explains all the variability in the data.

sentation did not entail the dynamic linkage like the simulations using Air MIDAS with the RFS. (In figure 5.4.13, Air MIDAS is represented in the process by the two blue octagons depicting the human brain and hands.)

The objective was to combine the significant advances in modeling of individual human performance and in understanding of human-performance factors (human factors in general and human cognitive behavior in particular) with an established risk-assessment tool to perform large-scale simulations for aviation-safety-risk assessment. The hypothesis was that the provision of the representations of individual human operators would improve the reliability and validity of aviation safety-risk assessment using TOPAZ. It was further hypothesized that such an integrated aviation safety-risk simulation model would illuminate the Air MIDAS model entities (either specific operators, functions within an operator, or interaction among operators) that are most critical in managing the aviation safety-risk level. Because the consequence of an identified hazard was assumed severe if left unattended, the focus of this risk analysis was on determining the likelihood of its occurrence.

Figure 5.4.13 depicts the process of development for the TOPAZ hazard model, and illustrates where data and model structure from the Air MIDAS model were incorporated into that model. The figure illustrates the entire process of TOPAZ model development and analysis. Beginning in the lower left corner, task decomposition for the specified scenario is undertaken (illustrated by the connected box graphic). In that step of the process, the Air MIDAS model is used to support the task decomposition for definition of the activities to be modeled in Air MIDAS. Moving up on the left, the task decomposition is presented to domain experts, who refine it. In the center top of the figure, a

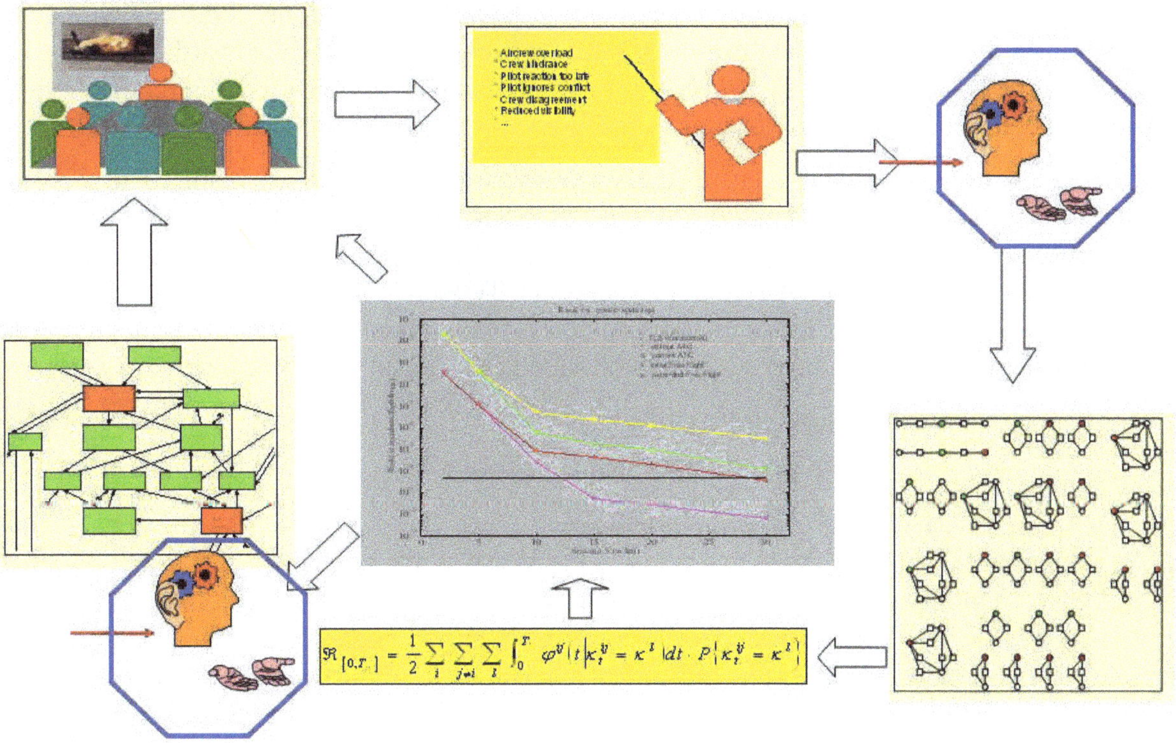

Figure 5.4.13. Air MIDAS-TOPAZ interlink.

list of possible hazards encountered in the specified scenario is developed, and these hazards are then implemented as procedures in the Air MIDAS model (shown in the upper right corner). The Air MIDAS model is then run, and distributions of performance times for specific parameters that are shared between MIDAS and TOPAZ are generated. These parameter distributions are then coded into TOPAZ as part of the dynamically colored Petri net process (in the lower right corner). The TOPAZ algorithms are then run and conditionally probabilistic likelihoods of hazard outcome are generated.

For effective simulation, the complementary modeling approaches of Air MIDAS and TOPAZ were integrated at all of the following three levels:

- *Context level:* This approach ensures that similar application contexts are being considered. Common understanding about issues such as boundaries, operations, and task analysis are required. In this level of integration, the assumptions about the operational context are coordinated. For example, in a study of runway operations, specification of the physical layout of the runways, operational procedures, and roles and responsibilities would be among the context-level coordination parameters.

- *Model level:* This approach ensures that the model integration is done appropriately; there was need for understanding and agreement about how Air MIDAS and TOPAZ models complement each other. Integration at the model level ensures that the simulation scenario is jointly represented in the two modeling systems, allowing identification of values for specific parameters of human performance in the TOPAZ simulation model to be supplied by the Air MIDAS simulation. These parameter values are generated in Monte Carlo runs of the Air MIDAS model and the results are subsequently supplied as input to the TOPAZ simulations. If the modeling paradigms allow similar representations, this parameter exchange is straightforward. For example, simulation of pilot reaction time to recognition of an incursion by the taxiing aircraft is represented in both models; hence exchange of reaction-time values is straightforward.

- *Parameter level:* This approach ensures that similar parameter values are being used; there was a need to be able to use detailed outputs of Air-MIDAS at the parameter level as inputs to TOPAZ simulations. Upon examination of the similarities and differences of the models used for the surface operation considered by Air MIDAS and by the TOPAZ in a similar study using a toolset called TOPAZ-TAXIR,[13] a list of model parameters to be affected by the joint runs was identified. These parameters can be grouped and provided as follows:

  - braking initiation times of pilots flying,

  - inter-monitoring time of pilot flying of taxiing aircraft, and

  - duration of visual observation of pilots flying.

Integration at the model level ensured that the simulation scenario under examination was consistently defined in the Air-MIDAS and TOPAZ models. Parameter values that were generated in Monte Carlo simulations of the human-performance model were subsequently supplied as input to the TOPAZ simulations. For example, the first parameter group listed included the braking initiation

---

[13] TAXIR (TAXI Risk) is the name of the TOPAZ simulation toolset designed specifically for assessing accident risks of runway operations.

times of pilots flying or taking-off or taxiing aircraft in either tactical or opportunistic mode, when they have become aware of a conflict with the other aircraft.

In an earlier study, known as TOPAZ-TAXIR, TOPAZ was used to evaluate a new operational concept for managing aircraft crossing an active runway with aircraft takeoffs. Figure 5.4.14 shows the simplified representation of the runway configuration used, where the taxiway crosses the runway at a distance of y_3^b from the runway start threshold, the taxiway has remotely controlled stop bars on both sides of the runway at a distance of x_3^x from the runway centerline, and the taxiway entrance is at a distance of x_3^1 from the runway centerline. Aircraft are crossing over the runway as they move from the apron to a second runway.

The involved human operators include the startup controller, the ground controller, a runway controller for each runway, the departure controller, and a pilot flying (PF) and a pilot not flying (PNF) for each aircraft. Two aircraft are shown in figure 5.4.14, one (a/c i) is in position to take off from the runway start and the other (a/c j) is in position to taxi and cross that runway.

Communication between controllers and pilots was by standard VHF receive/transmit (R/T) and communication between controllers was by telephone lines. Aircraft monitoring by the controllers was by direct visual observation with support from radar-track displays. Aircraft monitoring by the pilots was by visual observation with support by the VHF R/T party-line effect.

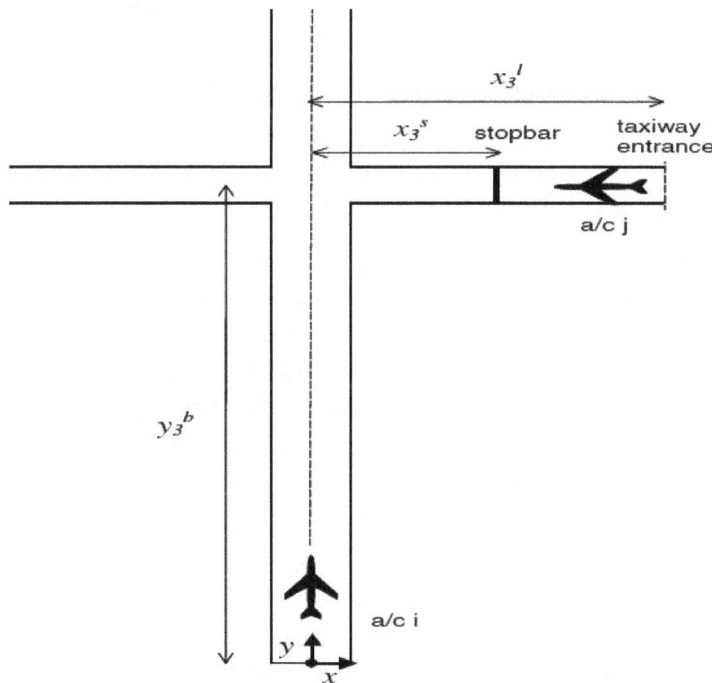

Figure 5.4.14. Configuration of active runway crossing operation considered.

In the runway-crossing operational concept, control over the crossing aircraft is transferred from the ground controller to the controller of the runway to be crossed. If the runway controller is aware that its runway is not used for a takeoff, the crew of an aircraft intending to cross is cleared to do so. The pilot not flying the crossing aircraft acknowledges the clearance and then the pilot flying initiates the runway crossing. After the crossing aircraft has vacated the runway, the pilot not flying reports this fact to the controller of that runway. Then control of the aircraft is transferred from this runway controller to either another runway controller or to the ground controller.

This scenario was used to evaluate the effects of modifying certain human-performance parameters used in the original TOPAZ-TAXIR model to correspond to those parameters as they are generated using Air MIDAS. The first issue was to find a common parameter shared by TOPAZ and Air MIDAS. In this example, "braking time" was that common parameter. The times at which braking is initiated by the pilot flying the aircraft that is taking off and the pilot of the taxiing aircraft (when they have become aware of the conflict with the other aircraft) is the parameter called "braking time." Both Air MIDAS and TOPAZ-TAXIR use braking time as a parameter that impacts the safety risk.

Both the Air MIDAS model and the TOPAZ-TAXIR model of the pilots have states or operating modes in which the pilots' actions are different (in terms of the parameter values used) depending on their operating modes. The operating modes modeled in Air MIDAS are strategic, tactical, and opportunistic (Blom, Corker, and Stoeve, (2005)). The models may change operating modes depending on many factors, of which one of the most important is situation awareness. The impact of situation awareness occurs in the following manner:

- If the pilot model is aware that its aircraft is on a taxiway that will cross an active runway or on an active runway, then the model remains in a strategic mode, the activities associated with potential surface conflict are assigned high priority, and the activities of detecting aircraft departing can be performed quickly without interruption.

- If the pilot model is not aware that its aircraft is on an active runway or on a taxiway that crosses an active runway, then the scheduling process of the model does not assign priority to those actions that are associated with possible conflict with another aircraft. It may take the model longer to schedule these activities because they do not have high priority and the model is performing other activities it considers to be of like or higher priority. In this case, when the model becomes aware of the situation, it changes to the tactical or opportunistic operating mode, depending on the contextual parameters.

In this scenario, response speed to a possible conflict is the variable that changes as a function of the operating mode for the human model.

Air MIDAS and the TOPAZ models use somewhat different mechanisms to switch between modes of operation, but fundamentally they both look at the current and near-term actions that need to be performed and the capacities available to perform them and exercise algorithmic transforms to pick set points that represent a level associated with that operating mode.

As stated previously, both Air MIDAS and TOPAZ-TAXIR use braking time as a parameter that impacts the safety risk. In TOPAZ, braking time is defined by a prescribed, empirically based Probability Density Function (PDF) that assigns a probability of occurrence to each value of the braking time. In Air MIDAS, braking time is considered composed of perceptual recognition of the need to break, a decision to brake, and the actual application of brakes to the aircraft dynamic model. A PDF is calculated for each of these components to arrive at the PDF for braking time.

In figure 5.4.15, an overview is provided of the PDF and related parameter values for Air MIDAS and the original and the modified TOPAZ-TAXIR models. In all three models, equal PDF types and parameter values for the braking initiation times are chosen for the pilots flying the taking-off and taxiing aircraft, regardless of their cognitive control modes. Figure 5.4.15 shows that, in comparison to Air-MIDAS, the original TOPAZ-TAXIR model has a smaller mean braking initiation time and a larger tail (i.e., probability of more than 5 seconds initiation time). In order to improve on these aspects, the Rayleigh PDF has been selected for the modified TOPAZ-TAXIR model. The improvements are:

- its shape better fits to the Air-MIDAS data,
- it supports positive values only, and
- it has a more realistic tail than Gaussian PDF.

The parameter value of the Rayleigh PDF has been chosen such that its standard deviation equals the standard deviation of the PDF chosen in Air-MIDAS.

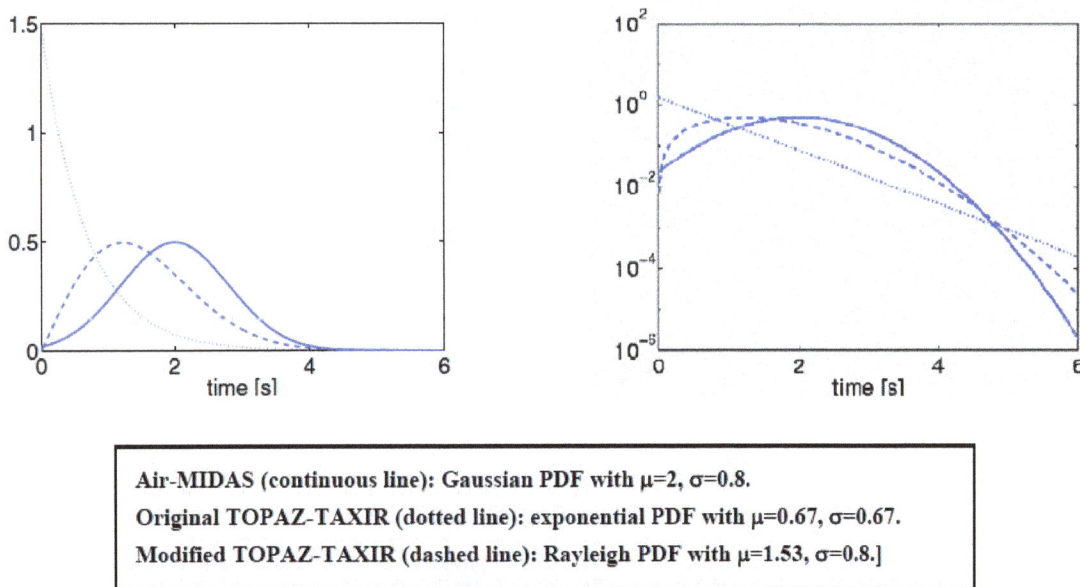

Air-MIDAS (continuous line): Gaussian PDF with $\mu=2$, $\sigma=0.8$.
Original TOPAZ-TAXIR (dotted line): exponential PDF with $\mu=0.67$, $\sigma=0.67$.
Modified TOPAZ-TAXIR (dashed line): Rayleigh PDF with $\mu=1.53$, $\sigma=0.8$.]

Figure 5.4.15. PDFs for braking initiation time; linear scale (left figure); logarithmic scale (right figure).

Table 5.4.1 provides the information about the parameter values integrated in the original TOPAZ model, Air MIDAS, and the modified TOPAZ simulation for braking initiation times.

It is assumed in TOPAZ-TAXIR that the inter-monitoring time of the pilot flying the taxiing aircraft is independent of the cognitive control mode of the pilot. In the original TOPAZ model, this time was represented by an exponential PDF. Simulations of Air MIDAS resulted in a dataset of 536 inter-monitoring times of the taxiing pilot flying. A histogram of this dataset is shown in figure 5.4.16, which shows that this histogram can be well represented by an exponential PDF. Therefore, in the modified model the inter-monitoring times of the pilot flying the taxiing aircraft are also chosen from an exponential PDF with a mean equal to the estimated mean of the Air-MIDAS data.

TABLE 5.4.1. PARAMETER VALUES FOR RUNWAY INCURSION SIMULATIONS

Source	PDF	Mean, μ	Std Dev, σ
4. Braking initiation time after detection of a conflict by taking-off PF in tactical mode			
Original TOPAZ-TAXIR	Exponential	0.67 s	0.67 s
Air MIDAS	Normal	2.0 s	0.8 s
Modified TOPAZ-TAXIR	Rayleigh	1.53 s	0.8 s
5. Braking initiation time after detection of a conflict by taxiing PF in tactical mode			
Original TOPAZ-TAXIR	Exponential	0.67 s	0.67 s
Air MIDAS	Normal	2.0 s	0.8 s
Modified TOPAZ-TAXIR	Rayleigh	1.53 s	0.8 s
6. Braking initiation time after detection of a conflict by taking-off PF in opportunistic mode			
Original TOPAZ-TAXIR	Exponential	0.67 s	0.67 s
Air MIDAS	Normal	2.0 s	0.8 s
Modified TOPAZ-TAXIR	Rayleigh	1.53 s	0.8 s
7. Braking initiation time after detection of a conflict by taxiing PF in opportunistic mode			
Original TOPAZ-TAXIR	Exponential	0.67 s	0.67 s
Air MIDAS	Normal	2.0 s	0.8 s
Modified TOPAZ-TAXIR	Rayleigh	1.53 s	0.8 s

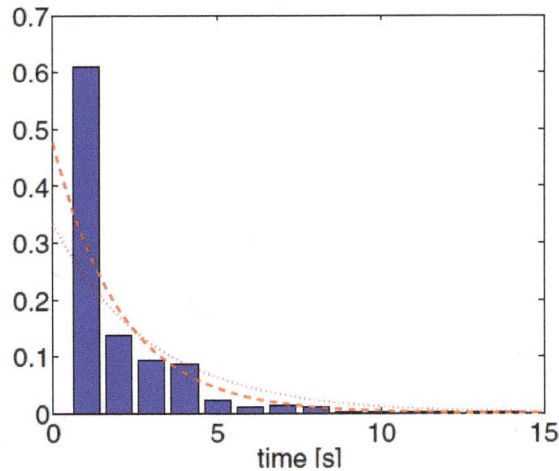

Figure 5.4.16. Data of inter-monitoring times of the PF of taxiing aircraft.

The third parameter group includes the visual observation times of pilots flying the taking-off or taxiing aircraft in either tactical or opportunistic mode. The PDFs of these times in the original TOPAZ model are exponential PDFs with a mean that is smaller in the opportunistic mode than in the tactical mode. Air-MIDAS simulations provided data on the duration of the tasks:

- "Monitor Out The Window" for the PF of the taking-off aircraft, and

- "Decide Action – Decide Take-off Spotted" for the PF of the taxiing aircraft.

These tasks were found to be in good agreement with the visual observation tasks of the pilots flying the taking-off and taxiing aircraft, respectively. These data were provided for the three control modes used in Air-MIDAS.

Figure 5.4.17 shows that the histograms of the visual observation time of the pilot flying the taking-off aircraft (figure 5.4.17 A, B, and C) and the pilot flying the taxiing aircraft (figure 5.4.17 D, E, and F).

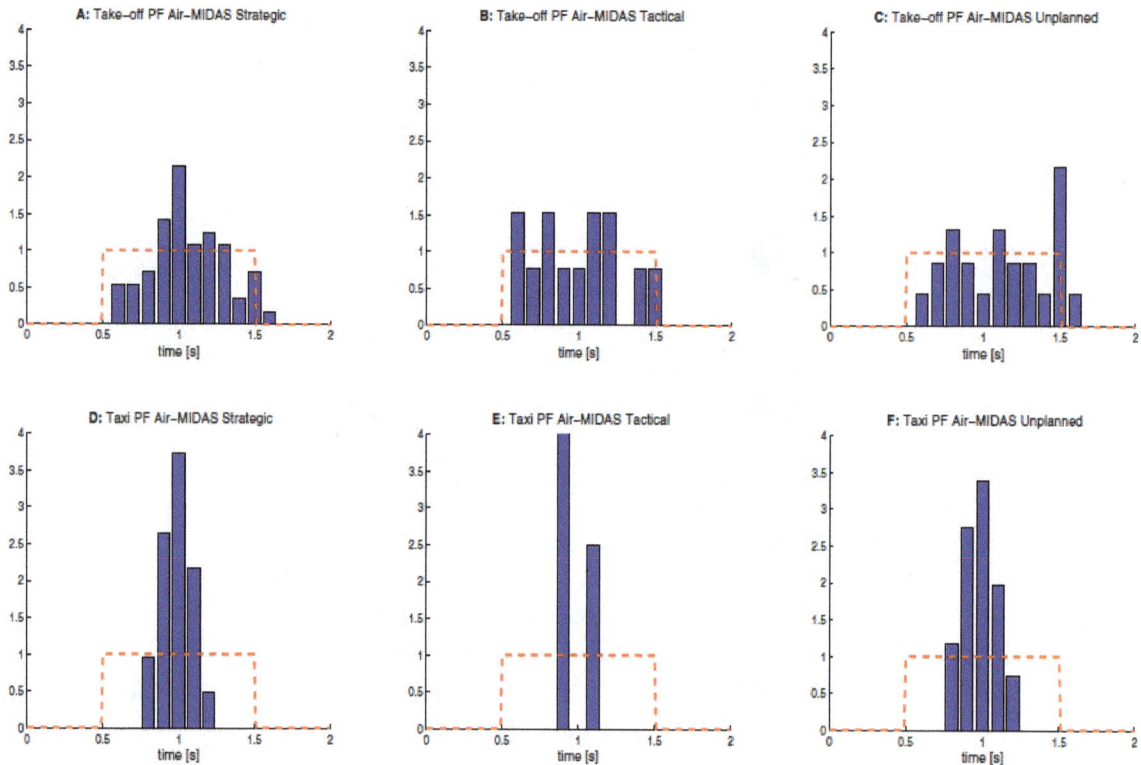

A: Take-off PF Air–MIDAS Strategic
B: Take-off PF Air–MIDAS Tactical
C: Take-off PF Air–MIDAS Unplanned
D: Taxi PF Air–MIDAS Strategic
E: Taxi PF Air–MIDAS Tactical
F: Taxi PF Air–MIDAS Unplanned

Air-MIDAS cognitive control mode is similar to strategic (left), tactical (center) or unplanned (right).
Air-MIDAS data (bars): normalized histograms of data and one uniform density fit (dashed lines).

Figure 5.4.17. Visual observation times of the pilot flying taking-off aircraft (upper figures A, B, and C) and taxiing aircraft (bottom figures D, E, and F).

It can be seen in figures 5.4.17 A, B, and C that the normalized histograms of the visual observation time of the pilot flying the taking-off aircraft can be reasonably approximated by uniform PDFs with a lower bound of 0.5 and an upper bound of 1.5 (Blom, Corker, and Stoeve, (2005)). Therefore, this simple representation was chosen in the modified TOPAZ-TAXIR model. The Air MIDAS data presented in figure 5.4.17 indicate that the variance of the visual observation duration for the pilot flying the taking-off aircraft is smaller than that for the pilot flying the taxiing aircraft (see figure 5.4.17 D, E, and F), whereas their means are about the same. Nevertheless, the PDF for the taxiing-aircraft pilot was chosen to be equal to the PDF for the taking-off-aircraft pilot so as to minimize the complication of the modified TOPAZ-TAXIR model and since there are no manifest reasons why the PDF of the visual observation duration should be different for pilots flying the taking-off and taxiing aircraft.

Figure 5.4.18 is an example of the results of calculating the probability of collision in the scenario of runway incursions using the original TOPAZ-TAXIR model and using the modified TOPAZ-TAXIR model with inputs from Air MIDAS. The Air MIDAS-based adaptation of TOPAZ-TAXIR showed as much as a factor two reduction in assessed likelihood of a collision.

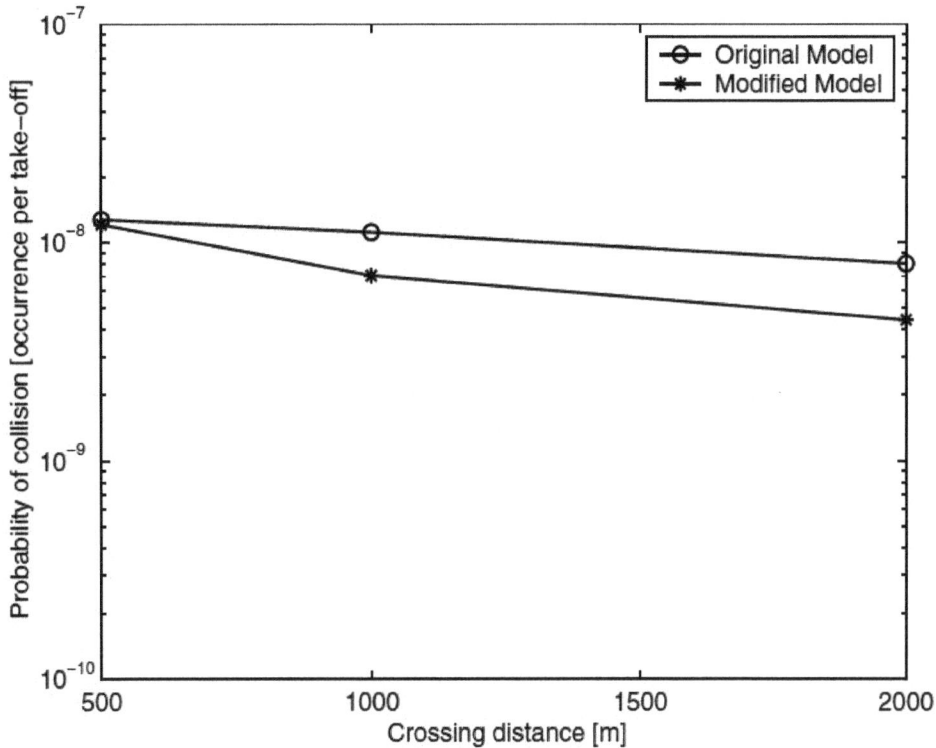

Figure 5.4.18. Comparison of collision risk as a function of crossing distance using TOPAZ with and without Air MIDAS inputs.

Table 5.4.2 provides the differences in the likelihood of a collision between the original and the modified TOPAZ models. There is no uncertainty in the conditional risk. The uncertainties in this mode are associated with the parameters of performance that are used to calculate the risk. Blom, Corker, and Stoeve (2001) discuss these details and the data that establish confidence in the calculated probabilities.

The results summarized here and presented in detail in Blom, Corker, and Stoeve (2001) demonstrated not only the feasibility, but also the value of coupling Air MIDAS and TOPAZ. More importantly, this experiment provided the unique opportunity to compare the model of human performance as it had been implemented in the original TOPAZ-TAXIR model with the representation from Air-MIDAS incorporated into TOPAZ-TAXIR model for the same scenario. This experiment also enabled further comparisons to be made between the two simulation approaches.

TABLE 5.4.2. LIKELIHOOD COMPARISON USING ORIGINAL AND MODIFIED TOPAZ

Crossing distance	Original collision likelihood (occurrence per takeoff)	Modified collision likelihood (occurrence per takeoff)
500 m	$1.3 \ 10^{-8}$	$1.2 \ 10^{-8}$
1000 m	$1.1 \ 10^{-8}$	$7.1 \ 10^{-9}$
2000 m	$8.0 \ 10^{-9}$	$4.4 \ 10^{-9}$

The probability of collision that resulted from the TOPAZ-TAXIR simulations is composed of contributions from many combinatorial event sequences. Stroeve et al. (2003) give further details on this study, the results of the analyses, and confidence levels. In particular, the event sequence classes include the status of technical systems, such as alerting systems, communication systems, aircraft types, and human-operator situation awareness. Since all the adaptations of TOPAZ-TAXIR inter-linked with Air-MIDAS incorporate assumptions describing the behavior of pilots flying, it is interesting to compare the decomposition of the probability of collision for a pilot flying in the original and modified models.

In particular, in figure 5.4.19, the probability of collision is shown for the situations in which (a) the pilot flying the taxiing aircraft believes he/she is on a regular taxiway with no runway crossing (Proceed taxiway) and (b) the pilot flying the taxiing aircraft believes he/she is on a regular taxiway and that runway crossing is allowed (Cross runway). In the first case, the pilot is lost. In the second case, the pilot's situation awareness corresponds well with the actual position of the aircraft. It can be seen in figure 5.4.19 that, in both the original TOPAZ-TAXIR model and the modified TOPAZ-TAXIR model using inputs from Air MIDAS, the contribution to the probability of collision for the situation in which the pilot believes he/she is on a regular taxiway exceeds the probability of collision for the situation in which the pilot believes a runway crossing is allowed. However, the difference between those probabilities is smaller in the modified model, in which the components of the braking time have been considered, than in the original model, in which the PDF for overall braking time was used.

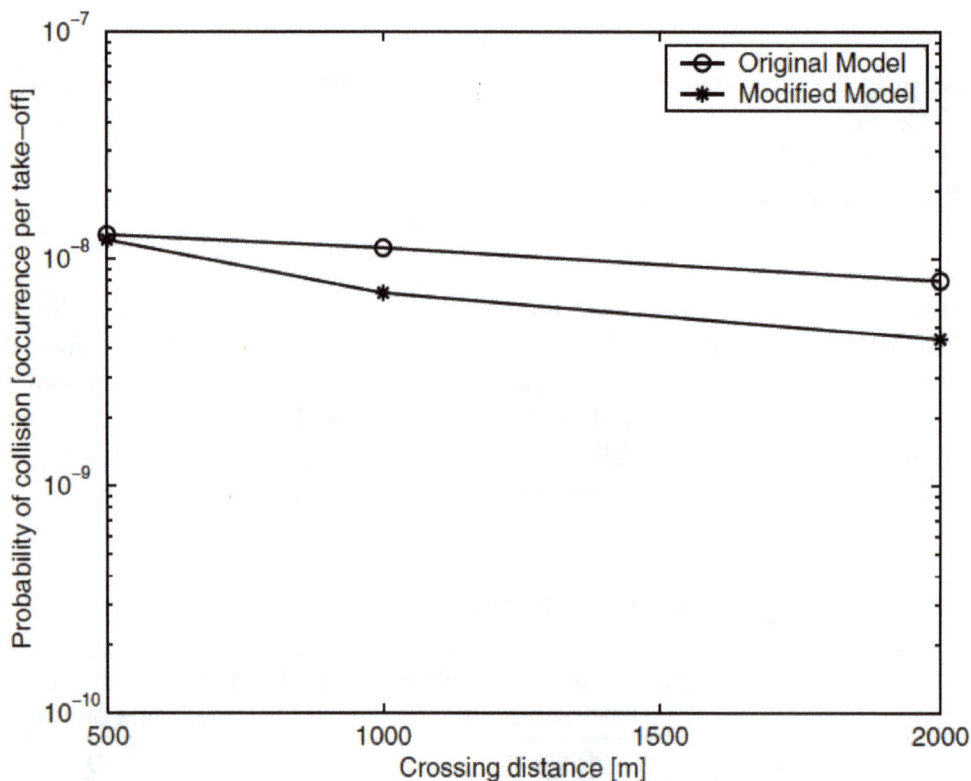

Figure 5.4.19. TOPAZ-Air MIDAS interlinked operation.

In the final experiment before the AvSP ended, another approach to risk assessment was explored. This approach was an extension of the study described previously of the workloads associated with TBM and MIT metering of high-density traffic on approach to LAX through the Southern California TRACON and the Los Angeles ARTCC (ZLA). As stated previously, the study of risk assessment was based on the predicted frequency of occurrence of the hazard, whereas the severity of its consequences was based on well-known existing analysis of potential consequences of loss of separation. This study entailed two phases. In phase 1, indicated in figure 5.4.20, the full high-fidelity human-performance model representing the controller behavior used in the studies described previously was used to perform a sufficient number (typically, hundreds) of simulations to obtain stable probability distributions of human behavior across the range of human activities associated with control. Each activity (for example, conflict detection, speed error detection, and flightpath deviations) has an associated probability distribution around the time of performance and the accuracy of performance. During these simulations, the model was subjected to parameter variations of "stressors," e.g., changes in the direction and speed of the wind or in the density of traffic or the visibility. The results of the full-model performance were then examined to identify the functions in the model that were critical in their contributions to potential hazardous performance (as measured, for example, by points of closest approach). Those critical functions were then made efficient in the code of a lower-fidelity human-performance model, integrated into the RFS, and embedded in a high-speed parallel simulation cluster. This process was necessary to make it practical to conduct the very large number of simulation runs needed for estimates of the occurrences of the very low-probability event of less-than-minimum separation in ATC. In phase 2, indicated in figure 5.4.20, a very large number of simulations (representing over 175,000 flight hours) were then performed in the cluster to look for the "rare event" in the system model and the system environmental "stress" contexts of those rare events. The phase 2 experiment on likelihood of occurrence is described in section 5.4.2

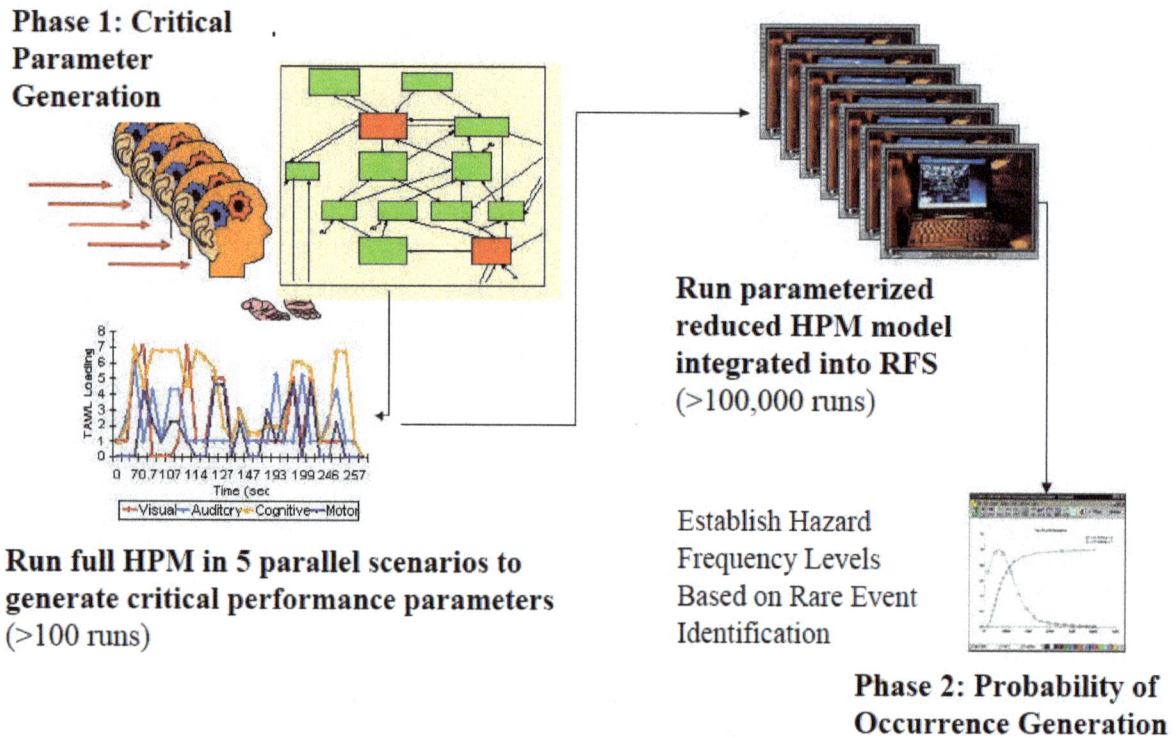

Phase 1: Critical Parameter Generation

Run parameterized reduced HPM model integrated into RFS (>100,000 runs)

Run full HPM in 5 parallel scenarios to generate critical performance parameters (>100 runs)

Establish Hazard Frequency Levels Based on Rare Event Identification

Phase 2: Probability of Occurrence Generation

Figure 5.4.20. The process for assessing frequency of occurrence of hazards.

Phase 1 of the process of generating the hazards, the human-performance parameters critical to the occurrence of those hazards, and the probability of the occurrence of the hazards was applied to two metering concepts for high-density air traffic (figure 5.4.20). This experiment was an extension of the second experiment described previously in which the differences in workload were explored that resulted from MIT metering and TBM of high-density traffic on approach to LAX through the Southern California TRACON and the Los Angeles ARTCC (ZLA). In this part of the study, procedures for both types of approach control were developed for four Center sectors and one TRACON sector. These sectors were sectors 29 and 30 in ZLA transition sectors and sectors 19 and 20 in ZLA approach sectors managing descent from 310 FL through 110 FL to the LFDR, as shown in figure 5.4.21.

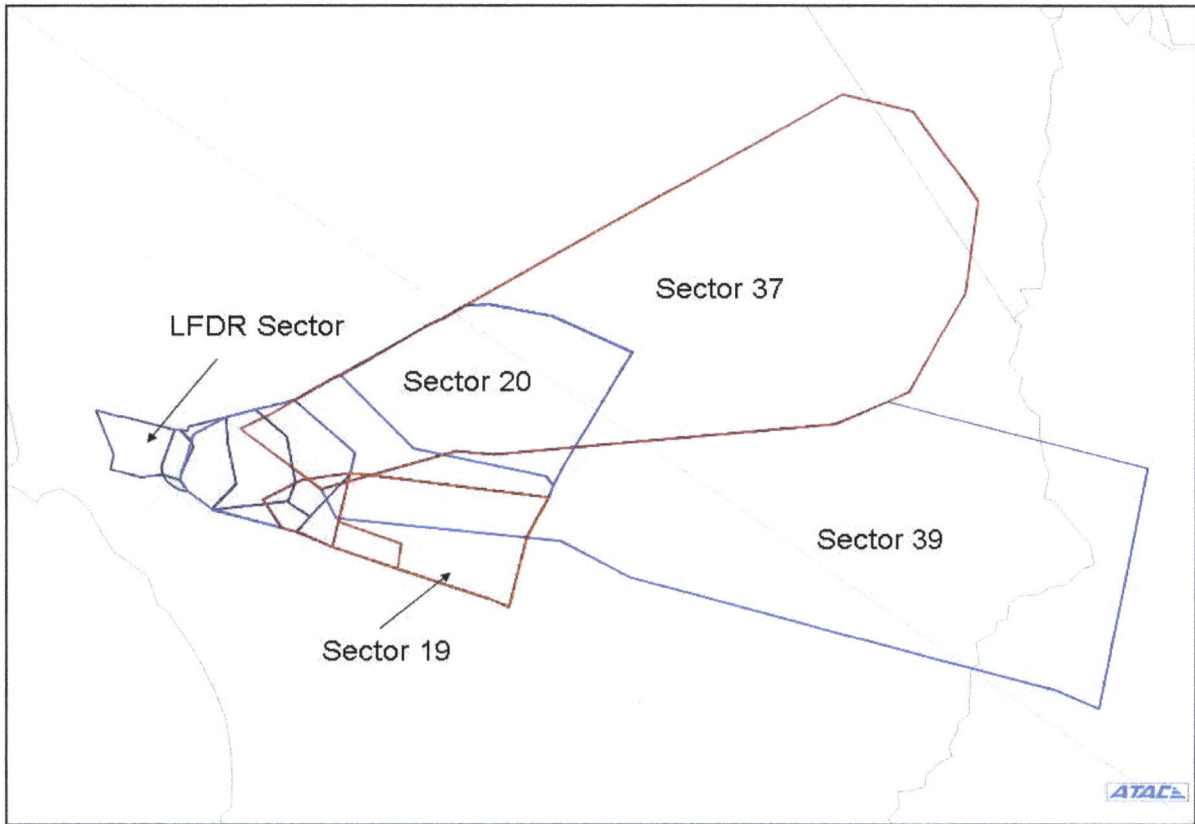

Figure 5.4.21. LAX Center and feeder sector.

Air traffic was recorded by PDARS under both MIT and TBM metering, and these data served as the validation traffic base. Human-performance models (Air MIDAS) and air-traffic models (RFS) were operated in integrated simulations to represent the performance of air-traffic controllers managing that traffic. The traffic-management performance of these models was compared with the performance of the human operators managing the same traffic. The metrics of comparison for each flight were:

- Time flown in sector,
- Distance flown in sector,
- Minimum indicated airspeed in sector,
- Maximum indicated airspeed in sector,
- Delta indicated airspeed (entry-exit) through sector,
- Average airspeed in sector,
- Number of altitude change commands in sector,
- Number of airspeed change commands in sector, and
- Total number of commands in sector.

Traffic samples were collected over a period of approximately one hour for each metering condition and for each wind (east/west direction and speed) condition, providing six one-hour validation samples for each condition. Verification of the human-performance model was based on statistical comparisons with the human performance of each of these metrics. Following are a few representative results.

Figure 5.4.22 shows the total distance flown by an aircraft to transit each of the five sectors under MIT metering. The four sets of data points in each sector represent the four different conditions under which data were collected with PDARS and the four conditions under which the simulations were performed. There were two conditions of wind direction (east or west), and two visibility conditions, or acceptance-rate conditions at the airport (high or low). In this figure, the approach to landing is from right to left (east to west) with the four Center sectors decreasing in size and accommodating merging traffic that moves into the TRACON in the LFDR sector. The total number of actual flights transiting each sector during the period (approximately one hour) of recorded PDARS data varies by sector and by day; however, all flights being sequenced to arrival at LAX eventually transition through the LFDR. The total number of arriving flights transiting the sectors varies between 20 and 25 in approximately one hour. The total number in any sector at any time is a subset of that total. The simulated data are generated by taking these same flights (the 20 to 25 transiting and being sequenced to the LAX arrival) and performing 20 simulations in each of the four scenarios of differing wind direction and visibility.

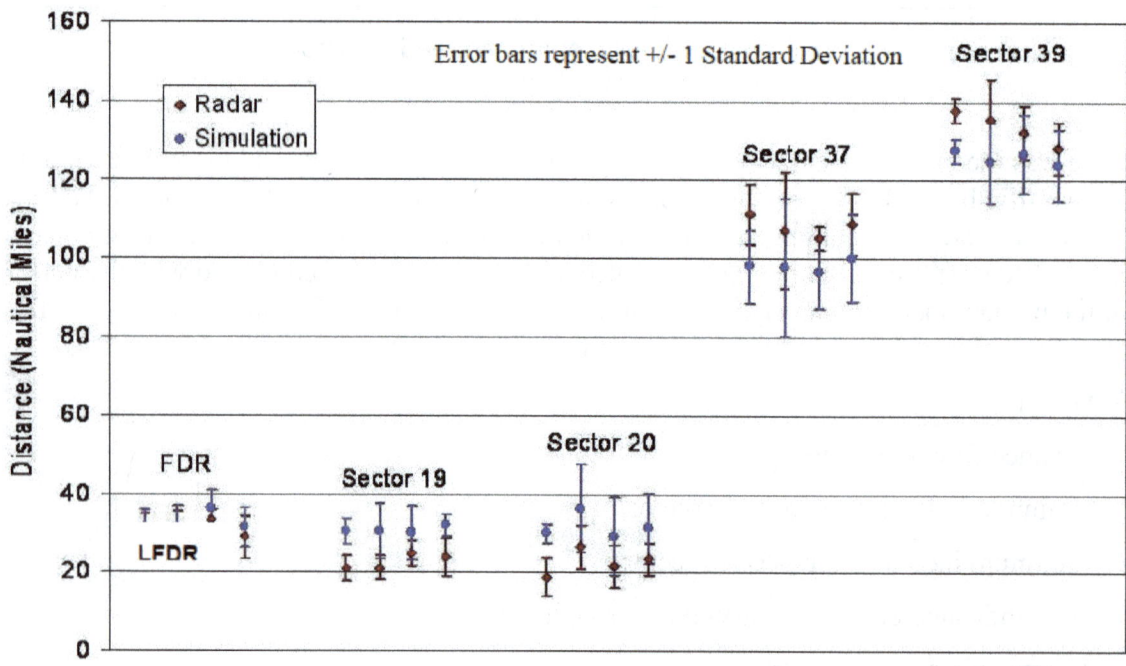

Figure 5.4.22. Comparison of distance in sector for MIT.

The distance flown in a sector is one aggregate measure of a controller's strategy in managing the traffic. Commands for heading changes and airspeed modifications will increase or decrease the time and distance flown in a sector. This comparison is one representative example of the nine metrics of performance of the human controller (as determined from the PDARS radar-track data) and the performance of the modeled controller (as determined from the simulations). The simulation data exhibit a slightly larger variability than the human performance in several of the scenarios. However, the close overlap of the means and standard errors provide a visual representation of their statistical equivalence.

The data of figure 5.4.23 represent the same distribution of distance flown for four scenarios and the five sectors with TBM rather than MIT metering. It shows a reduction in variability of both the actual data and the modeled data compared to the MIT metering data in figure 5.4.22. As in figure 5.4.22, there is an overlap of the standard-deviation bars and a close proximity of the means, indicating statistical identity between the simulated and actual performance. (The Pearson product moment correlation between the controller and the behavior of the model was R= 0.897.)

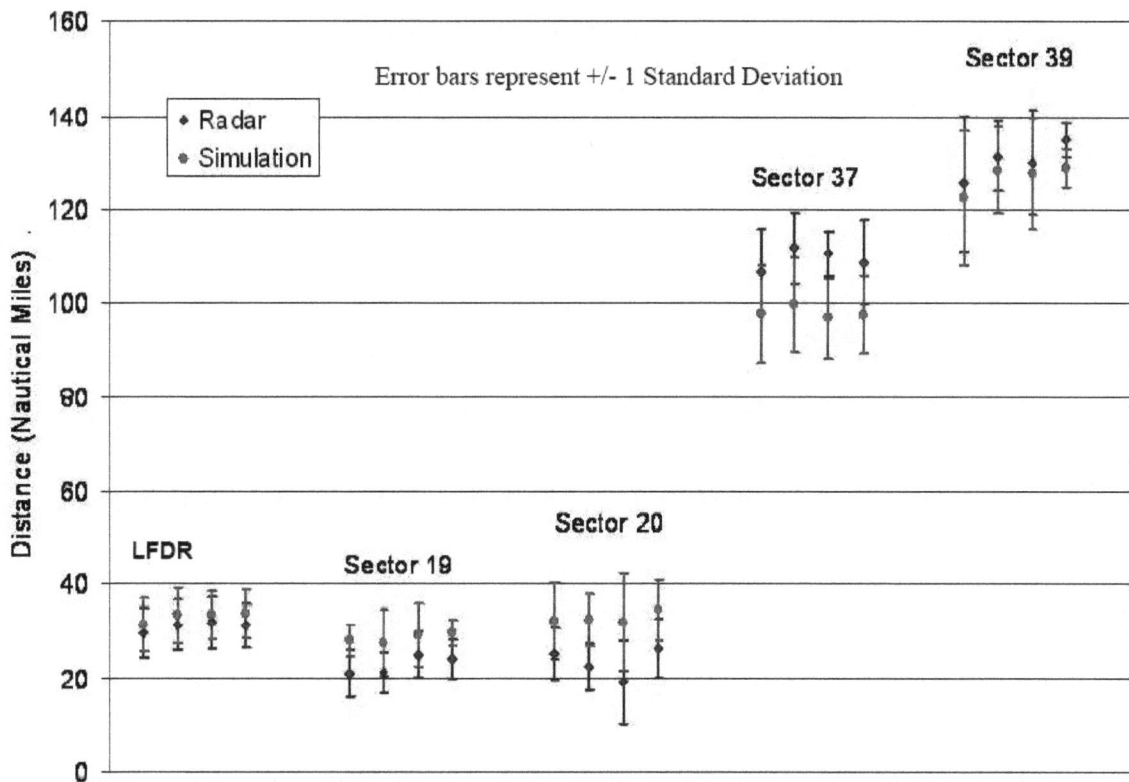

Figure 5.4.23. Comparison of distance in sector for TBM.

Figure 5.4.24 illustrates the performance of the model with respect to the aircraft meeting the time at which it is required to be over the fix as specified by the TBM. Each pair of bars represents the performance of one aircraft as measured by the delta between the actual and scheduled time over the fix. The zero axis represents the condition when the aircraft is over the fix at precisely the time specified by the TBM. A bar in the upward direction indicates that acceleration is needed by that aircraft to catch up on the time required to meet the specified time over the fix; a bar in the downward direction indicates that deceleration is required. The blue bar represents the delta from the simulation, and the purple bar represents the target delta, i.e., the required deceleration or acceleration needed to achieve the time specified by TBM. The simulated delta is a measure of the modeled controller's performance in achieving those times in the simulation by commanding the appropriate increase or decrease in speed. The difference in height between these two bars is the mean error in the model's making the required time. It was not possible to compare this measure directly to the human operator's performance because there were no comparable data on the human controller's actual achieved time over the fix.

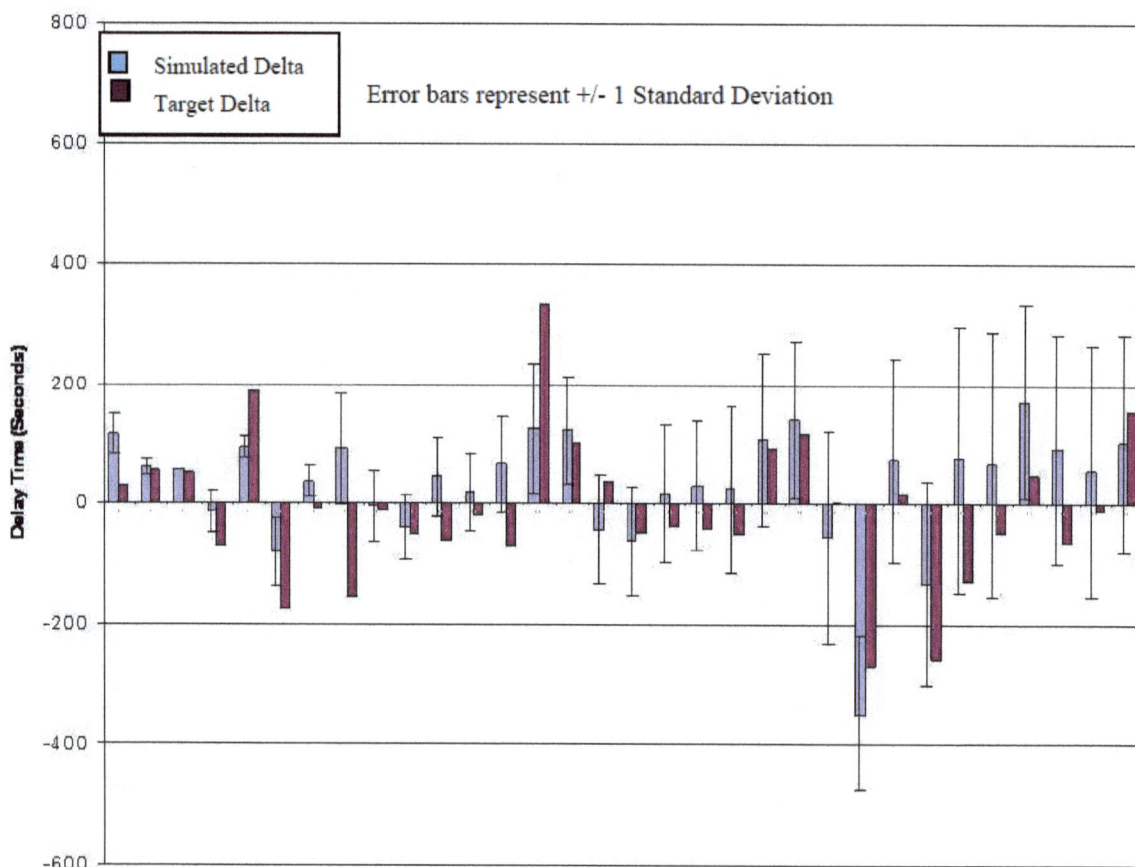

Figure 5.4.24. Mean delta relative to specified time over fix.

Figure 5.4.25 illustrates the mean and one-standard-deviation values of the three-dimensional separation distance between aircraft for the four scenarios across the five sectors under TBM. Virtually identical performance between the actual and simulated controller occurs in sector LFDR, and may be due, in part, to the limited airspace available as well as to the convergence of the human's and model's strategies of management. It is significant to note that the variation of the average separation distance at one standard error increment is over 20 miles in the outer sectors and decreases to approximately 10 miles in the inner sectors. Separation violations defined at 5 lateral miles and 1000 feet in altitude in the Center sectors represent a significant excursion from the norm and are appropriately rare events. This result of the phase 1 study is relevant to a result of phase 2 simulations discussed in the next section.

These representative results of the agreements that were found for all of the nine metrics shown previously provided confidence in the representation of the Air MIDAS model of the human controller's performance.

The human-performance characteristics that contribute to the potential for failing to meet minimum separation requirements were identified and were used to define an abbreviated, very efficient model that was incorporated into RFS for the phase 2 assessment study of probability of occurrences described in the next section.

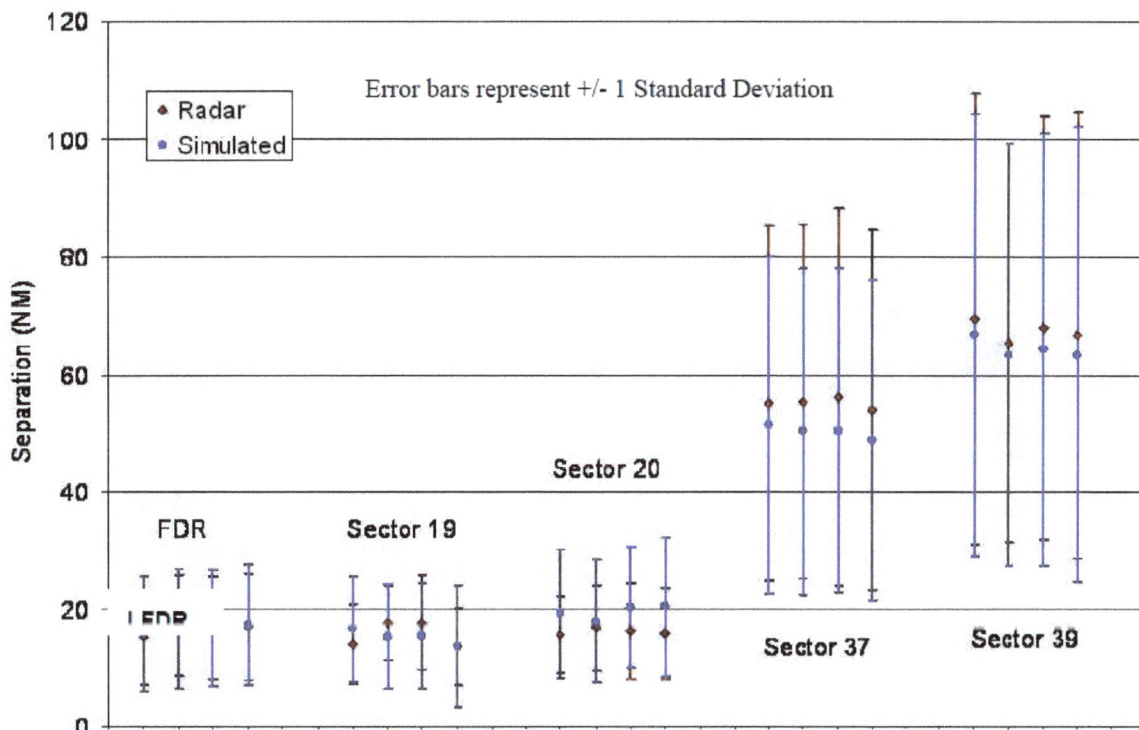

Figure 5.4.25. Minimum separation distance for TBM.

5.4.2 Systemwide Risk Assessment

Figure 2.0.1 presented a concept of proactive management of safety risk—an iterative cycle of identifying contributors to risk, evaluating their impact and causal relationships, and formulating methods of reducing risk that are intimately joined in with implementation. Another version of figure 2.0.1 is shown in figure 5.4.26 to show where risk assessment fits in this cycle. The functions of identification, evaluation, and formulation are all part of the process of safety-risk assessment. Although the implementation of an intervention to mitigate the possibility of the event or trend is the responsibility of the operators and the regulatory agencies, the systemic safety-risk assessment has strong linkages with the mechanisms used for implementation. Specifically, (1) the "identify" function should be a natural part of the implementation process that can be conducted both during and after implementation in order to monitor and assess effectiveness, and (2) the "evaluate" and "formulate" functions should provide directly applicable insights about the hazards and their causal factors to the implementation process in a language and form that are immediately understandable by the operators.

Table 5.4.3 addresses two perspectives on risk assessment: retrospective and predictive. Retrospective risk assessment entails analyses of data on past or current performances of an existing system, whereas predictive risk assessment is the evaluation of the unknown hazards that might be encountered in the operations of a new concept of system design, procedures, or personnel selection and training.

Retrospective risk assessment entails effective integration of information developed from multiple perspectives and data sources to identify existing hazards in the aviation transportation system and their causal factors, and, thereby, help formulate risk mitigations.

Figure 5.4.26. Risk assessment in the cycle of proactive management.

TABLE 5.4.3. PERSPECTIVES OF RISK ASSESSMENT

Retrospective (Diagnostic)	Predictive (Prognostic)

• Data exist on past safety incidents.
• Expertise and experience exist.
• Effective analysis tools have proven to:
  – identify hazards,
  – define their causal factors,
  – establish frequencies, and
  – provide learning.
• Validation of modeling and simulation have been proven by comparison to actual data.
• Implementation effectiveness is monitored.
	• No data exist on future safety incidents.
	• Expertise and experience may not exist.
	• Modeling and simulations are required to:
	  – predict likely hazards,
	  – identify probable causal factors,
	  – determine probable frequencies, and
	  – provide bases for focused studies using real-time, human-in-the loop simulations.
	• Implementation effectiveness may be difficult to monitor.

A process for predicting unknown hazards and assessing risks in the operations of proposed future systems builds upon a validated capability for assessing risks of the known hazards in the current system. Many tools are available to support retrospective risk assessment using existing operational data and expert knowledge. However, only fast-time simulations are available for identifying hazards, analyzing their causal factors, and providing consistent and reliable risk assessment of the operations of future aviation technologies and systems for which there are neither data nor experts. Although no predictive method is perfect, simulation can build on what is known, look at emergent behaviors, enable analysis of possible causal factors, and provide detailed assessments of safety risks.

A study discussed in section 5.4.1 and diagrammed in figure 5.4.17 showcased the predictive methods developed within ASMM, using fast-time simulation and human-performance models. Specifically, it was a study of the operational impact of different methods of metering traffic flow during high-density operations at that same facility. This study demonstrated the use of fast-time simulation to analyze current and new types of operations in a two-phase approach.

In the first phase, detailed, high-fidelity human-performance models were used in simulations to identify potentially hazardous scenarios and to provide insight into the behaviors relevant to the occurrence of these hazards. These results were used to identify the key human-performance characteristics critical to the scenarios of interest and the identified hazards that guided the development of the less complex, efficient human-performance models needed for sufficient computational efficiency to perform the high number of simulation runs needed in the second phase. A very large number of runs is required to identify the probability of occurrences of rare events with adequate degrees of confidence. In the second phase, hundreds of thousands of fast-time simulation runs were performed to determine the probability of occurrence of the rare hazardous events identified in the first phase. In each case, sufficient runs were performed until the probabilities of occurrences were determined to an acceptable level of reliability. This process was automated as is described later in figure 5.4.29 and its associated text.

Insight into rare hazardous events and their primary causal factors requires a model with a sufficient level of detail that reasonably accurate assessments are made of the interactions and dependencies of the primary variables, leading to the question of "What is the appropriate scope and decomposition of the model?" (See figure 5.4.27.) Historically, models that covered a broad scope had to use very aggregate models, suitable for looking at broad economic and policy issues, but not suitable for investigating the impact of new operational concepts. Likewise, many of these aggregate models were parameterized by retrospective analysis of current operations, further limiting their predictive ability. When it comes to identifying the rare hazard events of new operational concepts in which methods of interactions and the dynamics of traffic flows may be substantially different from current operations with which there has been experience or for which there are operational data, predictive simulations must be built directly on what is known: the "physics and cognition" aspects of technological and human performance acting within a specified future operational concept. By starting with these basic building blocks, it may be possible in the future to develop detailed models that can also span the scope of regions and nations. With increasing computational power this modeling will become possible, but will require the development of new models and simulation architectures beyond anything done before.

The NATS has multiple human and nonhuman agents acting both independently and interactively among themselves and with the environment. "Hazards" emerge out of their collective behavior in a manner that, for two reasons, cannot be predicted by examining agents in isolation. First, some of the causal factors of the hazardous situation arise because the agents are placed in a dynamic environment that promotes or induces risky behaviors. At one extreme an environment may overload the capabilities of an agent, such as occurred in the CAT simulations described previously in which the modeled controller simply did not perform certain tasks when the available time became too short.

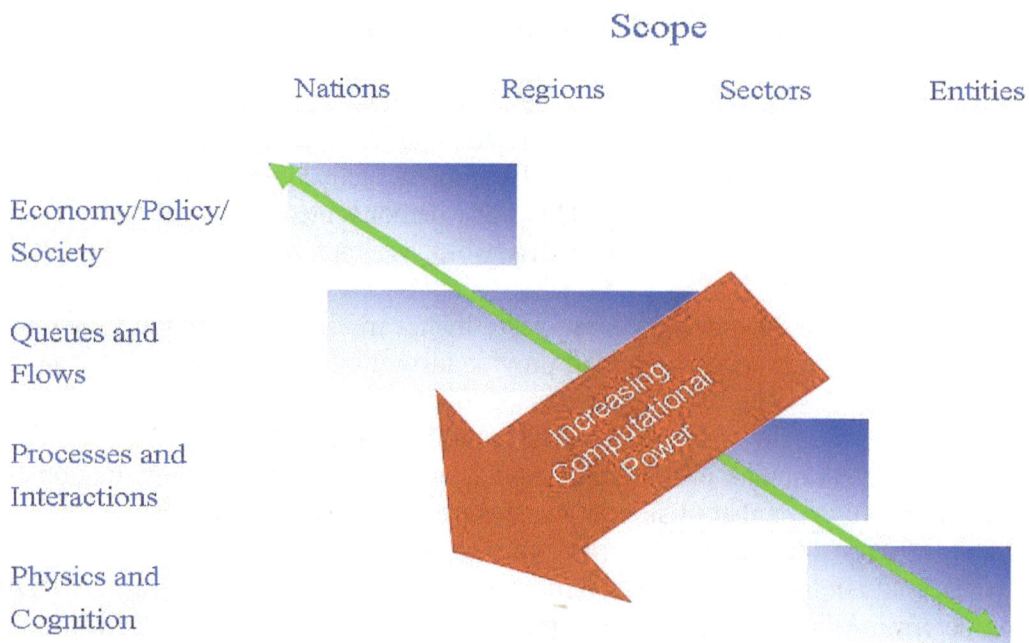

Figure 5.4.27. Decomposition and relationship between level of model detail and scope of simulation.

At the other extreme an environment may not provide the agent with sufficiently accurate and timely information to function in an effective manner, such as when a pilot encounters CAT with no advance warning. Second, additional contributors to the hazard arise that cannot be isolated to single agents, but instead are more accurately described as arising out of their collective behavior, such as multiple "upstream" sectors creating an overly dense traffic-flow condition downstream. Consequently, development of a hazardous situation and its risk is an emergent behavior.[14] Such behavior can be predicted with reasonable accuracy and confidence only by simulating these human and nonhuman agents all acting together and all interacting with their environment.

A related benefit of simulating all the agents acting together is that it enables "structure-preserving" modeling methods, in which the modeled agents are direct representations of the system components. In the NATS, new operational concepts are enabled by incorporating new technologies and establishing new procedures to describe and direct human activities and behavior. Thus, the ideal structure-preserving model of NATS would use agents to model technologies directly in terms of their immediate functions within the system, and would use agents to represent procedures directly using the same language and format as they give the humans who are expected to execute them while all are operating in representations of the environment appropriate to the scenario. Such an ideal structure-preserving model of the system minimizes the loss of fidelity and applicability that can be incurred when translating and abstracting from reality to a computational model and back to reality.

In these simulations of the NATS, the agents are established by technological functions, operating procedures, and coordination procedures that define the boundaries and interfaces by which agents interact with each other while operating in the environments represented in the scenario of the simulation. This concept is diagrammed in figure 5.4.28, where the technological functions, operating procedures, coordination procedures, and operational environment are indicated as known (i.e., specified in the models and simulation) and there are two unknown emergent behaviors to be identified by the simulations. First, human performance may be viewed as an emergent behavior, as the human agents will be acting within an environment that obligates much of their activities; analysis of their activities may provide additional inferences about risk, such as concerns about environments that overload an agent or do not provide the agent with sufficiently accurate and timely information. Second, the systemwide behavior may also be examined, including multi-agent dynamics that have implications for probability of occurrence of a hazardous event, such as aircraft losing safe separation. Therefore, the objective is to use simulations to predict the unknown hazardous events or trends that could compromise the safety of the operations of the NATS. These hazards are a consequence of unknown (possibly emergent) behavior of the human agents the causal factors of which are an unknown confluence of environmental factors, technological factors, procedural factors, coordination factors, and, possibly, cultural factors. The aim of simulations is to gain insight into all these unknowns.

---

[14] An emergent behavior of a system is defined here as a behavior at one level of abstraction that cannot be predicted from another level of abstraction.

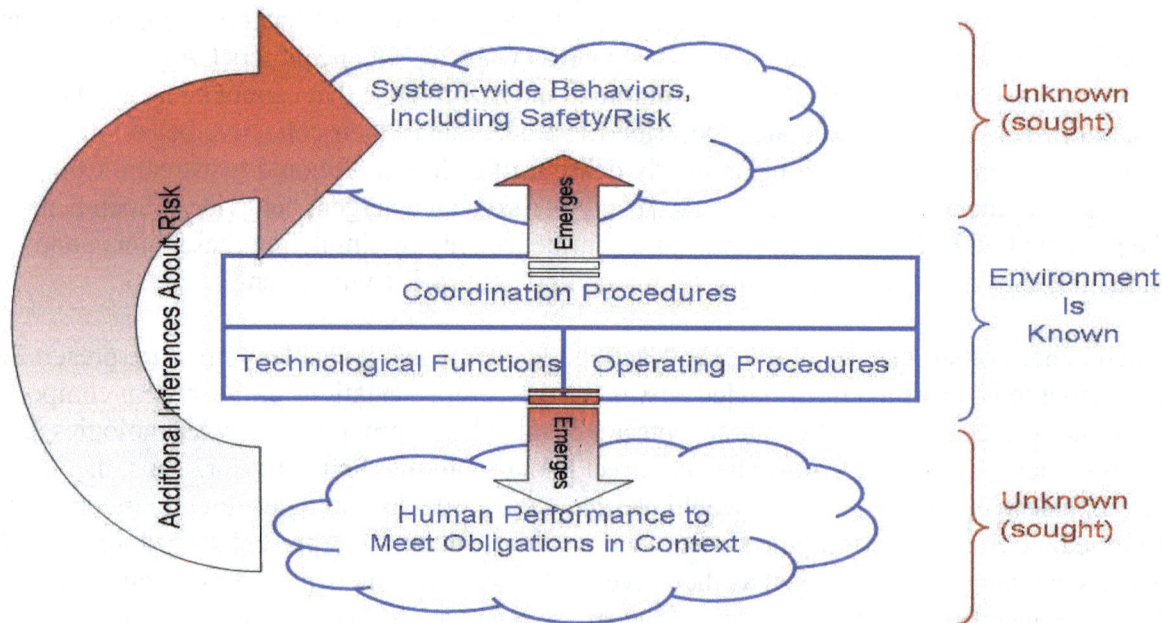

Figure 5.4.28. NATS knowns and unknowns.

The results of the simulations that identified the hazards and the scenarios of interest in phase 1 were described in section 5.4.1. In this section, the focus is on broader issues associated with maximizing the utility of large-scale simulations for a reliable assessment of the probabilities of occurrences of the hazards identified in phase 1. A very large number of simulation runs is required to develop high confidence levels in estimating the probabilities of the occurrences of rare events such as the hazards of the very safe NATS. Even a single simulation may involve a significant amount of computational time, depending on its scope and detail. Likewise, a single simulation can produce a vast amount of output documenting continuously the behavior of every individual agent. The time required for an analyst to examine these data can be prohibitive. Therefore, in the ASMM Project, mechanisms were developed to minimize computational run time by using COTS networks of workstations, where each workstation provides independent replications of a scenario of interest. Also, data-analysis tools were imbedded into this simulation architecture to analyze automatically the data output of each simulation run to provide statistics on the probabilities of hazards. Taking this concept further, a closed-loop control mechanism depicted in figure 5.4.29 was developed, by which the performance metrics resulting from each simulation run were automatically examined statistically to decide on the conditions for the next run. The statistical analysis may follow a classical Monte Carlo design of simulation experiments (although this method is not an efficient way to study rare events), or it may command more simulations runs of the same scenario until sufficient data have been generated for that scenario to state the probability of occurrence of a hazardous situation with the desired level of confidence, or it may command proceeding to the next scenario until sufficient data have been generated for each of the various scenarios to be able to compare and rank them with a desired degree of statistical confidence. This closed-loop control ensured that the number of simulations runs (observations) was minimized by commanding additional runs only when the desired statistical confidence level had not yet been reached.

146

Number of Observations/Configurations

Statistical Analysis
(Design of Experiments/
Confidence /
Ranking and Selection)

Simulation

Data Analysis

Performance Metrics

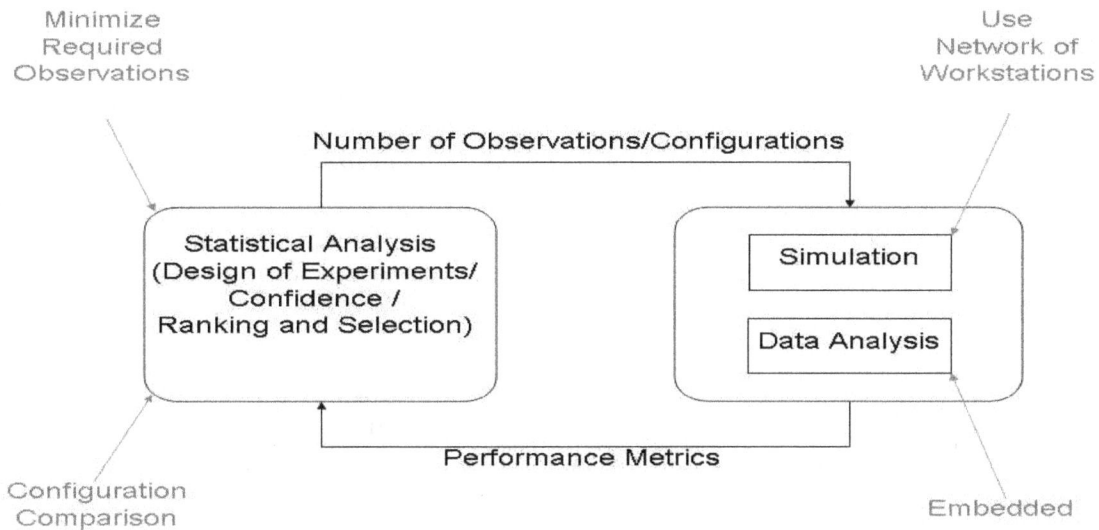

Figure 5.4.29. Automated simulation control: Closed-loop statistical control to maximize the utility of simulations.

It was not possible to analyze completely all the results that the phases 1 and 2 of this study generated before the AvSP ended. These results will very likely be reported in some subsequent paper(s) or presentation(s). However, one very interesting result can be reported here.

The phase 1 study determined the human-performance characteristics that were most important to meeting minimum separation requirements using two procedures for managing dense traffic under various scenarios of wind, traffic, and visibility by running hundreds of simulations. The validity of the human-performance model used in these scenarios was established by statistical comparison with radar data of the human controller's performance on nine observable metrics. Figure 5.4.25 showed the excellent agreement in the performance of the actual and the modeled controller in maintaining the three-dimensional separation distance between aircraft for the four scenarios of wind and visibility across the five sectors under TBM.

During the phase 2 study, about 175,000 flight hours were simulated to gain reliable estimates of the very low probability of the occurrence of less-than-minimum separation in air traffic control (i.e., operational errors). The data in figure 5.4.30 produced by the phase 2 simulations show that the model made more operational errors than did the human controller, especially in the LFDR sector. The model had five operational errors across the runs, whereas the human operator in the same period had none. There are likely many reasons for this result. The most important is probably the fact that the model was designed to follow procedure as its primary guidance for action selection, whereas the human controller has the goal of preventing conflicts. During the phase 1 study, it was discovered that, when the model followed the TBM procedures as designed, there was no assurance of conflict-free traffic downstream of the fix and violations were likely. Consequently, the phase 2 simulations identified the probability of the operational errors for the simulations, as shown in figure 5.4.30, revealing a flaw in modeling the controller using the goal of following procedures rather than the pragmatic goal of the human controller of maintaining separation. The model did not have the long-term goal of no separation violations regardless of efficiency costs as the controllers do.

Airspace and controller modeling heuristics and constraints are inaccurate, but perhaps more important was the discovery that TBM procedures are not infallible. (It is significant that this fallibility in TBM procedures was not being reported by the controllers, although the PDARS data showed clearly that, on frequent occasions, the controllers were compelled to maneuver aircraft so that they did not arrive over the fix at the specified time or did not fly over the fix at all in order to prevent conflict.) The human controller will not permit separation violations and will maneuver the aircraft as necessary to avoid a potential conflict. This fact was demonstrated in the results of phase 2 shown in figure 5.4.30, which shows that there were no separation violations in the Center sectors and only one in the LFDR identified by the radar data.

This discovery of the simulations that following the TBM procedures did not always ensure conflict-free air traffic was further validated by observing flight data as shown in figures 5.4.31 and 5.4.32. The simulated flight results for 1 hour of flight operations with TBM are shown in figure 5.4.31. Only the nominal flightpaths are shown in blue to minimize clutter. The short green segments are indications of vectoring that was deemed needed during that 1 hour of operations to avoid conflicts of the simulated flights at the merge point of Civet, reflecting the same information presented in figure 5.4.30 and portraying the potential for operational errors. A retrospective search using PDARS over 10 months of operation verified this prediction. In figure 5.4.32, each of the short green segments denotes the actual occurrence of an operational error during TBM. (Note that, as the airspace becomes more constrained in the feeder sector, the number of operational errors increases despite the reduction from 5 to 3 nautical miles as the separation standard.)

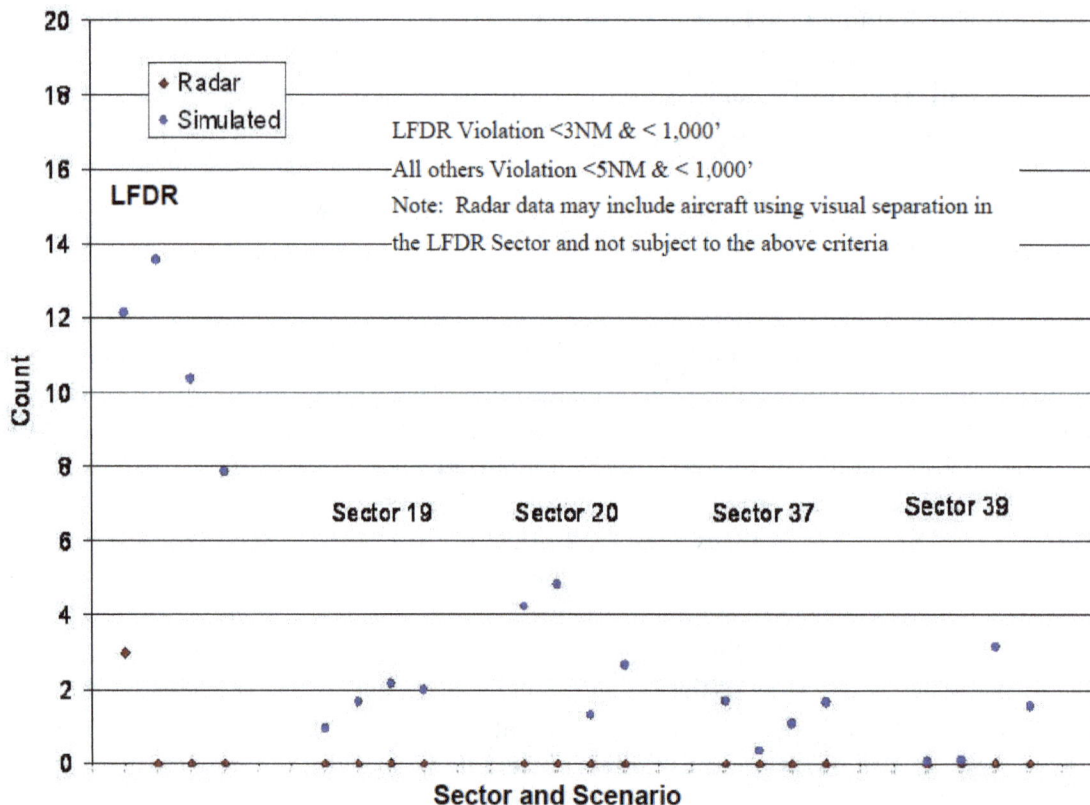

Figure 5.4.30. Separation violations for TBM.

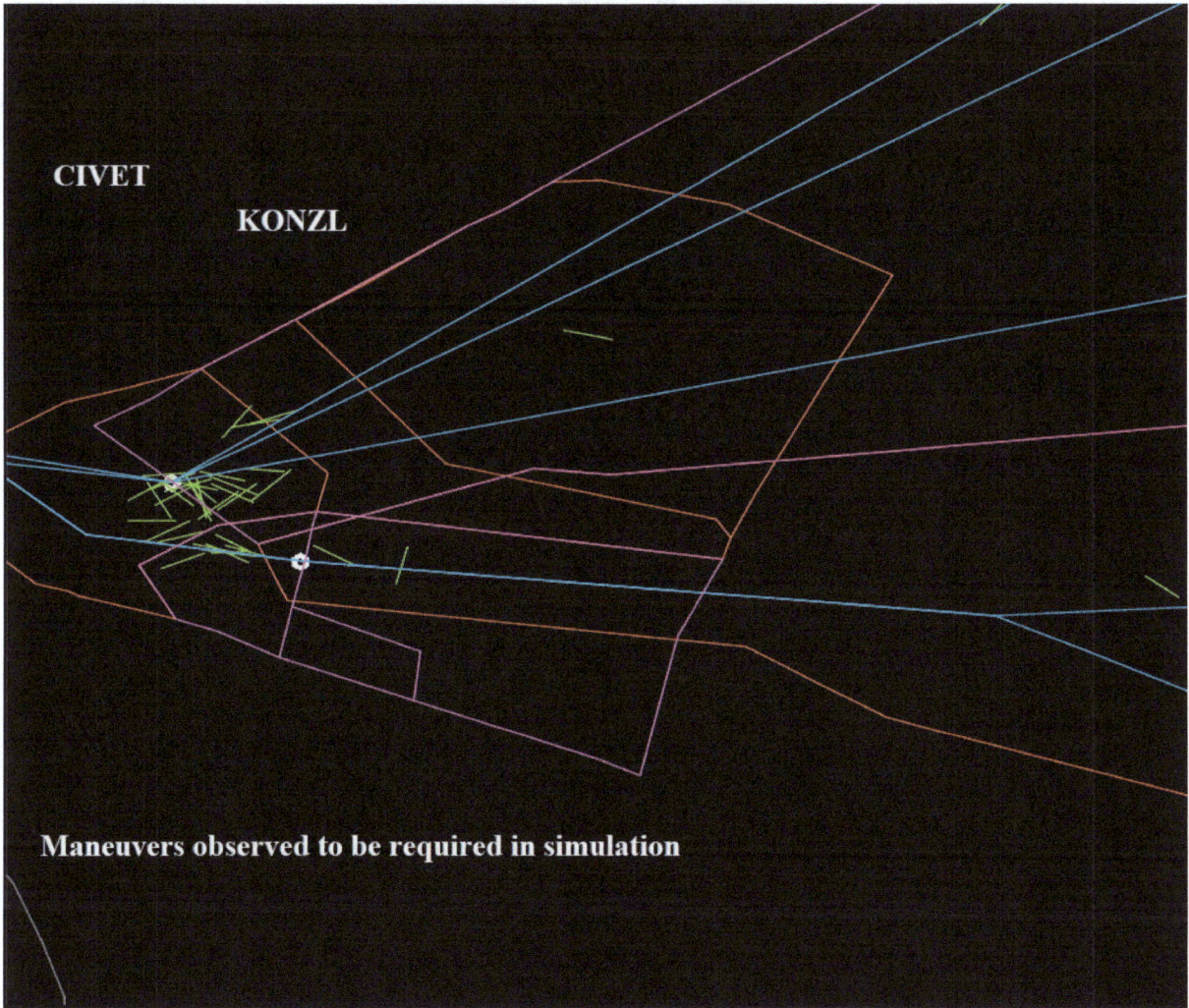

Figure 5.4.31. Results of TBM simulations of one-hour flights.

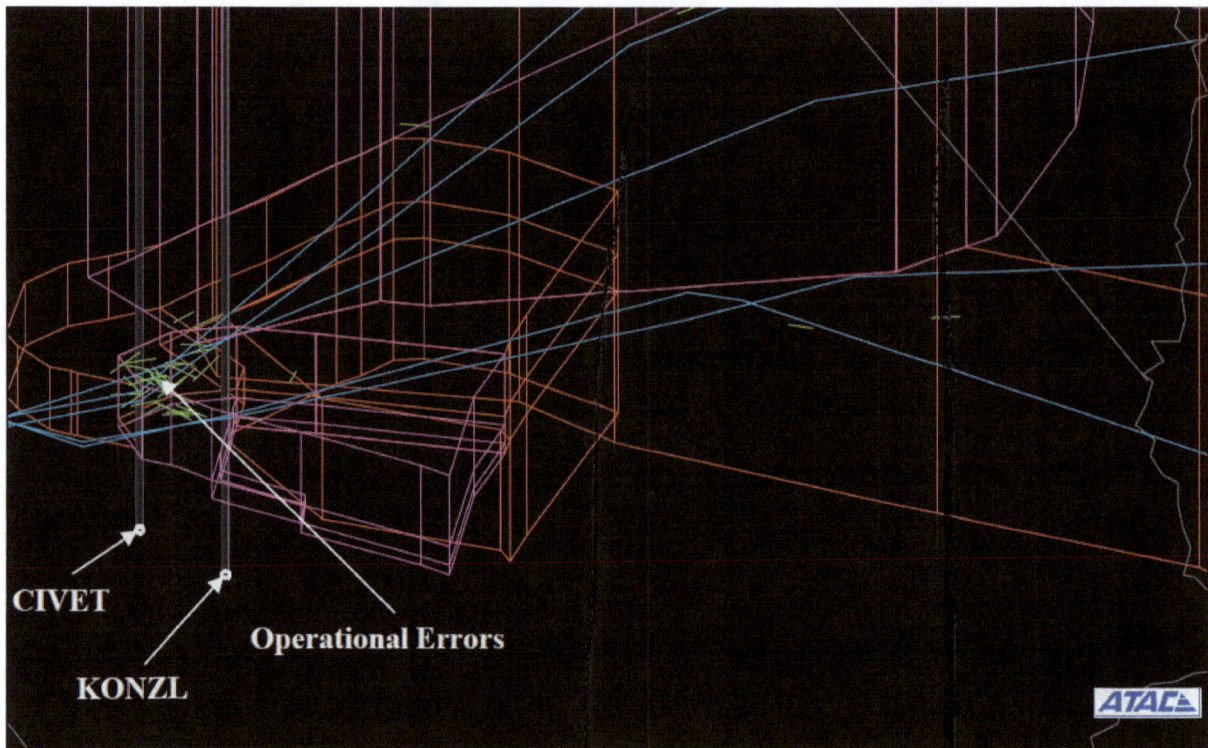

Figure 5.4.32. Operational errors during 10 months from PDARS.

The work performed under the element *Modeling and Simulations* demonstrated that:

- In the cycle of proactive management of system-safety risk, monitoring and fast-time simulation technologies play complementary roles. Monitoring of current operations provides the data needed for both verification and validation of fast-time simulation capabilities to identify hazards in current operations and their causal factors. This scenario establishes the credibility of fast-time simulations needed to trust the predictions of the systemwide impacts of the interventions that are proposed to mitigate the occurrences of the identified hazards.

- An integrated process of monitoring, modeling, and simulation has the potential to provide an important data-driven assessment of hazard likelihood.

- Fast-time simulations can be used for a reliable systemwide perspective on assessment of likelihood of hazard occurrence.

- The dynamic interaction among all the agents is the basis of unknown emergent behavior of the agents and, consequently, of the system performance that can be a critical aspect of system-safety-risk assessment.

- Fast-time simulations are needed for sensitivity studies and for prediction of hazards in new operations.

- The multi-tool approach supports different query types and analyses at various levels of system maturity and specificity.

150

- Customized model development, data analysis, and validation relevant to each new domain are now the primary limitations to the widespread use of fast-time simulations for predictive risk assessment.

5.4.3 Human-Performance Models

The *Modeling and Simulations* activity of the ASMM Project relied on an activity of another AvSP project named Systemwide Accident Prevention (SWAP). An element of SWAP named *Human-Performance Modeling* provided the models of human performance used in the fast-time simulations. The goal of the *Human-Performance Modeling* element of the SWAP project was to develop modeling capability for use in simulations like those described in the previous two sections to assess the efficacies of proposed technological and procedural solutions to operational problems and of potential mitigation strategies. The activities of the *Human-Performance Modeling* element of the SWAP project advanced the state of cognitive modeling with a focus on computational frameworks that facilitate the use of fast-time simulations for predictive analysis of human behavior in real-world environments. The work under ASMM associated with the development of fast-time simulations entailed a selection from among the hierarchy of models developed under *Human-Performance Modeling* and their adaptation to the requirements of the interactions among multiple human and nonhuman agents in each systemwide scenario.

The work of the *Human-Performance Modeling* element of the SWAP project is described in the two papers included in appendix E. The first report, an overview of the state-of-the-art of human-performance modeling, speaks of two development tracks: one of predictive human-performance models, the other of a prescriptive human-performance model. The second paper in appendix E reports on the results of an experiment in which five modeling teams from industry and academia developed human-performance models of pilots performing taxi operations and runway instrument approaches with and without advanced displays. The latter paper describes how the results of each of the five models are influenced by the architecture and structure of that model, the role of the external environment, and the specific modeling advances that these scenarios required.

The human-performance model selected for the experiments conducted under ASMM was the Air MIDAS predictive human-performance model developed at San Jose State University. Each of the five predictive models described in appendix E has unique characteristics. Air MIDAS was found by the staff of the *Human-Performance Modeling* element to be the most appropriate for the particular scenarios and the queries of the experiments described previously. Important to the selection was the fact that Air MIDAS was also the easiest to integrate with the RFS simulation of the rest of the system required in each scenario.

5.4.4 Human-Performance Modeling of Approach and Landing Operations

Other projects of the AvSP had been developing augmentative technologies for enhancing safety in flight operations. These developments included a synthetic-vision system (SVS) for commercial aviation as well as for business jets and general-aviation operations. The system was designed to generate a texture-mapped (or wire-frame) display of the terrain in proximity to the aircraft. Text and other symbology was intended to be overlaid onto the terrain display to show, for instance, the aircraft itself, its velocity, a "follow-me" aircraft, a "tunnel-in-the-sky" indication of the route, and indications of other nearby aircraft. In addition, flight information (i.e., airspeed, attitude, pitch, etc.) were to be overlaid on the display.

In its concept of operations, the existing display elements of current aircraft would be maintained in addition to these augmentations in an SVS-equipped aircraft. Inclusion of both current-presentation modes and SVS-presentation modes offers a challenge in research into the operational concept of their joint usage. It was recognized by the developers that providing both of these sources of information could be problematic. On one hand, they support cross-checking of flight-deck systems. On the other hand, two types of information that are similar in source and content but different in presentation mode may cause transformation workload for the pilot. SWAP managers suggested that computational human-performance models could be used in early design phases of systems such as those proposed for the SVS to predict performance effects of introducing such augmented technologies.

NASA's SWAP Human Performance Modeling leadership defined an experimental scenario in which five teams of modelers would generate predictions of human performance under three visualization scenarios and under three conditions of approach and landing (see table 5.4 4.). NASA provided the detailed cognitive task analysis with which each team of modelers developed a model of a "normative" approach and of landing with and without augmented SVS display. NASA also provided performance and eye-tracking data obtained during a pilot-in-the-loop simulation of these scenarios for validation of the model. The predictions produced using the Air MIDAS model of human-system performance are described here because of their relevance to the ASMM's *Modeling and Simulations*. Best, et al. (2004); Byrne and Kirlik (2004); Deutsch and Pew (2004); and Wickens, et al. (2005) give results using the other models.

Figure 5.4.33 depicts the approach chart for the simulation. A GPS-VNAV/LNAV approach to runway 33L at Santa Barbara, California, was assumed based on the existing GPS approach procedure to that airport. Initial position was located 5 nm north west of GAVIOTA; initial approach fix and flight using autopilot and auto throttle with VNAV and LNAV mode were assumed. Two Decision Altitudes (DAs) were assigned in order to examine the impact of SVS usage on DA selection for approach procedures. DA 650 feet is as high as the usual nonprecision approach decision altitude, and DA 200 feet is as high as the Category I instrument approach decision height. A detailed missed-approach pattern was not prepared because the scope of the go-around simulation focused on the phase of making the decision to go around and initiating a go-around climb. The go-around simulation terminated when the aircraft achieved positive climb with gear up and flaps-5 configuration. The landing simulation terminated when the aircraft touched down. No other aircraft were assumed to be in the area.

TABLE 5.4.4. APPROACH AND LANDING SCENARIOS

Display Configuration	Baseline	Baseline	SVS
Visibility	VMC	IMC	IMC
Nominal approach and landing	Scenario #1	Scenario #4	Scenario #7
Late reassignment – side step and land	Scenario #2		Scenario #8
Missed approach (go-around)	Scenario #3	Scenario #5	Scenario #9

VMC = Visual meteorological conditions
IMC = Instrument meteorological conditions

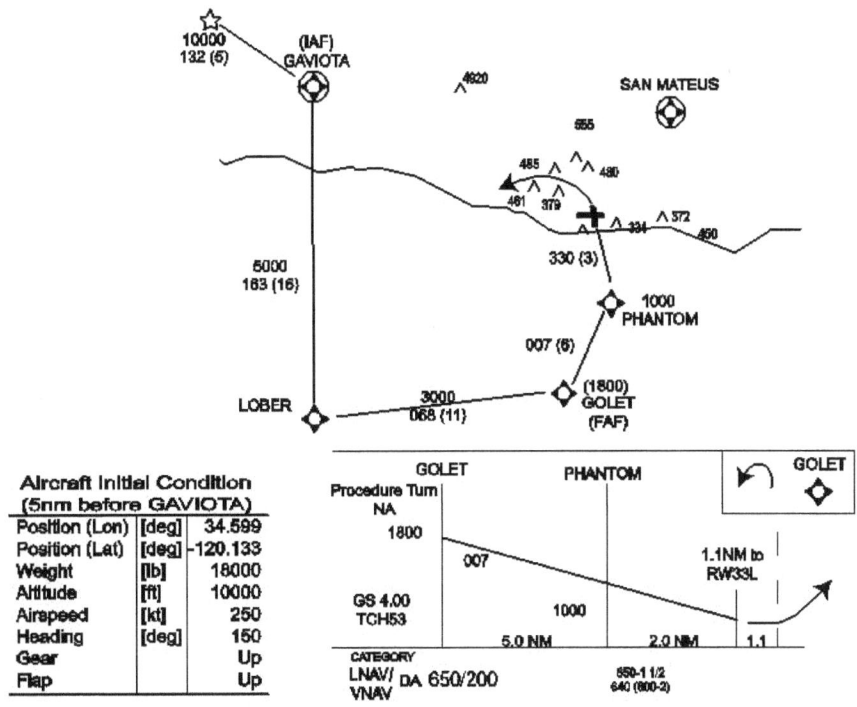

Figure 5.4.33. Approach chart for Santa Barbara simulation scenario.

The response of the Air MIDAS model and tests of the accuracy of the procedures implemented required calibration because this model was being used in the new environment imposed for the proposed SVS operation. An enhancement in the visual model within the Air MIDAS model gave the model a capability to perceive and respond to information from the environment that previously did not exist within the Air MIDAS structure. Specifically, this enhancement required the implementation of a scan-pattern policy (to be described later in this report) with which to model the flight-crew's information-seeking behavior. Data collected during human-in-the-loop simulations conducted by the Boeing Company were used as a basis for the scan pattern (see Mumaw et al. (2000)). The scenario for these human-in-the-loop simulations was an area navigation system (RNAV) (GPS) approach to 33L runway at Santa Barbara in daylight operations under calm winds. The aircraft was flown fully coupled to the autopilot down to DA at 650 feet afe. Experimenters served as flightcrew and air-traffic controllers. The accuracy of the scan-pattern model was checked against the Boeing source data in Mumaw et al. (2000) (using a split-halves validation methodology).

An informal verification process was followed during Air MIDAS model development with reference to the calibration data from Mumaw et al. (2000). In the Mumaw study, in-cockpit eye fixations were recorded on a series of approach-to-landing operations. The visual scan patterns for the model (dwell fixations and dwell durations) were developed using the scan-pattern data taken from the human-in-the-loop simulation data developed by Mumaw et al. (2000). The Mumaw research data were used to parameterize the procedures and information-seeking behavior of the Air MIDAS model in all three scenarios: namely, approach under baseline without SVS, baseline with SVS, and side-step with SVS. Because the Mumaw study was conducted without an SVS system, SVS fixation data were not available from that study. An assumption was made that, in cases in which the

Mumaw data included out-the-window fixations, those would be implemented as SVS scans in the SVS mode of operation. The fixations provided the aircraft state data that were used to populate the equipment representation in the model.

A strong correlation was found between the Mumaw flightcrew eye-fixation data and the Air MIDAS percent of visual fixations data across all scenarios. Correlations were R = 0.9936 for the baseline scenario and R = 0.9955 for the SVS scenario (figure 5.4.35). This result was viewed as evidence for verification that the model was accurately representing the data source for the initial scan pattern based on procedures. On the basis of the satisfactory statistical match between Boeing's human-in-the-loop simulation data and the model-generated data, the scan-pattern-policy model of information seeking was used during an approach to landing into Santa Barbara in the scenarios of table 5.4.4.

Figure 5.4.34 shows the fixation times on the respective elements within the flightcrew's scan pattern of the crew station and the external environment. This figure compares the results of the fixation times from the simulations using the Air MIDAS model to the data from the human-in-the-loop simulations reported in Mumaw et al. (2000). These results indicate that the procedural and visual sampling behavior that were encoded and run through the simulation largely replicate the data gathered on human performance from human-in-the-loop simulations. This result verifies that the model behaves as designed.

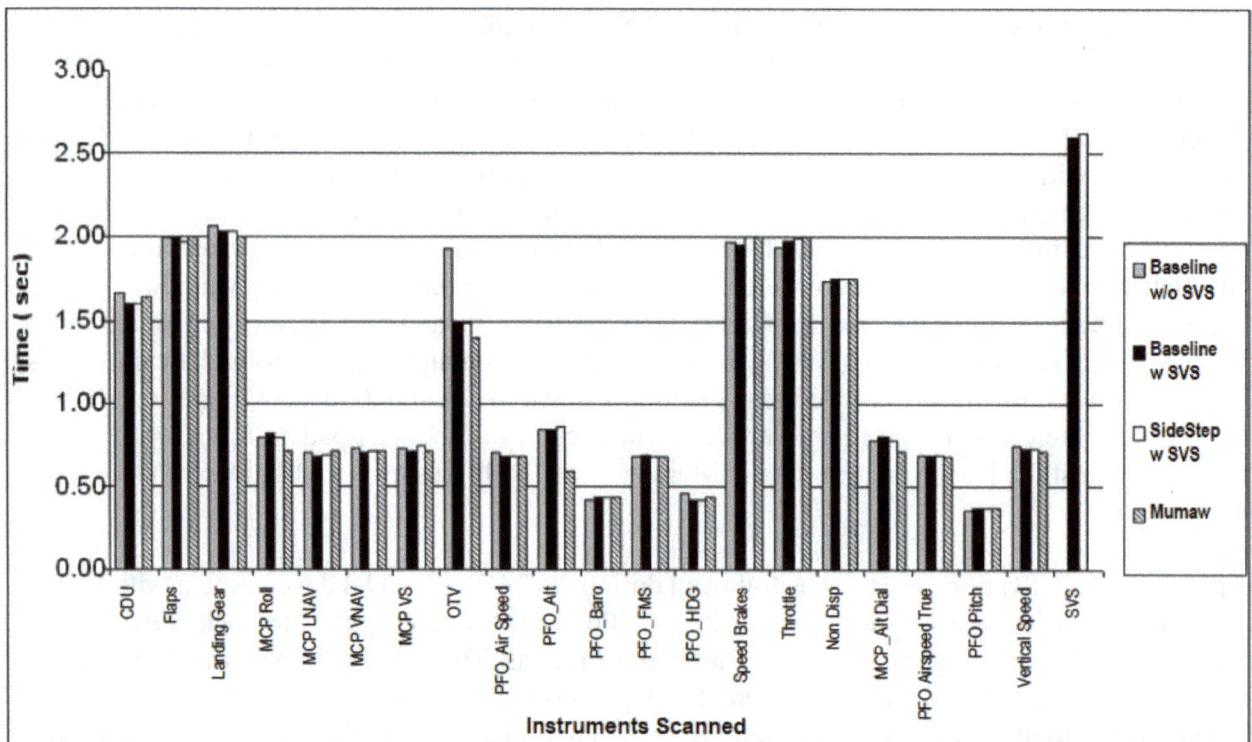

Figure 5.4.34. Pilot flying scan patterns: Air MIDAS model and Mumaw et al. (2000) results.

The approach to Santa Barbara was used in a study by NASA of SVS using human-in-the-loop simulation described in appendix E of the scenarios of table 5.4.4 to examine the use of SVS while performing a side-step maneuver late in approach. The predictive validity of the Air MIDAS model was tested by running the model through simulations of the same scenarios. Validation for appropriate visual-scanning behavior on the part of the model was examined by looking at the model-generated dwell frequency compared to the human flightcrew dwell-frequency patterns that were measured using eye-movement sensors in the human-in-the-loop simulation. It was found that "percent of fixations" required in completing an approach and landing during the human-in-the-loop simulation correlated with the Air MIDAS performance data. As seen in figure 5.4.35, the correlation between NASA's human-in-the-loop part-task simulation (Foyle, Goodman, and Hooey, (2003)) and the Air MIDAS percent of fixations was R = 0.7608 in the baseline scenario, R = 0.8782 in the baseline with SVS scenario, and R = 0.5538 in the SVS side-step scenario. The somewhat lower correlation of the model-generated side-step data to the human performance is likely due to the fact that no side-step conditions were available in the Mumaw data, which had been used as the source of the scan pattern. Therefore, the scan pattern of the model in the side-step scenario was predicted based on information requirements to achieve active goals.

In the simulations, an aircraft dynamic model, PC Plane©,[15] was integrated with the human-performance model so that the aircraft dynamics, which are critical in this approach phase of flight, were incorporated to impose the appropriate time pressures on the performance of the flightcrew model. Test scenarios for the simulation were developed and procedures were established that were based on established cockpit procedures for a current aircraft. Simulations were performed under the several conditions of table 5.4.4; i.e., approach and landing and go-around, both with "current day" technologies and with SVS cockpit configurations.

Percent Fixation Correlations[1]

Air MIDAS to Boeing Sim	Air MIDAS to NASA Sim
• Baseline: r = 0.9936	• Baseline: r = 0.7608
• With SVS: r = 0.9955	• With SVS: r = 0. 8782
• SVS with sidestep: r = 0.9948	• SVS with sidestep: r = 0. 5538
Verification	Validation

1 - Pearson Product Moment

Figure 5.4.35. Percent fixation correlations.

---

[15] PC Plane© was developed by NASA Langley Research Center and the NASA Ames Research Center.

The simulation scenarios follow:

- Normal approach without SVS case was used as a baseline.

- Two go-around situations: one go-around following controller's command, the other based on pilot's decision. In the case of the controller's command for go-around, two sets of the timing were used so that the interruption of pilot activities took place in both a busier and a less busy time during the approach. For the busier time scenarios, the controller's command was issued about 100 feet before DA when pilot activities included the tasks of "runway-in-sight" callout and "approaching-minimum" callout. In the less busy scenarios, the command for go-around was issued at 250 feet before DA, where no particular activity was expected.

- Two different visibilities, which switched at a specified altitude associated with cockpit activities "runway-in-sight" callout and pilot's go-around decision. Visibility was set so that runway came in sight at 150 feet before DA except in the pilot's go-around-decision cases. Visibility in other scenarios was set so that a go-around event happens because of an inability to see the runway at DA.

The system architecture of the Air MIDAS SVS simulation environment is shown in figure 5.4.36. Three modules, including PC Plane representation of the aircraft dynamics, Flight Management System/Control Display Unit (representing FMS/CDU), and Air MIDAS were integrated in the simulation. A set of Dynamic Link Library (DLL) functions generated cockpit-control input and time synchronization control to PC Plane through socket connection. DLL functions were invoked by the Air MIDAS module, which was written in List Processing (LISP) Programming Language, through JAVA network interface architecture. Time-synchronization control (functioning through coordination of the control function and the PC Plane response) maintained precise synchronization of Air MIDAS and PC Plane during simulation and enabled dynamic closed-loop simulation.

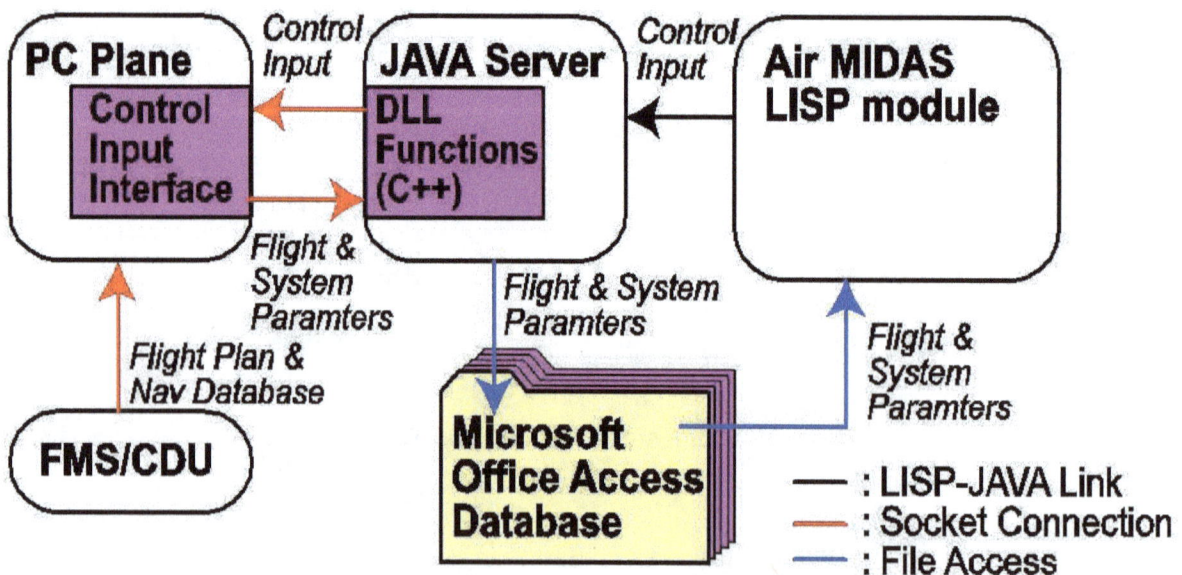

Figure 5.4.36. Air MIDAS system architecture for SVS study.

The Microsoft Office Access database architecture was used to share flight and aircraft system parameters between the Air MIDAS SOM and the representation of the scenario external to the human model. Figure 5.4.37 depicts the functional architecture of the entire Air MIDAS SVS simulation environment. As was described in section 5.4.1, the Air MIDAS model consists of a series of interlinked models. Working memory in the SOM contained domain knowledge and rules to invoke actions regarding cockpit procedures; namely, (1) approach and landing and go-around procedures, (2) standard callout, (3) checklist, (4) ATC communication, and (5) landing/go-around decisions were implemented. For this study, Working memory storages of Primary Flight Display (PFD), Navigation Display (ND), Engine Indication and Crew Alerting System (EICAS), Synthetic Vision System (SVS), and out-the-window (OTW) were prepared as input data sheets (as shown in figure 5.4.39) to be stored as the crew's information about the aircraft state. Visual scan of information provided by cockpit display models (shown in figure 5.4.38) was passed into the SOM through its visual perception models. The Air MIDAS visual-scan-pattern model reads data from a particular data sheet (figure 5.4.39) when the agent fixates on a corresponding display. When this information was perceived through the scan pattern, it was passed into the UWR and it is used to trigger cockpit motor activities through the scheduler (figure 5.4.37). Because the subject of this research effort is an evaluation of a visual display concept, perceptual processes associated with the SVS system and/or OTW observations are critically important.

On the left side of figure 5.4.37 are the "domain" models that represent the world, the equipment with which the operator interacts, and the mission/world characteristics that are required for the scenario of the simulation. These models can either be contained within the Air MIDAS model framework itself, or they can be an independent model of the domain that is linked to the air MIDAS human-performance model. In this study, PC Plane was used, and it updates all the parameters in the database as time proceeds. The right side of this framework represents the functions of the human operator. The system representation is a closed-loop control model with inputs coming from the external world representation on the left and action being taken by the human operator model that changes that world. The model provides psychological plausibility in the cognitive constructs of

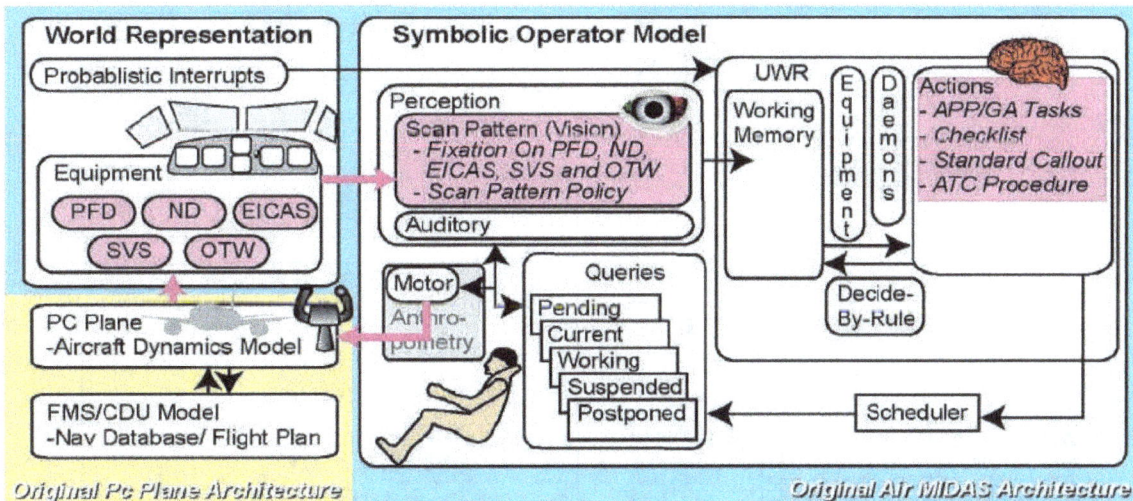

Note: Magenta Color implies implementation for SVS application.

Figure 5.4.37. Elaborations on Air MIDAS for study of SVS.

long-term, working memories (with articulation into spatial and verbal components of the theses models) and has sensory/perceptual and attentional components that focus, identify, and filter simulation world information for the operator, action, and control. The cognitive function is provided by the interaction of context and action. Context is a combination of declarative memory structures and incoming world information that is mapped to the agenda manager that is taking the plan (overall mission). The agenda manager acts in conjunction with the plan interpreter to provide a series of procedures to be performed in order to meet mission goals and to handle contingent activities (such as interruption or plan repair). The human-operator model is resource constrained, i.e., it is limited in the cognitive, perceptual, and motor behaviors that can be performed simultaneously. Output of action on the external world is effected through the models of the operator linked to anthropometric representations (if they are invoked by the analyst). The action changes the external world and the cycle begins.

Figure 5.4.39 illustrates the data and the parameters of those data that are stored as the information about the aircraft state and entered into the operator's UWR as a consequence of visually scanning the displays shown in figure 5.4.38.

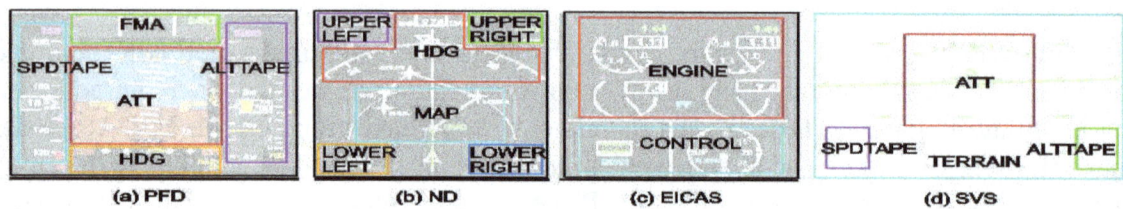

Figure 5.4.38. Visual scan information sources.

PFD

Parameter	Description	UNIT	VALUE (ex)	AREA
thedg	Pitch Angle	(deg)	5.20	ATT
phidg	Bank Angle	(deg)	10.1	ATT
easkt	IAS	(kt)	213	SPDTAPE
selias	Speed Command	(kt)	200	SPDTAPE
altft	Press. Altitude	(ft)	3,235	ALTTAPE
selalt	Altitude Command	(ft)	3,000	ALTTAPE
roc	Rate of Climb	(fpm)	500	ALTTAPE
apth_e01	Autothrottle Mode		SPD	FMA
appt_e01	Altopilot Pitch Mode		VNAV	FMA
aprl_e01	Autopilot Roll Mode		LNAV	FMA

EICAS

Parameter	Description	UNIT	VALUE (ex)	AREA
flap	Flap Angle	(deg)	20.0	CONTROL
nsgear	Gear Position		1	CONTROL
sbrk	Speed Brake Angle	(ratio)	0.1	CONTROL

OTW

Parameter	Description	UNIT	VALUE (ex)	AREA
thedg	Pitch Angle	(deg)	5.20	ATT
phidg	Bank Angle	(deg)	10.1	ATT
visibility	Visibility	(sml)	5.0	TRR
rpos_tw_dme	DME to Runway	(nm)	20.1	NAV
rpos_rw_brg	Bearing to Runway	(deg)	32.0	NAV

ND

Parameter	Description	UNIT	VALUE (ex)	AREA
psidg	Heading Angle	(deg)	276.0	HDG
track	Track Angle	(deg)	269.0	HDG
selhdg	Heading Command	(deg)	300.0	HDG
to_wpt	Name of To Waypoint		GOLET	MAP
rpos_to_dme	DME to To WPT	(nm)	11.2	MAP
rpos_to_brg	Bearing to To WPT	(deg)	125.0	MAP
rpos_tw_dme	DME to Runway	(nm)	20.1	MAP
rpos_rw_brg	Bearing to Runway	(deg)	32.0	MAP

SVS

Parameter	Description	UNIT	VALUE (ex)	AREA
thedg	Pitch Angle	(deg)	5.20	ATT
phidg	Bank Angle	(deg)	10.1	ATT
easkt	IAS	(kt)	213	SPDTAPE
selias	Speed Command	(kt)	200	SPDTAPE
altft	Press. Altitude	(ft)	3,235	ALTTAPE
selalt	Altitude Command	(ft)	3,000	ALTTAPE
roc	Rate of Climb	(fpm)	500	ALTTAPE
rpos_tw_dme	DME to Runway	(nm)	20.1	OTW
rpos_rw_brg	Bearing to Runway	(deg)	32.0	OTW

Note) Altitude and Speed on SVS was not used for the trigger of procedural tasks.

Figure 5.4.39. Display information source.

Figure 5.4.40 illustrates the logic of the Air MIDAS scan-pattern model. Scan pattern selects a display, selects its area where the Air MIDAS human operator model fixates its eyes, aggregates the values of displayed parameters, and then updates UWR. Failure of data aggregation was also simulated before updating the UWR.

A display, its area, and duration of fixation were randomly determined based on eye-fixation data obtained from human-in-the-loop simulations (Hooey and Foyle (2001)). After the display, its area, and duration of fixation were determined, then the fixation success/failure filter evaluated whether the fixation period was sufficient to aggregate the data successfully. It was assumed that all the data included in the area can be successfully perceived when the fixation duration was beyond the reading-time threshold and no data are perceived when the fixation duration was less than that threshold time. Consequently, Air MIDAS updated the parameters in UWR when the fixation duration on the corresponding display was longer than a specified threshold, but if shorter, it did not perform any UWR update.

The cockpit was divided into viewing planes (i.e., areas of the cockpit that contain information). These areas are represented on the "x" axis of figures 5.4.41 and 5.4.42. The total number of fixations (i.e., movement from one point of regard (defined by a dwell time of larger than 500 ms) to another) was collected. The percentage of fixations on each of the viewing planes was calculated for the human flightcrew in their information-seeking and for the model as it too sought information from specific parts of the cockpit. Figures 5.4.41 and 5.4.42 show the correlations between the human and model with respect to information-seeking fixations as a function of the scenario and the cockpit equipment configurations.

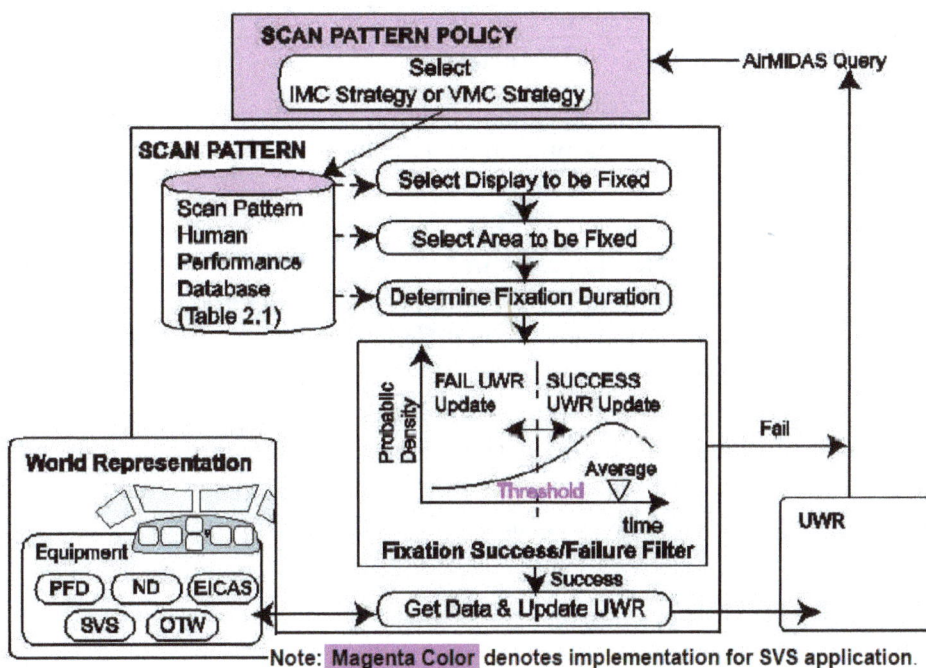

Figure 5.4.40. Scan pattern policy.

159

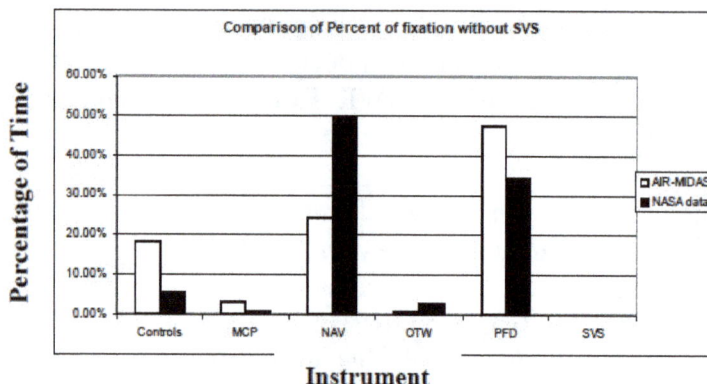

Figure 5.4.41. Comparison of percentage fixation time on specific instruments between the Air MIDAS and the human pilots (without SVS).

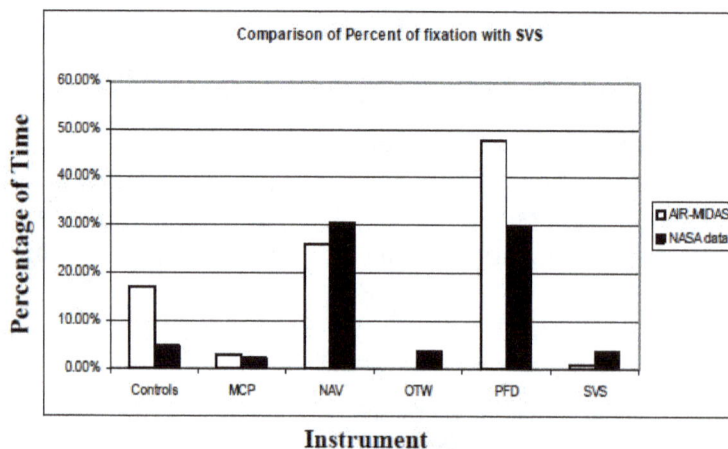

Figure 5.4.42. Comparison of percentage fixation time on specific instruments between Air MIDAS and the human pilots (with SVS).

The benefit of the visual augmentation implemented in the Air MIDAS model was that it appears to have been sensitive to the introduction of the SVS technologies. Given this information-source difference, the flightcrew is predicted to engage in different scanning patterns and engage in different behaviors, depending on the phase of flight that they are completing. The modeled flightcrew focuses on scanning the portions of the display that will serve the flightcrew's immediate goal of safely landing the aircraft as reflected by the performance-characteristics output from the approach and landing phases of Air MIDAS predictions. The information that is provided to the flightcrew from the SVS during these critical phases of flight is time-critical information.

Air MIDAS predicted higher fixation rate on the controls and the PFD than did the human data produced by the human-in-the-loop simulation, suggesting that the rules guiding human performance are different from those guiding the performance of the model. Based on their fixation patterns, the human pilot used the information on the ND to a larger extent than does the Air MIDAS pilot, while the Air MIDAS pilot fixated on the PFD to a larger extent than did the human pilots. Figure 5.4.41 illustrates that mismatch for the case of approach to Santa Barbara under "low" (IMC) visibility conditions without SVS and without any request for side-step.

The Air MIDAS model also predicted lower dwell times on the ND, the OTW scene, and the SVS displays than did the human simulation, suggesting that, when flying with the SVS display, the flightcrews looked at the SVS information to a greater extent than did the Air MIDAS model. It was postulated that the human flightcrew received PFD information from overlays in the SVS and the Air MIDAS model required looking at the PFD (not the SVS) for that information. Figure 5.4.42 illustrates that difference. This figure showed the approach to Santa Barbara under "low" (IMC) visibility conditions with the availability of an SVS and without any request for side-step.

Figure 5.4.43 shows flight and task sequence during one of the normal approach cases (scenario #1 in table 5.4.4). Speed command setting, flap lever position, and gear position are the system parameters, which were manipulated by Air MIDAS pilots. Since the aircraft was flown with autopilot (vertical navigation (VNAV) and lateral navigation (LNAV) modes) and auto-throttle (VNAV mode), control-surface parameters do not show any Air MIDAS manipulations. Figure 5.4.43 (a) shows the time histories of the key flight parameters during the approach and figure 5.4.43 (b) is the horizontal view of the flightpath.

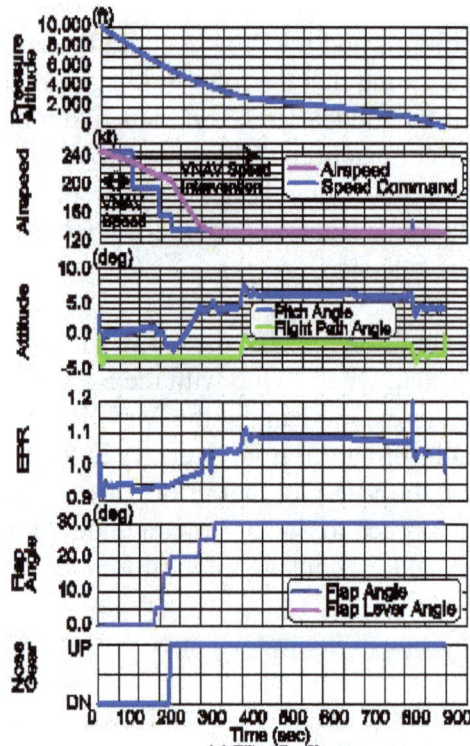

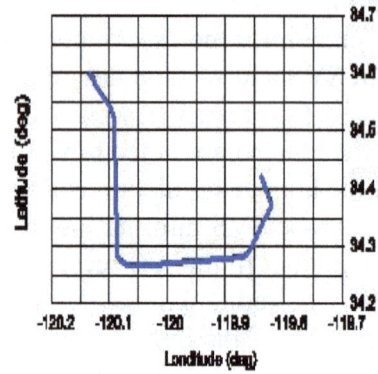

(b) Horizontal Profile

(a) Flihgt Profile

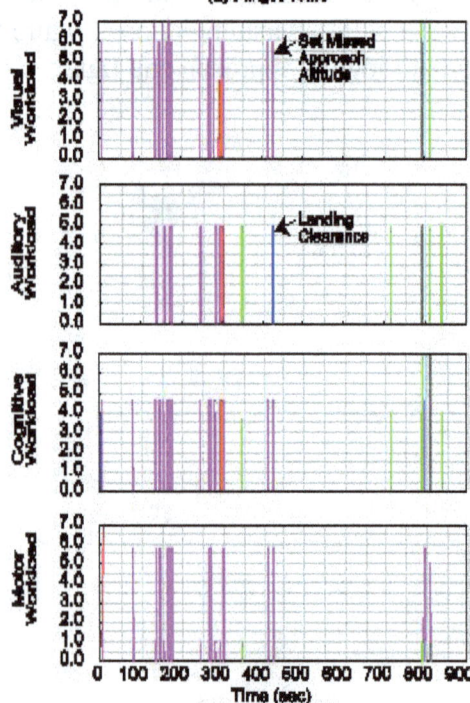

(c) Workload (PF)

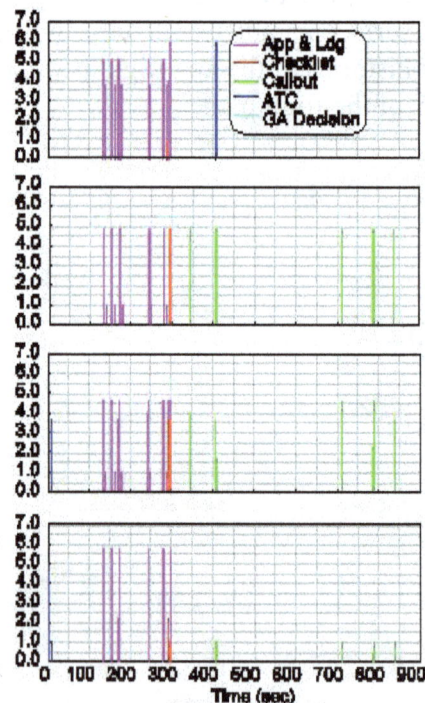

(d) Workload (PNF)

Figure 5.4.43. Flight and task sequences of normal approach (Scenario #1 in table 5.4.4).

The information presented in figure 5.4.43 conveys the story of a normal approach (Scenario #1 in table 5.4.4). After initiation of the simulation (i.e., start of approach), airspeed was reduced by VNAV-programmed airspeed command regardless of the setting of the speed knob. When the airspeed was reduced to 218 knots. at 131.7 seconds, the pilot-flying (PF) model ordered flap deployment and the pilot-not-flying model (PNF) set the flap lever to 5 degrees. At 134.5 seconds, the PF model decreased the airspeed to 160 knots, requiring further flap extension. This further airspeed reduction occurred before the aircraft had achieved the previous airspeed commanded. After passing 800 feet afe, which is a boundary altitude of low- and high-visibility area, the PNF model called out "runway-in-sight" at 772 feet afe and 784.7 seconds. At that moment, the PNF model fixated on OTW. At 796.3 seconds, after passing DA (= 650 feet), the PF model decided to land and called out "Landing." The aircraft touched down on the runway at 845.4 seconds.

Average workload, which was used to examine the contribution of each procedure to the overall workload, was defined by

$$\overline{WL_i}\mid_{total} = \sum_{j}^{j_{all}} \sum_{n}^{n_{all}} WL_{i,j,n} \cdot \Delta t_{i,j,n} / t_{total} \quad :\text{Total average workload}$$

$$\overline{WL_{i,j}} = \sum_{n}^{n_{all}} WL_{i,n} \cdot \Delta t_{i,n} / t_{total} \quad : \text{Average workload (each procedure)}$$

where

    i: V(Visual), A(Auditory), C(Cognitive), or M(Motor)

    j: Procedures (Approach and landing, Go-Around, Scan Pattern, Checklist, Standard Callout, ATC, GA Decision, and Others)

    n: Number of accomplished activities

    $WL_{i,j,n}$: Workload of each task item

    $\Delta t_{i,j,n}$: Duration of each task item (seconds)

    t_{total} : Flight Time (seconds)

Figures 5.4.43 (c) and (d) are time histories of the workloads of the PF and the PNF models during the normal approach. Both the PF and PNF models had a period of higher density of workload at around 130 seconds before they completed the configuration of landing flaps, airspeed, and gear. The PF model also had another period of higher density of workload period from around DA to touchdown. The PF model had maximum visual workloads of 7.0 when it performed speed reduction and ordered flap extension, and when it performed a landing/go-around decision where it also had maximum cognitive workload at 7.0. Maximum auditory and motor workloads were 5 in the simulated flight. The PNF model did not have a moment when its workload value reached 7.0.

Figure 5.4.44 is an example of Scenario #8 in table 5.4.2 in which a sidestep maneuver resulted in a go-around. As in the previous example, this figure shows the flight and task sequence of this simulation run. Figures 5.4.44 (a) and (b) show that at 812.6 seconds, a go-around command was issued by the air-traffic-controller model. Both pilot models heard it and the PF model called out "go around" at 812.9 seconds. Then the PF model pushed the go-around lever and set pitch attitude to 10 degrees. The PNF model set the flap lever to 5 degrees at 818.6 seconds, following the order from the PF model. After confirming positive climb, the PF model ordered gear up and the PNF set the gear lever to the up position at 821.5 seconds.

Figures 5.4.44 (c) and (d) show the time histories of the workloads of the PF and the PNF models. These figures show that the PF model in this scenario had a higher level of visual, cognitive, and motor workload after hearing the go-around command (with auditory workload at 5.0) compared with the workload around DA in the normal approach flight (figures 5.4.43 (c) and (d)). This level was caused by the time criticality of the tasks required to perform the go-around. Time-critical tasks (i.e., tasks whose required completion time is equal to or less than the current simulation time plus the time required to perform the activities) are increased in the priority levels and scheduled in a noninterruptible mode (i.e., once initiated no other activity will be able to interrupt them). This result conveys the impact of an externally ordered go-around as compared with the lower workload density at DA in normal approach flight.

The total visual, auditory, and cognitive average workloads of both the PF and PNF were 4.0, 0.2, and 1.0, respectively, in both normal approach and go-around cases. Total motor workload was about 2.5 for PF and 1.6 for PNF in both the normal approach and go-around simulations. The major differences in activities across the two scenarios were only in the short final phase, with no significant impact on the total average workload.

A summary of results follows:

- Small delays of action initiation in flight control were observed in the approach phase with SVS operation. These delays occurred because the chance of fixation on each instrument, given a fixed scan time, was decreased by adding SVS to conventional display configuration, resulting in fewer fixations on other instruments.

- No degradation in human performance or in delay of task initiation were observed in landing and go-around phases with SVS, though there were time shifts when the necessary actions were performed in the approach phase.

- A scan-pattern model that simulates the pilot's instrument scan was validated by using the data of human-in-the-loop simulation.

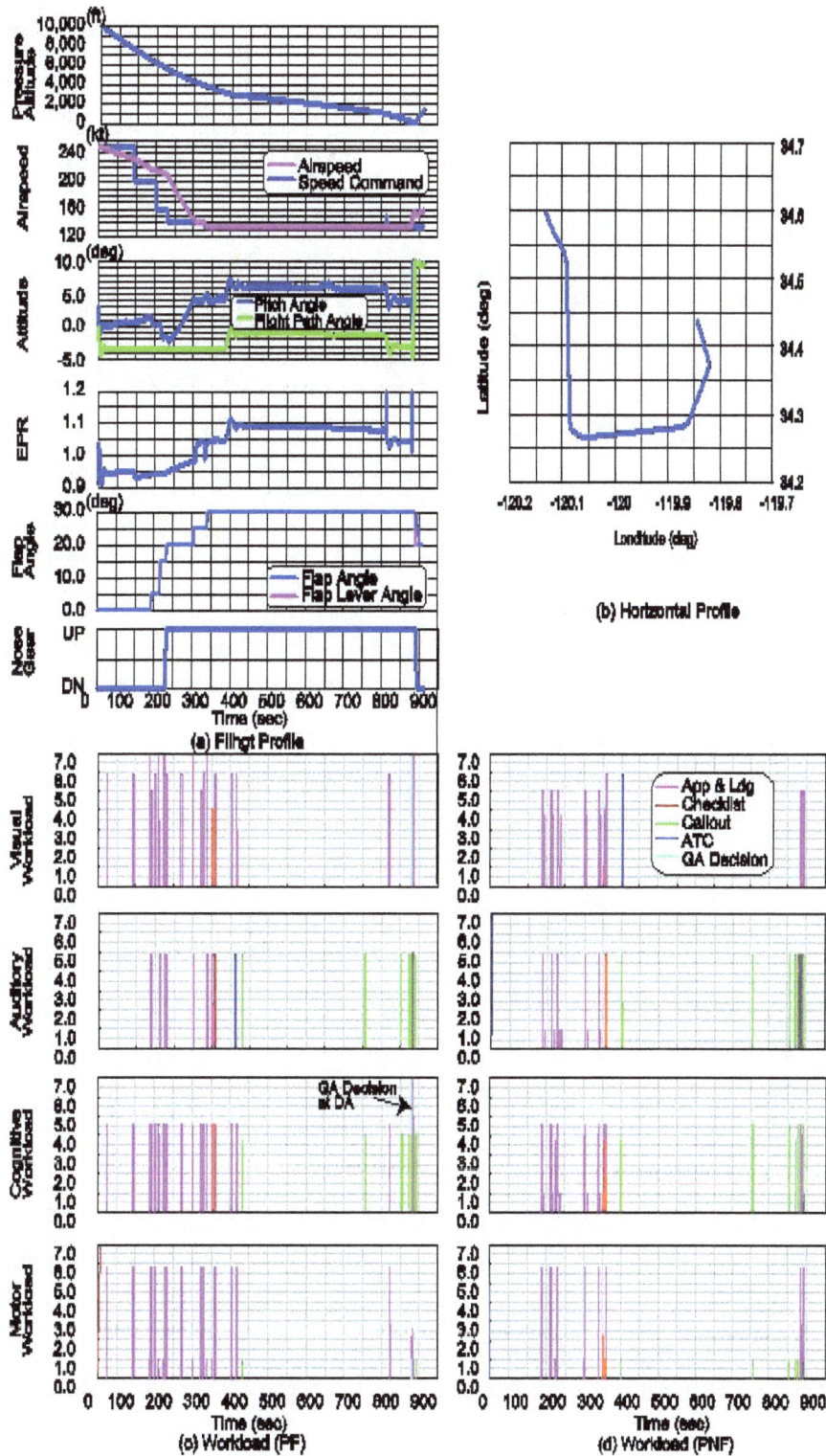

Figure 5.4.44. Flight and task sequence of go around (Run 1, Scenario #8, in table 5.4.4).

6.0 SUMMARY AND CONCLUDING THOUGHTS

The aviation community has recognized that accident rate alone is an unreliable safety indicator. Fortunately, the NATS, especially in the U.S., is very safe and, although the ultimate metric is fatal accidents, it is not a practical metric against which to evaluate the efficacy of an intervention to mitigate the cause of an accident. It takes many years to verify a change in the probability of occurrence of this rare event because so few accidents occur each year. Furthermore, the circumstances of accidents vary greatly. This fact was recognized by the Aviation Safety Investment Strategy Team and, consequently, it was suggested that the challenge to NASA was to develop technologies and methodologies to identify accident precursors from the incident reports and from operational flight-recorded data acquired. As stated previously, a precursor is an incident, event, or trend, that is identified as a hazard when it is the symptom of a systemic problem that, if left unresolved, has the potential to result in increased probability of an accident.

NASA, under its Aviation Safety Program, established the Aviation System Monitoring and Modeling Project to provide the technologies and methodologies that would make it feasible to monitor the NATS continuously, convert the collected data into reliable information, and share that information among the stakeholders for collaborative decision-making so as to enable a revolutionary, proactive approach to managing the aviation system for prevention of accidents. The concept was that, by identifying and understanding the causal factors of the precursors of the next accident, total accident risk would be minimized by:

- reducing the likelihood with which anomalous incidents occur,

- reducing the likelihood that anomalous incidents will pose a severe hazard,

- reducing the system exposure to hazards, and

- reducing the likelihood that exposure to hazard will result in an accident.

When the ASMM Project ended in 2005, it had produced:
- validated automated capabilities to extract operationally useful information from any numerical data sources
  - that were tested and evaluated operationally on flight-recorded data and on radar-track data;
- validated automated capabilities to extract useful and consistent information from any textual data sources and linked to numerical data
  - that were tested and evaluated operationally on ASRS and ASAP reports and linked to FOQA data;
- a statistically sound, top-down survey methodology;
- Measurement capabilities for ATC performance and information-sharing among ATC facilities over a secure network; and
- Fast-time, system simulations incorporating human performance
  - that were validated against operational data.

The transfer of ASMM products to providers of current commercial data-analysis systems was a requirement of the project so as to extend the use of these safety-enhancing capabilities across a large sector of the aviation community. As previously indicated in table 5.3.1 regarding APMS, many of the individual ASMM analysis tools were commercialized, including:

- Full-flight data storage, pattern search, and routine events became vendor standards.

- Automated data integration was commercialized through a nonexclusive license to SAGEM Avionics, Inc.

- ADIS weather archive was transferred to the FAA with Website access to all air carriers with FOQA programs.

- The Morning Report of atypical operations was commercialized through a nonexclusive license to SAGEM Avionics, Inc.

- The PDARS network operating at 20 Centers and 13 TRACONs was transferred to the FAA.

In retrospect, the ASMM Project can be deemed highly successful. However, it did not achieve its original goal, because the capability to integrate information extracted from all of the diverse distributed data sources across the entire industry to identify, assess, and analyze systemic-level issues could not be demonstrated.

Each of the ASMM products such as APMS, PDARS, and NAOMS demonstrated standalone capabilities that will continue to evolve as the Data Analysis Tools continue to be adopted and commercialized by the industry and continue to be adapted to meet the evolving needs of the stakeholders. However, while these standalone capabilities have advanced the analysis of individual data sources, mitigations designed from the perspective of the owner of any single data source may create risks to the others. The true and overriding value of the ASMM products is their use as an integrated suite of tools to enable a *systemwide* perspective on proactive management of the safety risk of the NATS. The AvSP ended before the ASMM products could be demonstrated as a capability to enable information extracted from diverse data sources to be integrated and used for safety-risk assessment from a systemic-issue perspective.

From the outset of the ASMM Project, the concept of operations was always that the tools and methodologies it developed are converged into an operating, integrated capability for sharing information across the aviation community to enable continuous systemwide safety-risk assessment for the benefit of, and access by, the entire aviation industry.

The criticality of this existing need has been recognized by the FAA, industry, and the Commercial Aviation Safety Team (CAST). They designated the Aviation Safety Information Sharing task (ASIST)[16] Team to define an ideal future state for safety information-sharing in the commercial aviation community and to agree on a roadmap to implement the future state. ASIST developed a Concept Paper in response to its charter that is provided in appendix D. The concern about the current inadequate sharing of safety information is expressed, for example, in the following statement taken from the ASIST Team's Concept Paper:

---

[16] This Aviation Safety Information Sharing Task (ASIST) should not be confused with the previously mentioned NASA's Aviation Safety Investment Strategy Team (ASIST).

"Why Change is Urgent

Recent studies have suggested that in many fatal accidents, information existed somewhere in the system that could have been used to prevent the accidents, had it been available to the right people at the right time.[17] That is, existing information could have been used to identify the need to mitigate risks and thus prevent accidents. This is unacceptable – it means that we are not using existing safety data and information optimally in today's environment."

The following is another statement from that Concept Paper that addresses the need to facilitate sharing of aviation-safety information:

"A key element of future safety management systems will be a dramatic change in the way we collect, manage, and share data and information. We envision a future state in which safety information from stakeholders is accessed through a safety information sharing process and is available to decision-makers for managing risks."

The ASIST team's Concept Paper was endorsed by the CAST. It was also submitted to the JPDO's Safety IPT and the need was reiterated in that team's Action Plan.

The following statement taken from the paper presented by the FAA Administrator, Marion Blakely, to the AIAA Air and Space Conference in Washington, D.C., on April 20, 2004, recognized the need for merging information from diverse data sources:

"So far, our safety improvements have been limited to specific operators or specific locations. But now, we are on the threshold of being able to marry—to merge—the data that we gather through ASAP and FOQA. And that will be a major improvement for safety. You take the subjective— ASAP—and view it in tandem with the objective—FOQA—and you have taken an important step toward increasing the overall safety of the system."

---

[17] *Accident Precursor Analysis and Management, Reducing Technological Risk through Diligence.* National Academy of Engineering of the National Academies, The National Academies Press, Washington, D.C., 2004.

Roelen, A. L. C.; Wever, R.; and Verbeek, M. J.: *Improving Aviation safety by better understanding and handling of interfaces, a pilot study.* NLR-CR-2004-025, National Aerospace Laboratory, Amsterdam, 2004.

Commercial Airplane Process Study, An Evaluation of Selected Aircraft Certification, Operations, and Maintenance Processes. The Report of the FAA Associate Administrator for Regulation and Certification, 2002.

"Continuing Airworthiness Risk Evaluation: An Exploratory Study." Flight Safety Foundation, *Flight Safety Digest* September–October, 1999.

Improving the Continued Airworthiness of Civil Aircraft, A Strategy for the FAA's Aircraft Certification Service. Committee on Aircraft Certification Safety Management, Aeronautics and Space Engineering Board, Commission on Engineering and Technical Systems, National Research Council, National Academy Press, 1998.

As was said, the ASMM Project was not able to meet this need. Implementation of various ASMM products as separate entities will improve ways to handle data and facilitate understanding from the perspective represented in a single data source, but automated integration of the information from these products was not demonstrated. Even with the advanced capabilities of ASMM, when a problem is discovered, analysts must laboriously seek out and examine information extracted from other data sources to gain the essential system perspective of the problem. Although concepts for linking information from diverse sources had been demonstrated, when it ended in 2005, the ASMM Project had not tested at operational levels a capability for merging information extracted from flight data, ATC data, survey data, simulations, experiential reports, and other heterogeneous data sources and, therefore, failed to achieve an operational capability for gaining an integrated *systemwide* and systems-deep perspective of risks. There remained a recognized need to converge the tools and methodologies developed under ASMM into an operating, integrated capability for systemwide safety-risk assessment for the benefit of, and access by, the entire aviation industry.

7.0 TOWARD THE FUTURE VISION

Several studies conducted under the ASMM Project demonstrated the value of integrating information from diverse sources of aviation-safety data. These sources showed that when the data-analysis capabilities developed in the ASMM Project can be applied across airlines, across ATC facilities, and across diverse data sources, it will be possible to gain insight into systemic operational issues.

The study of In-close Approach Changes (ICACs) described in appendix B started with recognition by the analysts in the ASRS office that, based on reports, ICACs were causing problems for some pilots. The NAOMS survey revealed that the frequency of problematic ICACs was, indeed, much higher than had been suspected (or reported). The radar-track data from PDARS showed that flights that had been given a late runway change crossed the threshold somewhat higher on average. The flight data from APMS showed that ICACs frequently resulted in less stable approaches. All of the information extracted from these sources combined to show that problematic ICACs were frequent events that caused an increased safety risk sufficiently high to take some corrective action, but not so high as to prevent controllers from using this valuable procedure for maintaining smooth traffic flow. Analyses of the ASRS database and the results of the NAOMS survey suggested some of the possible causal factors of when an ICAC becomes a problem and led to the preliminary observations on changes that might mitigate anomalous consequences of ICACs (see appendix B).

Another experiment under the ASMM Project was a comparative study of time-based metering (TBM) with miles-in-trail metering (MITM), both of which are methods used by controllers for sequencing aircraft into approach airspace during high-traffic-flow periods. (This study was described in section 5.4.) With the older technique, MITM, the controller manages the distance between arriving aircraft; with the new method, TBM, the controller manages the time of entry into the approach airspace. Radar-track data from PDARS were used to compare traffic patterns. Flight data from APMS were used to compare effects on flight operations, and fast-time simulations were used to compare controller workloads. The PDARS data showed that TBM resulted in advantage over MITM in time and distance flown in a sector and in more consistent descent profiles. The APMS data showed an advantage of TBM over MITM with regard to cockpit navigation workload and confirmed more consistent descent profiles shown by PDARS. The fast-time simulation study showed how these advantages could be realized as a consequence of a redistribution of workload between the Center controller and the TRACON controller.

The analyses conducted of the high-energy arrivals described in section 5.3 were another example of the value of being able to examine an issue from various different perspectives.

Many groups representing the industry have documented the need for integrating information from data sources across the NATS. See, for example, figure 7.0.1, taken from the report of the ASIST provided in appendix D, which was endorsed by the CAST, FAA leadership, and the JPDO Safety IPT. The challenge remaining after the ASMM Project was integrating information.

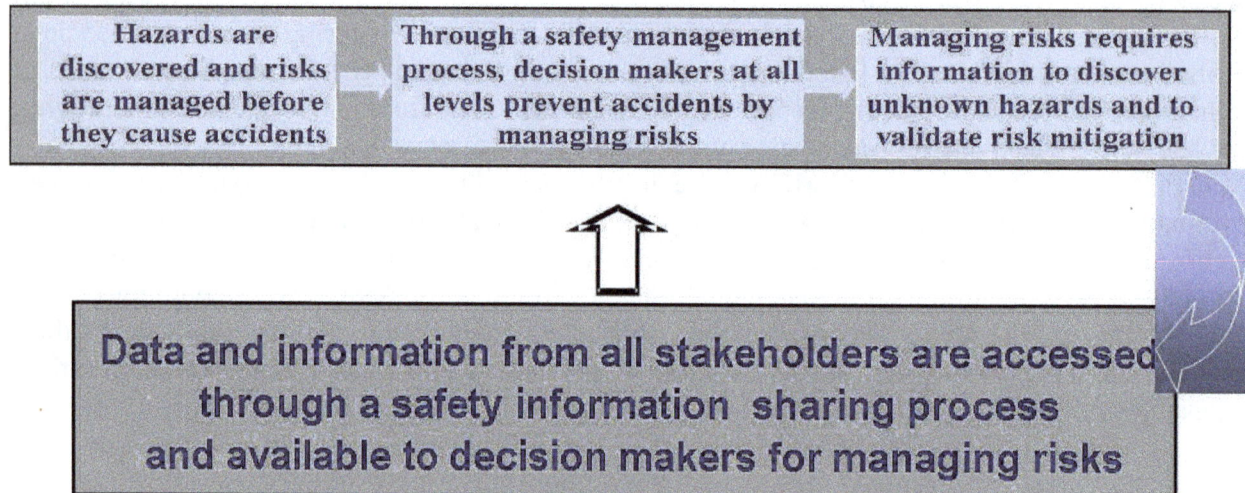

The future air transportation system is <u>accident-free</u>

HOW?

| Hazards are discovered and risks are managed before they cause accidents | Through a safety management process, decision makers at all levels prevent accidents by managing risks | Managing risks requires information to discover unknown hazards and to validate risk mitigation |

Data and information from all stakeholders are accessed through a safety information sharing process and available to decision makers for managing risks

Figure 7.0.1. The vision of the future state (from ASIST report in appendix D).

This challenge entails both political and technical aspects requiring an approach that:

- builds up from sources that owners are willing to share and will require:
  - building trust,
  - demonstrating that sharing can do national good without individual or corporate harm, and
  - adding sources as trust and technology allow; and
- builds tools for making data sources interoperable that:
  - build from individual sources of information,
  - do not force standardization on data owners, and
  - derive knowledge by merging information extracted from across distributed, heterogeneous data.

In response to this challenge, NASA's AvSP undertook two new projects: the Information Sharing Initiative (ISI) and Integrated Safety Data for Strategic Response (ISDSR), as the ASMM Project came to a close in FY05.

7.1 Information Sharing Initiative

The Information Sharing Initiative (ISI) was initiated by NASA under the Aviation Safety and Security Program in response to the industry's request to develop the technology and infrastructure to enable a distributed archive of FOQA and ASAP data across U.S. air carriers.

In 2003, the commercial aviation industry established the Voluntary Aviation Safety Information-sharing Process (VASIP) to provide a means for the commercial aviation industry and the FAA to collect and share safety-related information and to use that information to proactively identify, analyze, and correct safety issues that affect commercial aviation. VASIP participants include more than a dozen air carriers, organized labor, the FAA, and NASA.

The key to VASIP is the development of a technical process to extract de-identified safety information from any participating airline Flight Operations Quality Assurance (FOQA) or Aviation Safety Action Program (ASAP) program, aggregate it with comparable information from across a distributed database, and make it accessible to appropriate industry stakeholders for analysis. Initially, the process is envisioned to focus on issues that have been identified through individual airline programs. Ultimately, it must be capable of analyzing those issues that are identified from the aggregate safety database as well.

The FAA established the Voluntary Safety Information Sharing Aviation Rulemaking Committee (VSIS ARC) to serve as a forum for interaction between industry representatives and the FAA, to provide advice to FAA on safety-information sharing, and to prepare recommendations for rulemaking. The distributed national archives were designed to be in compliance with the VASIP and to operate under the rules of engagement established by the VSIS ARC and the VASIP Executive Steering Committee (ESC). The Distributed National FOQA Archives (DNFA) and the Distributed National ASAP Archives (DNAA) data collection processes were to be distributed and secure. Air carriers would retain ownership of the data they provide to the national archives. These data would be fully de-identified before being placed in secure repositories located on the respective air carriers' property. De-identification of these data was deemed essential to ensure the protection of proprietary information. Furthermore, analyses of the distributed archives were to be conducted only as directed by the VASIP ESC and the VSIS ARC, and would be overseen by the VASIP Rules and Procedures Subcommittee.

The stakeholders in VASIP identified NASA as having the institutional background, resources, and expertise required to develop this technical aggregation framework, as well as the analytical tools to support the process. NASA's AvSP agreed to undertake the development of these national archives in collaboration with, and acceptable to, all of the participants, including the owners of the FOQA and ASAP data, the vendors of the FOQA data-analysis software, and the FAA. Upon urging from the FAA, NASA began development of the ISI in 2004. ISI was to develop and implement the hardware, software, and networking to enable a demonstration within two years of an archive of distributed FOQA data with sufficient reliability that it could be handed off to an operating organization at the end of that time. NASA further agreed to implement and demonstrate an extension of the FOQA archive for an archive of ASAP safety reports. ISI was accomplished in a collaborative partnership with participating airlines, employee organizations, and the FAA to demonstrate operational distributed national archives of FOQA and ASAP data by 30 September 2006. The FOQA repositories are collectively known as the DNFA and the ASAP Archives are collectively known as the DNAA.

The DNFA and DNAA physical infrastructure consists of servers located at each participating air carrier as well as DNFA-DNAA Central at NASA ARC in Mountain View, California. DNFA and DNAA employ a service-oriented, messaging architecture that allows DNFA-DNAA Central to communicate with air-carrier sites over a secure, private network to cooperatively accomplish safety analyses. DNFA and DNAA use methods and algorithms that accomplish system-level aviation-safety assessments without moving raw FOQA or ASAP data from air-carrier locations. Instead, separate analyses are conducted at each participating air-carrier location. The results of these separate, air-carrier-level analyses are de-identified and sent to DNFA-DNAA Central, where they are combined into a unified, system-level picture of the aviation-safety issue being examined.

Figure 7.1.1 represents the infrastructure and process developed for the DNFA. Each gray rectangle at the top represents a participating airline. At the top of each of theses boxes is shown the local FOQA processing system for that airline. NASA has entered into contracts with Austin Digital, Inc. (ADI) and SAGEM Avionics, Inc. (who, between the two, provide FOQA processing services to all currently participating airlines) to transfer FOQA data. Each vendor was provided with the software to transform the data into a standard format across a NASA-supplied one-way firewall onto a NASA-supplied Local Archive Server (LAS). The LAS at each airline is shown in figure 7.1.1 at the bottom of each of the grey rectangles representing an airline. At each participating airline, all of the flight-recorded data from all aircraft equipped for FOQA are de-identified, processed into standard,

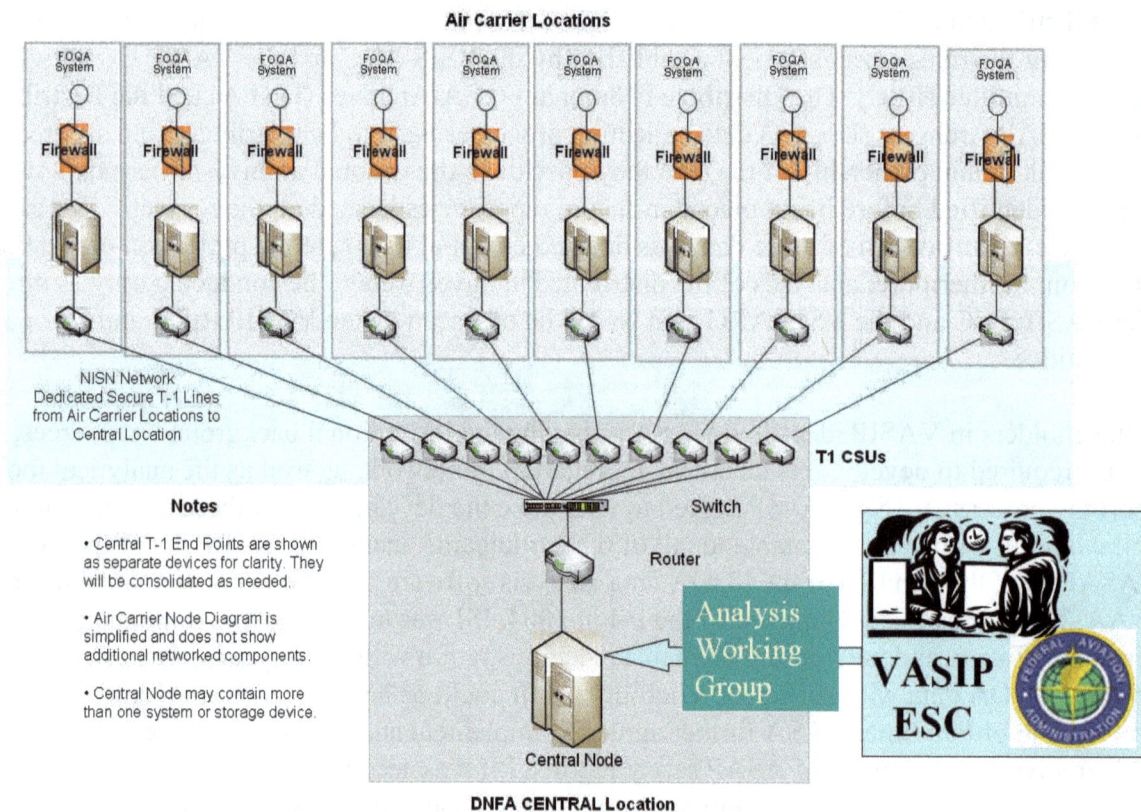

Figure 7.1.1. The concept of distributed archives.

compressed files, and stored on each LAS as those data are acquired by at that airline by its FOQA vendor. De-identification, performed by the local vendor, entails encryption of airline and flight number and changing the date of all flights in a month to the first of that month.

Each LAS communicates over a secure, private network to a Central Server housed at NASA ARC to respond to queries from the Central Server. These queries are made only of the archived data in support of an investigation of an issue that has been identified by the ESC appointed by the VASIP requiring access to, and research on, the archive. The ESC appoints a working group to study the issue, and that group defines specific queries to be processed by the archive. The Central Server issues commands for each query (indicated by the green arrows in figure 7.1.2) to each LAS, where searches are automatically conducted by software installed in the LAS for only those events that are relevant to the query and where calculations of summary statistics are accomplished.

After these analyses are completed on each LAS, de-identified lists of flights meeting the search criteria and the summary statistics are forwarded to the central server, where they are aggregated. This response is indicated by the red arrows in figure 7.1.3. No raw data ever leave the premises of their owner. Added de-identification is imposed automatically if the results of a query could identify an airline because it is the sole operator of a fleet of aircraft or the only airline servicing an airport. Such flights are automatically deleted from the particular study, but not from the archive because they may be used in some future study.

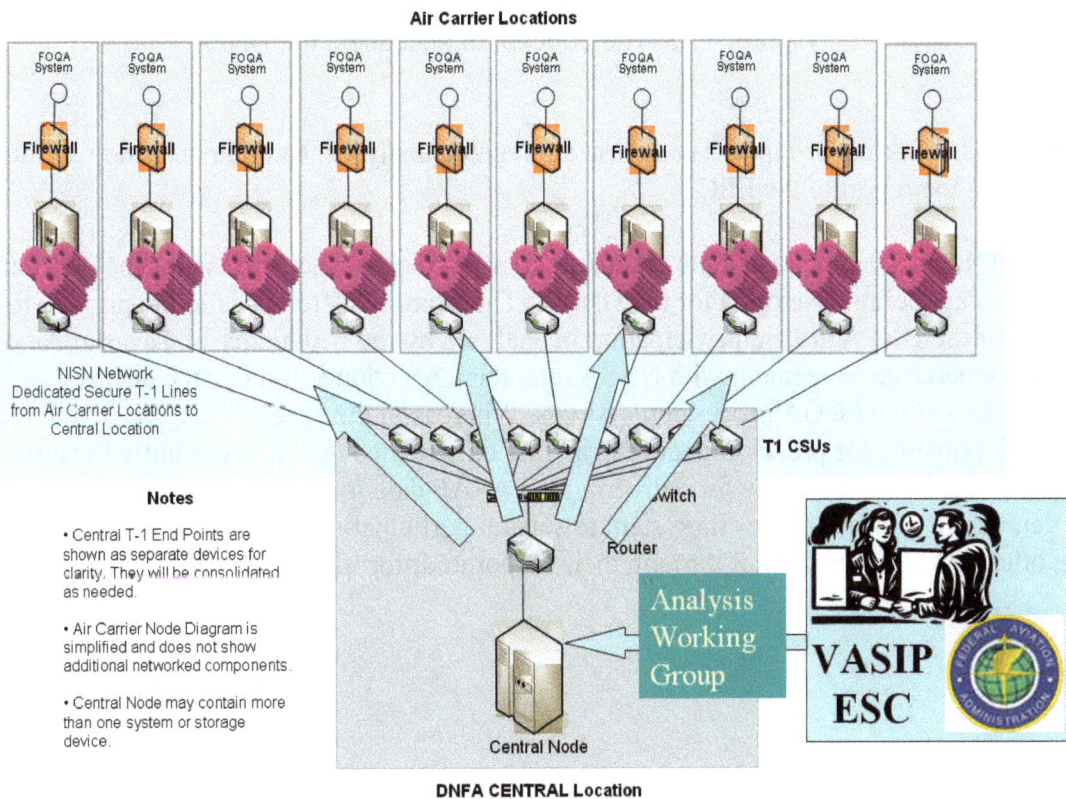

Figure 7.1.2. The concept of submitting a query to the DNFA.

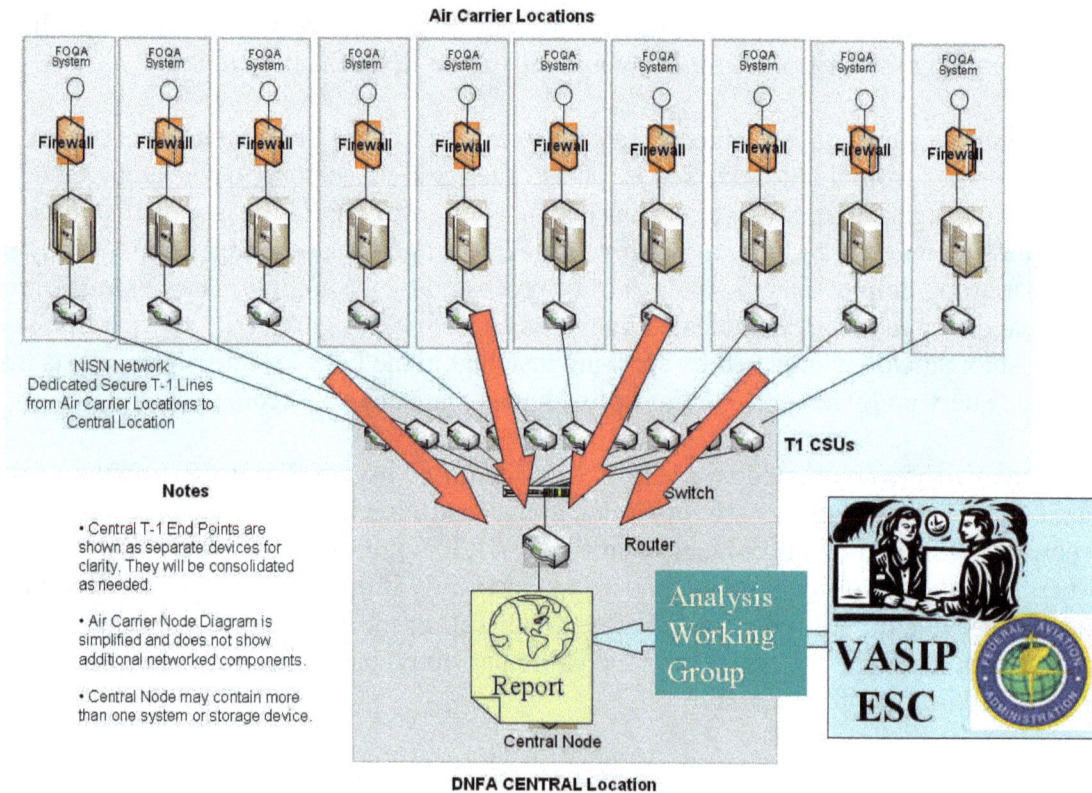

Figure 7.1.3. The concept of accessing the DNFA.

The Analysis Working Group reviews these aggregated analyses, summarizes findings into a report, and provides the report to the ESC.

Figures 7.1.4 and 7.1.5 represent the comparable infrastructure and process for the distributed-archive concept being developed for the DNAA. The primary differences in the process for DNAA from that of the DNFA are the participation in analyses by the University of Texas at Austin (UT) under a Cooperative Agreement with NASA, and the LAS connection to each airlines' ASAP servers rather than to the FOQA processing systems. Whereas in the case of FOQA data the local vendors are responsible for processing and storing the data in the LAS, it is currently the airlines themselves that are responsible for delivering the ASAP data to the DNAA. Some of these airlines have vendors (e.g., UT supplies its system to several participating airlines) to assist them in this work; others must accomplish it through their corporate Information Technology departments.

176

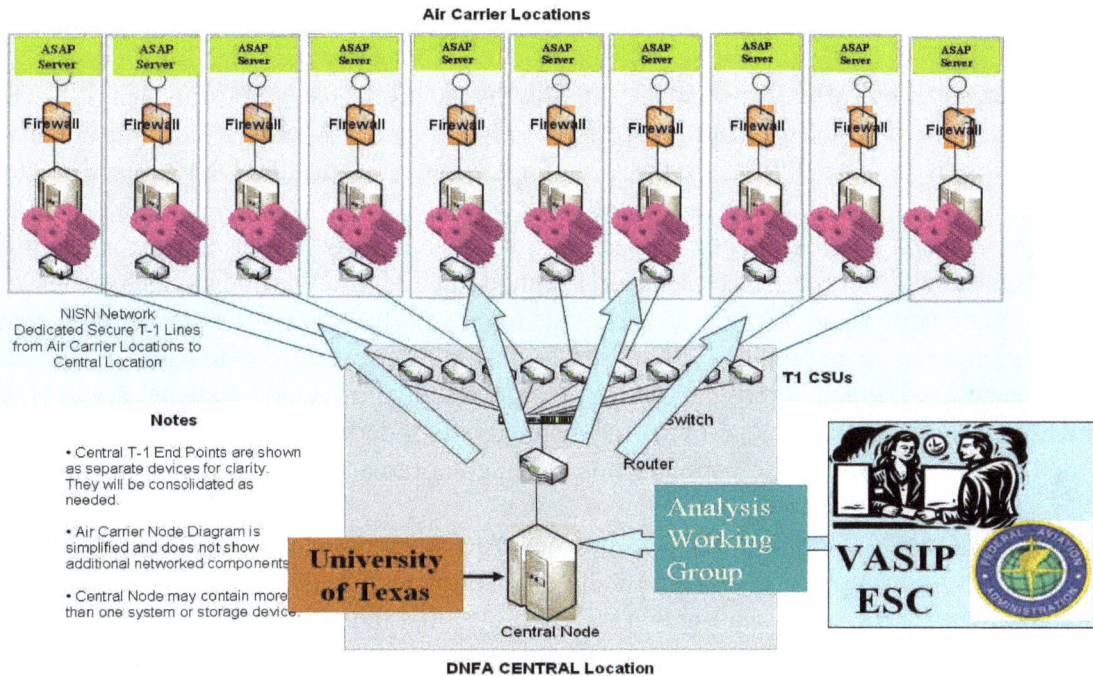

Figure 7.1.4. The concept of submitting a query to the DNAA.

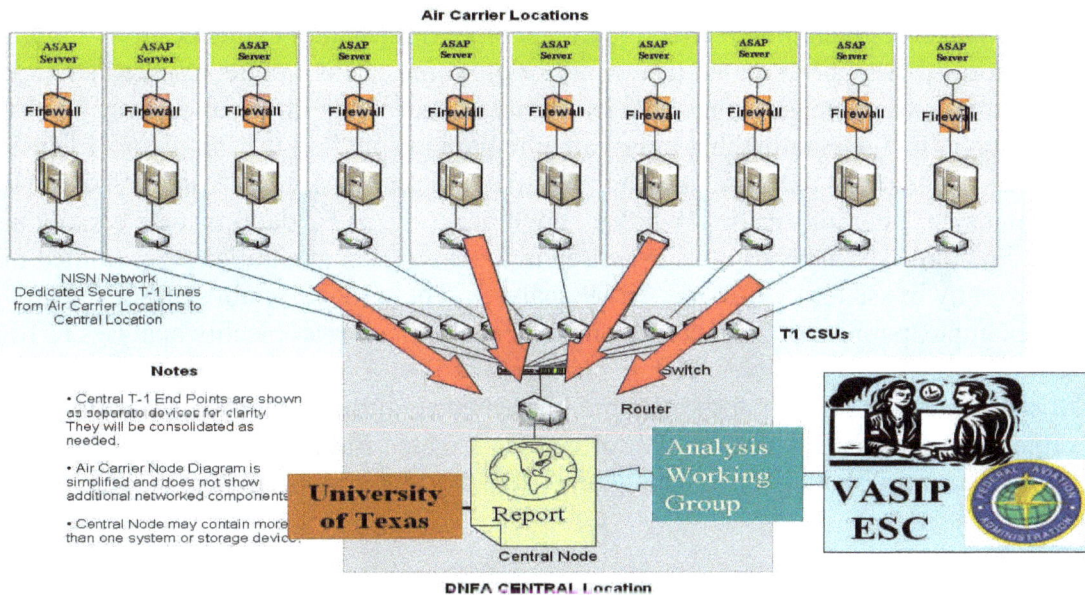

Figure 7.1.5. The concept of accessing the DNAA.

The ESC and Analysis Working Group roles in the DNAA are the same as they are in the DNFA, and the expectation is that each working group will, in practice, work with both archives to respond to ESC research issues. The FOQA data and the ASAP data are sources of complementary information relative to most queries. The quantitative FOQA data provide the information about what happened, while the ASAP incident reports are the best available sources of information about why it

177

happened. Because the former is numerical data and the latter is textual data, the automated tools for extracting information from each are very different. A challenge that was discussed previously in this report is to be able to reliably automatically link information extracted from each that is relevant to any query. Figure 7.1.5 shows that the creation of reports by the Analysis Working Groups and their distribution to the VASIP ESC are the same for the DNAA as for the DNFA.

In the current concept, queries can be initiated only from the Central Server. However, the infrastructure diagrammed in figure 7.1.1 has been designed to be highly flexible so that queries could be initiated from any node if the VASIP should decide to do so. The capability to issue the command functions currently generated by the Central Server for information relative to each query could be installed in each LAS. In that configuration, issues could be identified and analyses could be conducted within any node, and each participating airline could independently compare itself to the aggregated results of all the participating airlines (though not to any individual airline).

The ISI in conjunction with the VASIP is as much a process for developing trust in data-sharing as it is a technology for accomplishing sharing functions. The airlines, their employees, and their representatives need experience with sharing that demonstrates that industry-level good can be accomplished without individual or corporate harm. That demonstration requires procedural controls on when and how analyses are conducted, which are reflected in the policy documents governing the archive operation and, in some cases, written into software. Analyses and reports will be run only as directed by the VASIP ESC.

The ISI project had the potential of enabling integration of information extracted from FOQA-data sources and from ASAP reports across a large sector of airlines. This step represented an important first step toward, and a demonstration of the feasibility of, extracting information from all of the safety data sources and integrating that information to gain insight into the full picture of systemic issues. However, this initial demonstration has significant limitations. FOQA and ASAP are two very important sources of information, but there are others (e.g., ATC data) that are equally important for a full picture. The analysis tools exercised in this initial demonstration were limited to those that were currently in use for FOQA and ASAP analyses. These tools are not state-of-the-art. They do not, for example, permit the advanced automated capabilities for atypicality analyses of flight data or text mining of free narratives. The architectures of DNFA and DNAA were designed to support the use of these tools, but they were not implemented during the two-year demonstration of the ISI project. Further, the integration of information extracted from the DNFA with that from the DNAA will be accomplished by the human labor of the working groups and not by automated analysis tools.

The ISI two-year plan for the implementation and demonstration of the DNFA and the DNAA called for the Wide-Area Network to be connected and Local Archive Servers deployed by January, 2006, to those airlines that had agreed in 2004 to participate. Queries and analyses could then be accomplished on data from those airlines. By October 2006, hardware, software, and networking were scheduled to be deployed to all airlines that have agreed to participate by October 2005. NASA was scheduled to hand off the operation of the DNFA and DNAA to another agency on 30 September 2006.

7.2 Integrated Safety Data for Strategic Response

Another challenge in making diverse, heterogeneous data sources interoperable was undertaken in October 2005 as an AvSP five-year sub-project called Integrated Safety Data for Strategic Response (ISDSR) just prior to the end of the AvSP. The ISDSR was expected to build upon the work of the ASMM Project to develop and demonstrate the technology necessary to assess aviation-system safety risk by integrating information extracted from distributed, diverse types of safety data.

To develop and demonstrate analysis tools that make heterogeneous data sources interoperable, NASA planned to build a working prototype with representative samples of data, as diagrammed in figure 7.2.1. NASA expected to enter into agreements with owners of different types of aviation-safety data to obtain de-identified samples to be stored on computer workstations networked within the NASA firewall. Each computer was to contain a sample of de-identified data as a representation of the data source held or owned by an organization.

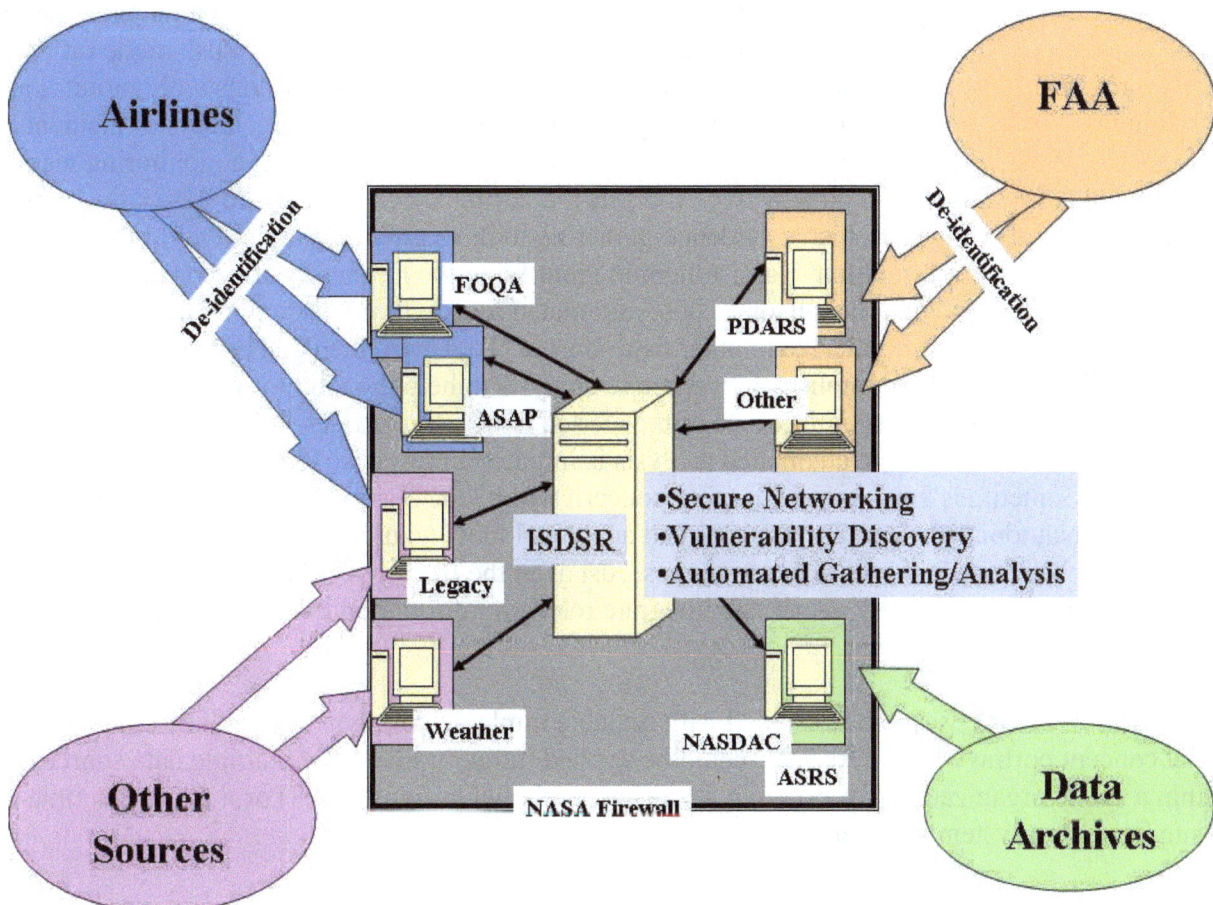

Figure 7.2.1. The ISDSR prototype.

The objectives of the ISDSR were to develop automated capabilities to:

- ensure acceptable levels of security when accessing full populations of de-identified data,

- monitor each data source for aviation-system safety issues to discover a potential vulnerability,

- identify other sources of information relevant to that potential systemic issue and gather that additional evidence,

- integrate all of the information relevant to the identified systemic issue and display a summary of the issue with explanation,

- use the same capabilities to respond to any query from an authorized stakeholder regarding a specific safety issue by extracting relevant information from across all data sources and displaying a summary of results, and

- produce reports of such investigations of diverse data sources in integrative, linked analysis reports.

Three general operating concepts envisioned for the prototype that are consistent with demonstrating the achievement of these objectives are illustrated in figure 7.2.2. In the System-Push mode (at the left of figure 7.2.2), automated tools for vulnerability discovery would nearly constantly monitor all data sources as they are being updated, looking for anomalies that might be precursors of incidents or accidents. When such an event or trend is discovered, these tools would alert a monitoring team to the potential safety issue. In the Interactive-Analysis mode (in the center of figure 7.2.2), the monitoring team would use the automated evidence-gathering tools to guide a search of all of the data sources for additional information to assist them in identifying the frequency of occurrence and the causal factors of the potential precursors. These automated tools will translate the issue that was alerted into search criteria applicable to other data sources. Tools will then gather relevant information from all sources and accomplish an integrative analysis to help the monitoring team understand and determine the operational significance of the alerted issues. In the User-Pull mode (at the right of figure 7.2.2), the integrative automated analysis is initiated in response to a query from the monitoring team. Sometimes the source of a safety concern will be the observations of industry experts rather than an automated alert. These automated analytical tools would translate the issue in question into search criteria applicable to all data sources. Just as in the Interactive-Analysis mode, these tools would then automatically gather and integrate relevant information to help the monitoring team understand the issue they raised.

The ISDSR prototype was to be designed with scalable implementation in mind; that is, the operational concept portrayed in figure 7.2.2 could be applied intramurally to the multiple data sources within a single organization (for example, an airline as shown in figure 7.2.3) or it could be implemented to gain a systemwide perspective.

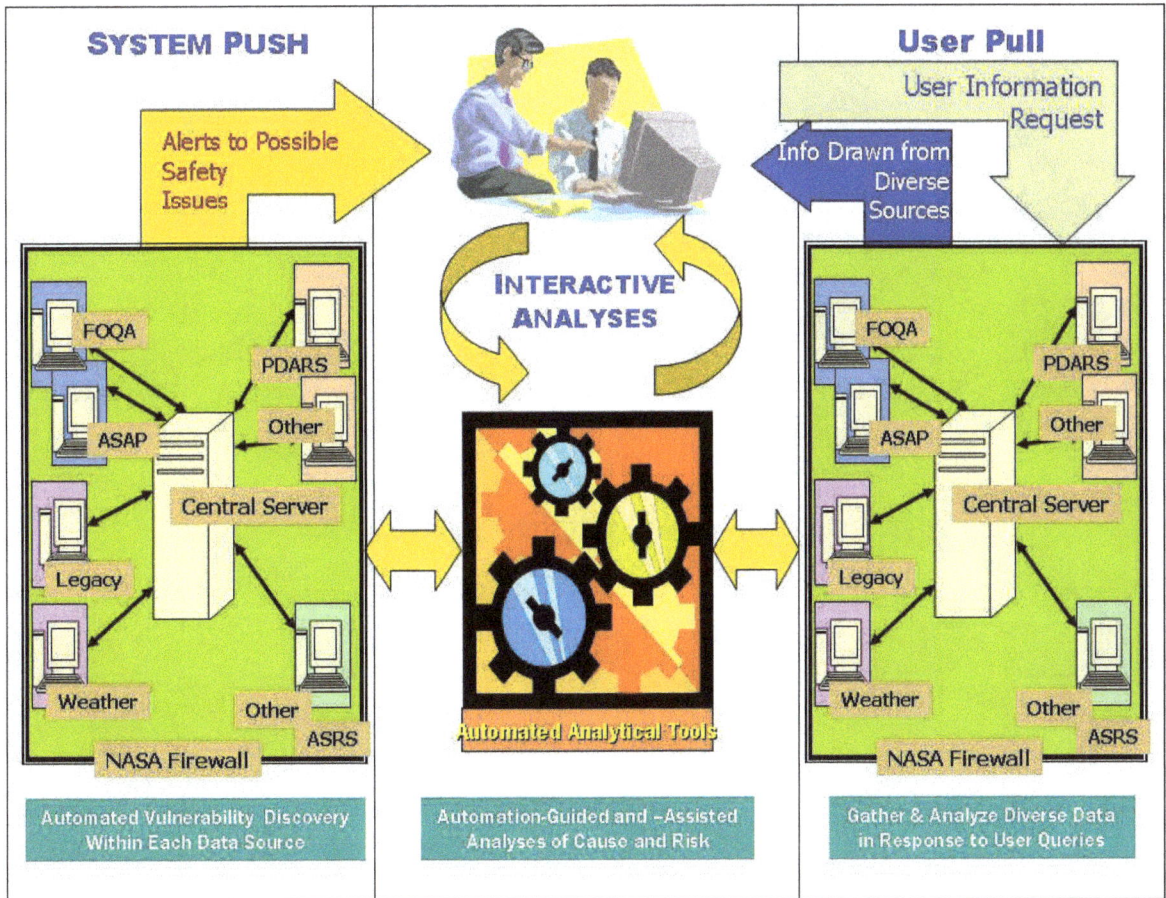

Figure 7.2.2. The operating concept of the ISDSR.

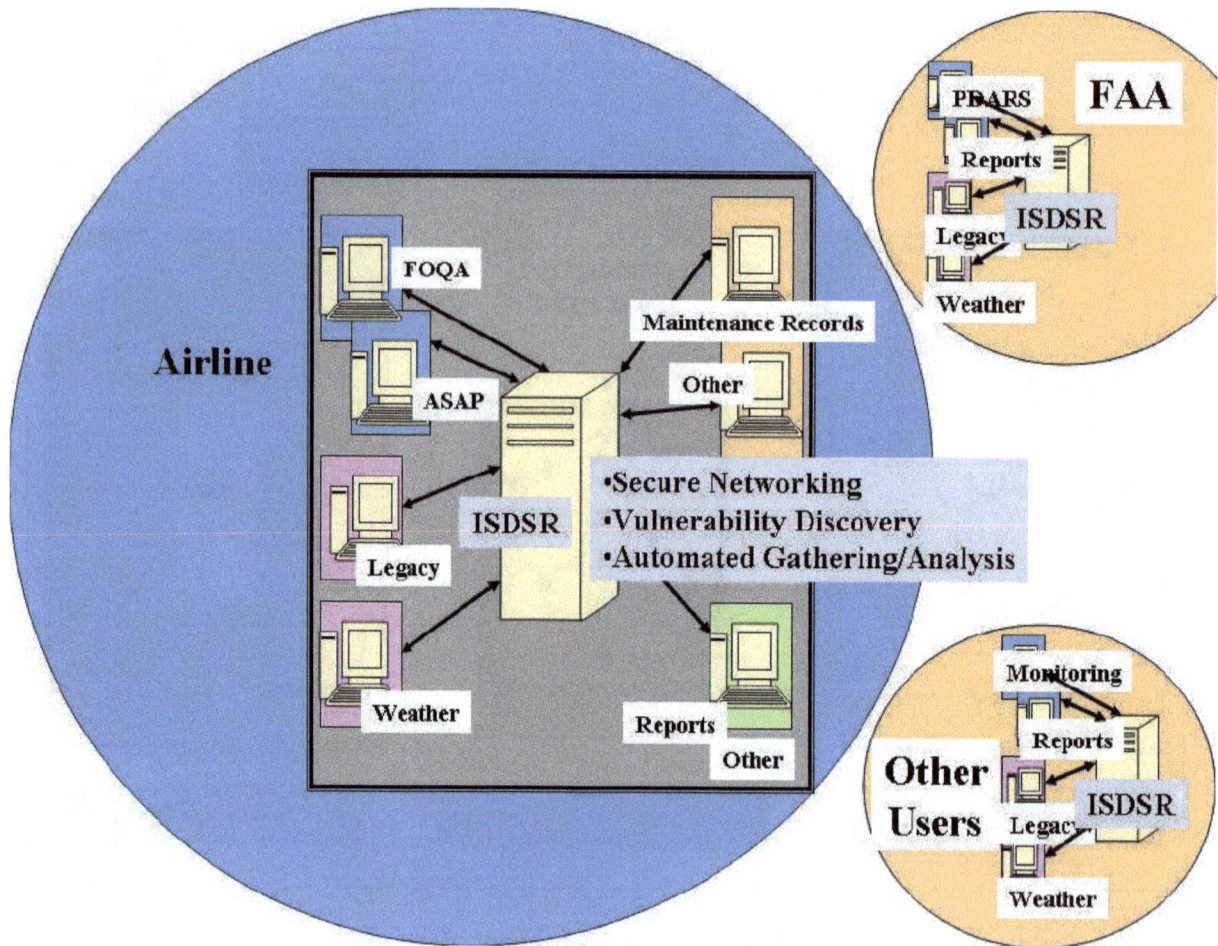

Figure 7.2.3. The intramural use of the ISDSR.

If the VASIP process and other initiatives overcome the significant barriers to information-sharing among all these organizations, the ISDSR prototype could become operational by connecting directly to copies of de-identified data at each participating organization, just as VASIP/ISI is doing with FOQA and ASAP data at participating airlines (see figure 7.2.4.).

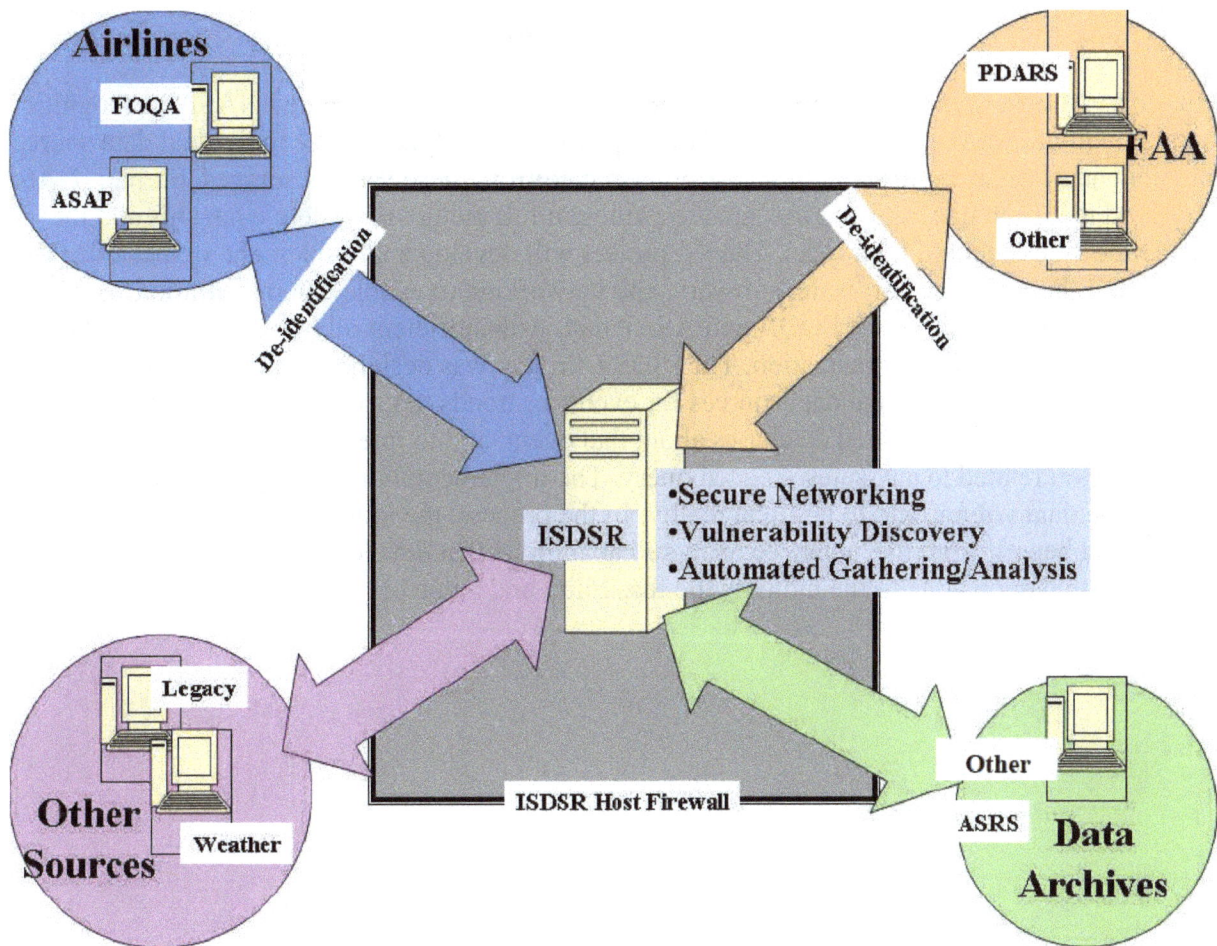

Figure 7.2.4. The systemwide concept of the future ISDSR.

The ASMM products have made the operational concept described previously for the ISDSR prototype feasible. However, although the advancements in technology required for the implementation of this concept appear reasonably probable, the risks are sufficiently high to discourage commercial vendors from undertaking their development. There is the political risk that some owners of essential data may be unwilling to grant a commercial vendor access to, and sharing of, their trade-sensitive, proprietary data. There is also the market risk of a limited customer base for this capability. Consequently, there is an effective role for NASA in the development of a demonstration of the feasibility and the value of an ISDSR prototype.

7.3 Conclusions

In summary, the work initiated under ISI and proposed for ISDSR represented NASA's recognition that, although the ASMM Project advanced the capability to analyze diverse individual data sources, it did not enable the integration of information extracted from these various sources. Further, NASA recognized that such integration is essential to gaining a full picture of systemic hazards and the systemwide assessment of their risks. The ISI project will develop and implement, on behalf of the industry and the FAA, the hardware, software, and networking to enable a demonstration within two years of a distributed archive of FOQA and ASAP data with sufficient reliability that it may be handed off to an operating organization. The ISDSR Project was designed to build a prototype of a capability to monitor individual data sources for events or trends that could compromise safety, to gather evidence from distributed databases about that event, and to interrogate across all databases for information related to any safety-related query. The implementation of the prototype would be scaled to the data volume and to political reality. In the process, the industry partners that enter into SAAs would benefit from the advanced analysis capabilities that develop. The goal of the ISDSR Project was to demonstrate to the industry the feasibility and value of identifying systemic issues.

8.0 ACKNOWLEDGEMENTS

The accomplishments of the ASMM Project are attributed to the integration of contributions from an exceptional group of experts representing:

- Aviation operations
- Causal analyses and risk assessment
- Computational linguistics
- Computer sciences
- Elicitation and modeling of expert knowledge
- Human performance
- Human-performance modeling
- Kernel methods for pattern analysis
- Non-real-time simulation
- Statistical analysis
- Survey sciences
- Visualization aids

During the six years of the project, its personnel produced 51 publications and made 174 presentations at international conferences and symposia, most of which were by invitation. Table 8-1 lists the members of the ASMM Team and their organizations.

The ASMM team expresses its appreciation to our partners, without whose collaboration the testing and evaluation of the ASMM products would not have been possible. These partners include:

- Alaska Airlines
- American Airlines
- Delta Airlines
- TransWorld Airlines
- United Airlines
- Joint Implementation Measurement Data Analysis Team (JIMDAT)

The ASMM Team also acknowledges the significant contributions of George Finelli, the Director of the Aviation Safety Program for most of its life, whose exceptional leadership, deep understanding, and continuous support during a time of severe fiscal constraints were essential to the survival of the ASMM Project.

TABLE 8-1. THE ASMM TEAM AND THEIR ORGANIZATIONS

NASA Headquarters
 Mr. George Finelli

NASA Ames Research Center
 Dr. Thomas R. Chidester
 Ms. Linda Connell
 Dr. Mary M. Connors
 Dr. Rich Keller
 Dr. Deepak Kulkarni
 Dr. David Maluf
 Dr. Michael McGreevy
 Mr. Robert Padilla
 Dr. Michael Shafto
 Mr. Brian Smith
 Dr. Ashok Srivastava
 Dr. Irving C. Statler
 Dr. Yao Wang

QSS
 Mr. David Hall
 Mr. Hemil Patel
 Ms. May Windrem

Battelle
 Mr. Brett G. Amidan
 Ms. Cathy Anderson
 Ms. Lee Bretoi
 Mr. Scott Cooley
 Dr. Joan Cwi
 Dr. Thomas Ferryman
 Ms. Louise Glezen
 Mr. Daniel Haber
 Mr. Griff Jay
 Ms. Pam Kaifer
 Captain Robert Lawrence
 Mr. Tim Leslie
 Captain Robert Lynch
 Mr. Brett Matzke
 Dr. Rowena Morrison
 Dr. Christian Posse
 Ms. Andrea Swickard
 Mr. Loren Rosenthal
 Ms. Amanda White
 Dr. Paul Whitney
 Dr. Alan Willse

ProWorks
 Mr. Adi Andrei
 Mr. Rob Birdwell
 Mr. Chris Mosbrucker
 Mr. Gary Prothero
 Mr. Jason Prothero
 Mr. Dan Robin
 Mr. Tim Romanowski

Intrinsyx Technologies
 Mr. Allen Carter
 Mr. Darin Foreman
 Mr. William Mortimor

SimAuthor
 Mr. Steve Lakowske

Expert Microsystems
 Mr. Randy Bickford

Dodd Associates
 Dr. Robert Dodd

ATAC
 Dr. Mike Abkin
 Mr. Raymond Bea
 Ms. Kennis Chan
 Mr. Don Crisp
 Mr. Minh Dang
 Mr. Brian Davis
 Mr. Wim den Braven
 Mr. David Holl
 Mr. Mike Kiniry
 Mr. Bryon Li
 Mr. John Schade
 Mr. Elliott Smith
 Mr. Eric Wang
 Mr. Corey Warner
 Mr. Phil Welliver
 Mr. David Xu
 Mr. Jae Yu
 Ms. Maggie Zhang

TABLE 8-1. THE ASMM TEAM AND THEIR ORGANIZATIONS (Concluded)

San Jose State University Dr. Kevin Corker Mr. Anthony Tang Mr. Amit Jadhav	**RIACS** Mr. Eugene Turkov
	UCSC Dr. Ram Akella Mr. Suratna Budalakoti
Georgia Tech Dr. Amy Pritchett Mr. Anuj P. Shah Mr. Sachidanand A. Kalaver Dr. Seung Man Lee Lt. Col.Kirk Benson, PhD Dr.. Margaret Loper Dr. David Goldsman Dr. Christos Alexopoulos Dr. Richard Fujimoto	**ONERA** Dr. Claude Barrouil Dr. Laurent Chaudron Mr. Patrick Le Blaye Dr. Nicolas Maille
	Cinq Demi General Jean-Claude Wanner Mme. Nicole Wanner General Pierre Lecomte
FAA Dr. Tom Longridge Dr. James McMahon Mr. Rich Nehl	**British CAA** Mr. David Wright
Ohio State University Dr. Jon A. Krosnick Dr. Michael Silver	**National Aerospace Laboratory NLR** Dr. Henk Blom

Also,
University of Alabama **Sandia National Labs**	**Virginia Polytechnic Institute** **Rannoch Corp**

9.0 MAJOR AWARDS

APMS received the NASA Group Achievement Award in 1999.

PDARS received the Administrator's Award for Transferring Goals into Reality in 2003.

ADIS received the NASA Group Achievement Award in 2005.

The Morning Report of Atypical Flights received an R&D 100 Award for 2005 and also the Editors' Choice Award for the product most likely to impact public safety.

10.0 REFERENCES

Aldrich, T. B.; Szabo, S. M.; and Bierbaum, C. R.: The Development and Application of Models to Predict Operator Workload During System Design. In G. MacMillan, D. Beevis, E. Salas, M. Strub, R. Sutton & L. Van Breda (Eds.), Applications of Human Performance Models to System Design, New York: Plenum Press, 1989, pp. 65–80.

Amidan, B. G.; Cooley, S. K.; and Ferryman, T. A.: Identifying In-Close-Approach-Changes in Aviation Performance Measuring System (APMS) Data. PNWD-3209, Battelle Pacific Northwest National Laboratory, Richland, WA, 2002.

Amidan, B. G.; Swickard, A. R.; Allen, R. E.; and Ferryman, T. A.: Identifying In-Close-Approach-Changes in Air Traffic Control (ATC) Data. PNWD-3210. Battelle Pacific Northwest National Laboratory: Richland, WA, 2002.

Amidan, B. G.; and Ferryman, T. A.: Atypical Event and Typical Pattern Detection within Complex Systems. In Proceedings of 2005 IEEE Aerospace Conference, art. no. 11.0701, Manhattan Beach, CA, 2005, pp. 1–12.

ASRS Internal Request Number 524, ASRS Primary Problem Coding Counts, 2005.

Best, B.; Lebiere, C.; Schunk, D.; Johnson, I.; and Archer, R.: Validating a Cognitive Model of Approach Based on the ACT-R Architecture. Contractor Report, Micro Analysis and Design, Inc., Boulder, CO, July 2004. Available at http://humanfactors.arc.nasa.gov/ihi/hcsl/2004.

Blom, H. A. P.; Bakker, G. J.; Blanker, P. J. G.; Daams, J.; Everdidj, M. H. C.; and Klompstra, M. B.: Accident Risk Assessment for Advanced ATM. Proceedings of the 2nd USA/Europe ATM R&D Seminar, FAA/EUROCONTROL, Orlando, FL, 1998.

Blom, H. A. P.; Corker, K.; Stroeve, S. H.; and Klompstra, M. B.: Study on the Integration of Air-MIDAS and TOPAZ: Phase 2 Final Report. Report NLR-CR-2001-528, Nov. 2001.

Blom, H.; Corker, K.; and Stoeve, S.: Study on the integration of human performance and ACCIDENT Risk assessment models: Air-MIDAS & TOPAZ. 5th FAA Eurocontrol International Symposium on Air Traffic Management, Baltimore, Maryland, June 2005.

Blom, H.; Stoeve, S.; Everidj, M.; and van der Park, M.: Human Cognitive Performance Model to Evaluate Safe Spacing in Air Traffic. Human Factors and Aerospace Safety J., vol. 3, no. 1, 2003, pp. 59–82.

Budalakoti, S.; Srivastava, A. N.; and Akella, R.: Discovering Atypical Flights in Sequences of Discrete Flight Parameters. 2006 Proceedings of the IEEE Aerospace Conference, Mar. 4–11, 2006, pp. 1–8.

Byrne, M. D.; and Kirlik, A.: Integrated Modeling of Cognition and the Information Environment: A Closed-loop, ACT-R Approach to Modeling Approach and Landing with and without Synthetic Vision Systems (SVS) Technology. In D. C. Foyle, A. Goodman, and B. L. Hooey (Eds.), Proceedings of the 2003 NASA Aviation Safety Program Conference on Human Performance Modeling of Approach and Landing with Augmented Displays, NASA/CP-2003-212267, 2003, pp. 91–117. Available at http://humanfactors.arc.nasa.gov/ihi/hcsl/2004.

Cacciabue, P. C.: Modeling and Simulation of Human Behavior in System Control. In Advances in Industrial Control, Amsterdam: Springer, 1998.

Card, S.; Moran, T.; and Newell, A.: The Model Human Processor: An Engineering Model of Human Performance. In K. Boff, L. Kaufman, and J. Thomas (eds.), Handbook of perception and human performance (vol. 2). New York: Wiley, 1986, pp. 1–35.

Chidester, T. R.: An Overview of the Enhanced Aviation Performance Measuring System. In Fifth GAIN World Conference Proceedings and Products, Miami, FL, December 5–6, 2001.

Chidester, T. R.: Understanding Normal and Atypical Operations through Analysis of Flight Data. In Proceedings of the 12th International Symposium on Aviation Psychology, Dayton, Ohio, 2003, pp. 239–242.

Corker, K.; Gore, B.; Fleming, K.; and Lane, J.: Free Flight and the Context of Control: Experiments and Modeling to Determine the Impact of Distributed Air-Ground Air Traffic Management on Safety and Procedures. In Proceedings of the 3rd FAA Eurocontrol International Symposium on Air Traffic Management. Naples, Italy, 2000.

Cormen, T.; Leiserson, C.; Rivest, R.; and Stein, C.: Introduction to Algorithms, 2nd Edition. The MIT Press and McGraw-Hill, 2001.

Cunningham, H.; Maynard, D.; Bontcheva, K.; and Tablan, V.: GATE: an Architecture for Development of Robust HLT Applications. In Proceedings of the 40th Meeting of the Association for Computational Linguistics (ACL'2002), Philadelphia, PA, July, 2001, pp. 168–175.

den Braven, W.; and Schade, J.: Concept and Operation of the Performance Data Analysis and Reporting System (PDARS). In Proceedings of SAE Advances in Aviation Safety Conference (ACE), Paper # 2003-01-2976, Montreal, QC, Sept. 2003.

Deutsch, S.; and Pew, R.: Modeling the NASA Baseline and SVS Equipped Approach and Landing Scenarios in D-OMAR. BBN Report No. 8399, Cambridge, MA: BBN Technologies, 2003. Also NASA/CP-2003-212267, 2003. Available at http://humanfactors.arc.nasa.gov/ihi/hcsl/2004.

Ferryman, T. A.; Willse A.; and Cooley, S.: Cluster Analysis of Digital Flight Data for the Aviation Performance Measurement System (APMS). In Proceedings of American Statistics Association, Atlanta, GA, August 5–9, 2001.

Ferryman, T. A.; Posse, C.; Rosenthal, L. J.; Srivastava, A. N.; and Statler, I. C.: What Happened and Why: Towards an Understanding of Human Error Based on Automated Analyses of Incident Reports—Vol. II. NASA TP-2006-213490, Dec. 2006.

Foyle, D. C.; Goodman, A.; and Hooey, B. L.: An Overview of the NASA Aviation Safety Program (AvSP) Systemwide Accident Prevention (SWAP) Human Performance Modeling (HPM) Element. In D. C. Foyle, A. Goodman, and B. L. Hooey (Eds.), Proceedings of the NASA Aviation Safety Program Conference on Human Performance Modeling of Approach and Landing with Augmented Displays, NASA/CP-2003-212267. Available at http://humanfactors.arc.nasa.gov/ihi/hcsl/2003.

Foyle, D. C.; Hooey, B. L.; Byrne, M. D.; Corker, K. M.; Deutsch, S.; Lebiere, C.; Leiden, K.; and Wickens, C. D.: Human Performance Models of Pilot Behavior. In Proceedings of the Human Factors and Ergonomics Society 49th Annual Meeting. Santa Monica, CA, 2005. Available at: http://humanfactors.arc.nasa.gov/ihi/hcsl/2005.

GAO Report to the Subcommittee on Aviation of the Committee on Transportation and Infrastructure Aviation Safety, GAO/RCED-00-111, July 2000. Safer Skies Initiative has Taken Initial Steps to Reduce Accident Rates by 2007, 2000.

Hooey, B. L.; and Foyle, D. C.: A Post-Hoc Analysis of Navigation Errors During Surface Operations: Identification of Contributing Factors and Mitigating Solutions. In Proceedings of the 11th International Symposium on Aviation Psychology, Ohio State University, Columbus, Ohio, 2001.

Hunt, J. W.; and Szymanski, T. G.: A Fast Algorithm for Computing Longest Common Subsequences. Communications of the ACM, vol. 20, no. 5, May 1977, pp. 350–353.

Landy, M. S.: Vision and attention for Air Midas. New York University, Dept. of Psychology and Center for Neural Science, NASA Final Report NCC2-5472, Ames Research Center, 2002.

Leveson, N.: Safeware: System Safety and Computers. New York, NY: Addison-Wesley Professional, 1995.

Maille, N. P.; Ferryman, T. A.; Rosenthal, L. J.; Shafto, M. G.; and Statler, I. C.: What Happened and Why: Towards an Understanding of Human Error Based on Automated Analyses of Incident Reports—Vol. I. NASA TP-2006-213466, March 2006.

Manning, C. D.; and Schütze, H.: Foundations of Statistical Natural Language Processing. Cambridge, MA: The MIT Press, 2000.

Mumaw, R.; Sarter, N.; Wickens, C.; Kimball, S.; Nikolic, M.; Marsh, R.;, Xu, W.; and Xu, X.: Analysis of Pilot Monitoring and Performance on Highly Automated Flight Decks. NASA Final Project Report: NAS2-99074, Ames Research Center, 2000.

Posse, C.;, Matzke, B.; Anderson, C.; Brothers, A.; Matzke, M.; and Ferryman, T.: Extracting Information from Narratives: An Application to Aviation Safety Reports. IEEEAC paper #1182, IEEE Conference, 2005.

Posse, C.; Matzke, B.; Anderson, C.; Brothers, A.; Matzke, M.; and Ferryman, T.A.: Extracting Human Performance Factors from Aviation Safety Reports using a Common Pattern Specification Language. In Conference Publications of 2005 IEEE Aerospace Conference, art. no. 11.0707, pp. 1–14, Manhattan Beach, CA, 2005.

Pritchett, A. R.: Human-Computer Interaction in Aerospace. In Handbook of Human Computer Interaction, 3rd ed., eds. J. Jacko and A. Sears. Mahwah, NJ: Erlbaum, 2002.

Rasmussen, J.: Skills, Rules, and Knowledge; Signals, Signs and Symbols, and Other Distinctions in Human Performance Models. In IEEE Transactions on Systems, Man, and Cybernetics, vol. SMC-13, no. 3, May/June, 1983, pp. 257–266.

Shah, A. R.; Pritchett, K. M.; Feigh, K. C.; Kalaver, S. A.; Jadhav, A.; Corker, K. C.; Holl, D. M.; and Bea, R. C.: Analyzing Air Traffic Management Systems Using Agent-Based Modeling and Simulation. 6th USA/Europe International Symposium on Air Traffic Control, Baltimore, Maryland, June 2005.

Srivastava, A. N.: Discovering Anomalies in Sequences with Applications to System Health. In Proceedings of the 2005 Joint Army Navy NASA Air Force Interagency Conference on Propulsion, Charleston, South Carolina, 2005.

Srivastava, A. N.; Akella, R.; Diev, V.; Kumaresan, S. P.; McIntosh, D. M.; Pontikakis, E. D.; Xu, Z.; and Zhang, Y.: Enabling the Discovery of Recurring Anomalies in Aerospace System Problem Reports using High-Dimensional Clustering Techniques. In 2006 Proceedings of the IEEE Aerospace Conference, Big Sky, Montana, March 4–11, 2006,

Statler, I. C.; Morrison, R.; and Rosenthal, L. J. Beyond Error Reporting Toward Risk Assessment. In Proceedings of the 12th Biennial International Symposium on Aviation Psychology, Dayton, Ohio, April 14–17, 2003.

Stroeve, S. H.; Blom, H. A. P.; and Van der Park, M. N. J.: Multi-agent Situation Awareness Error Evolution in Accident Risk Modeling. In Proc. 5th USA/Europe Air Traffic Management R&D Seminar, Budapest, Hungary, 2003.

Swain, A. D.; and Guttman, H. E.: Handbook of Human Reliability Analysis with Emphasis on Nuclear Power Plant Operations: Final Report. Report No. NURGE/CR 1278, Sandia National Laboratories, 1983.

Verma, S.; and Corker, K.: Introduction of Context in Human Performance Model as Applied to Dynamic Resectorization. In Proceedings of the 11th International Symposium on Aviation Psychology, Columbus, Ohio, March 5–8, 2001.

Wickens, C. D.: Processing Resources in Attention. In R. Parasuraman and R. Davies (eds.) Varieties of Attention. New York: Academic Press, 1984, pp. 63–102.

Wickens, C. D.; McCarley, J. S.; Alexander, A. L.; Thomas, L. C.; Ambinder, M.; and Zheng, S.: Attention-Situation Awareness (A-SA) Model of Pilot Error. Technical Report AHFD-04-15/NASA-04-5. Univ. of Illinois at Urbana-Champaign, Savoy, IL, Jan. 2005. Available at http://humanfactors.arc.nasa.gov/ihi/hcsl/2005.

Willse, A. R.; Cooley, S. K.; Chappell, A. R.; Daly, D. S.; and Ferryman, T. A.: Analysis of Rotary-wing Aircraft Incident and Accident Reports. PNWD-3070, Battelle Pacific Northwest National Laboratory, Richland, WA, 2002.

Woods, D. D.; Johannesen, L. J.; Cook, R. I.; and Sarter, N. B.: Behind Human Error: Cognitive Systems, Computers, and Hindsight. CSERIAC SOAR 94-01, Wright-Patterson Air Force Base, Information Analysis Center, Ohio, 1994.

APPENDIX A

THE FLIGHT OPERATIONAL QUALITY ASSURANCE (FOQA) PROGRAMS

FOQA is the acronym for Flight Operational Quality Assurance. The term describes the proactive, routine use of flight-recorded data to create a true quality assurance program in the area of flight operations. Worldwide, FOQA programs are given a variety of names and acronyms by airlines, but the goals and methods of their programs do not differ widely.

FOQA is a program to improve flight safety through the routine collection and analysis of digital flight data generated during line operations. The benefit of FOQA is that it provides a greater understanding of the total flight operations environment. The development of newer data-acquisition devices that allow quick, inexpensive access to flight data, along with sophisticated analysis software, have now made it possible to routinely collect and analyze flight data. The information and insights provided by FOQA analyses can be used to reduce operational costs and significantly enhance training effectiveness and operational procedures. FOQA data are unique because they provide objective information not available through any other methods.

Aircraft operators and airlines operate their own internal FOQA programs. They are supported by regulations and forums sponsored by the FAA. The airlines usually retrieve the data from a Quick Access Recorder (QAR) that operates in parallel with the Digital Flight Data Recorder (DFDR) on board their aircraft. Data from the data bus are recorded on Quick Access Recorders installed on board the aircraft. These data are downloaded at periodic intervals on the ground and sent to the airline's FOQA office. The data are processed and analyzed using one of the commercially available systems for processing FOQA data. The automated search focuses on identifying prescribed exceedances. Exceedances are flight parameters that fall outside of normal operating ranges during some phase of flight as determined by the organization's operational and training standards, as well as the manufacturer's aircraft operating limitations. The flight data for flights with identified exceedances are available to a FOQA monitoring team designated by the airline, which consists of personnel from various flight operations and maintenance disciplines. Line pilots are typically included in this process because of the operational insights they can provide. This group identifies areas where corrective action is needed, determines what corrective action should be taken, and then closes the loop by tracking the effectiveness of the corrective action.

Some airlines have had FOQA-type programs in operation for more than 40 years. A pioneer in this arena is British Airways, which started its program in 1962 to validate airworthiness criteria through the use of flight-recorder data. In 1991, the FAA contracted with the Flight Safety Foundation (FSF) to examine the technology, benefits, and other issues arising from the FOQA-type programs used by non-U.S. airlines. As a result of this study, the FSF recommended that the FAA initiate a program to examine the benefits that FOQA programs might provide to U.S. airline operators. At the January 1995 Department of Transportation (DOT) Aviation Safety Conference, it was recommended that the FAA encourage and help U.S. airlines voluntarily establish FOQA programs. In particular, it was recommended that the FAA sponsor FOQA demonstration studies, in cooperation with the airline industry, to develop guidelines for such programs, and to assuage the concerns of airlines and flight-crews regarding the appropriate use and protection of recorded flight data.

Today, over 40 airlines worldwide have FOQA programs. In the U.S., there are now 12 major air carriers and 1 regional carrier with FAA-approved FOQA programs. In addition, several other major and regional air carriers have FOQA programs under development. The FAA instituted a program to assist airlines in establishing their FOQA programs with both financial and technical assistance. UTRS, Inc. was selected by the FAA to provide services to U.S. air carriers under this program. Airlines may voluntarily agree to participate and share their lessons learned about FOQA with other carriers participating in the program. UTRS serves as a clearinghouse for this shared knowledge, and provides resources and technical assistance to participating airlines. As airlines see both the safety and economic benefits of such a program, it is expected that many more will participate.

Airlines carefully consider their decision to implement a FOQA program for several reasons. First, there has been concern about whether the government or the media would gain access to data gathered under a FOQA program. Second, a considerable investment of time, human, and financial resources is necessary to create a successful program. Third, pilot associations have been concerned about privacy and the possibility that data would be misused against their members. Finally, airlines are concerned about FOQA information being obtained and used in civil litigation.

In the U.S., a significant change occurred as a result of the January 1995 Joint Industry-Government Aviation Safety Conference. One of the recommendations of this conference was that the FAA should encourage airlines to establish voluntary FOQA programs, and provide assistance in getting such programs established. Further, the FAA Administrator issued a letter in February 1995 stating that the FAA would not use data gathered under an approved FOQA program in enforcement actions against either airlines or their employees. In December 1998, the FAA Administrator issued a FOQA Policy Statement announcing that the FAA will not use de-identified FOQA information to undertake enforcement actions except under special circumstances, such as in the conduct of a criminal investigation, or if there were an intentional disregard of safety. These protections have now been embodied in aviation regulations, giving them the force of law (see refs. 1 and 2). In competition for scarce resources within an airline, FOQA programs need to go through the same cost-justification process as any other program. Although there are clear and compelling benefits for a FOQA program to identify and reduce operational risks, they are often difficult to quantify. As U.S. airlines have become more familiar with FOQA, they have discovered uses of the data that have resulted in extended engine life, more efficient routings, and other areas that save money. Airlines with FOQA programs have found that these improvements, coupled with safety enhancements, have more than justified their cost. Pilot associations still have concerns about the collection and use of FOQA data, but have been agreeable to a FOQA program when appropriate protections are in place. Both labor and management also share a common interest in promoting safety. They have entered into binding agreements that cover the uses of FOQA data and the de-identification of individual crewmembers. With these agreements in place, there has been no misuse of FOQA data, and new bonds of trust have been built within the airlines. (For additional information about the FOQA program, see refs. 3 and 4.)

Appendix A References

1. FAA ORDERS: 8300.10 and 8400.10 BULLETIN NUMBER: HBAT 00-11, HBAW 00-10
   http://www.access.gpo.gov/nara/cfr/waisidx_02/14cfr193_02.html
   http://www.asy.faa.gov/gain/FOQA_&_ASAP/HBAT00-11_HBAW00-10.pdf

2. FAA 14 CFR Part 13 FOQA

3. Flight Safety Foundation FOQA.HTM

4. The Next Generation FOQA Programs Capt M H Brandt Director FOQA Allied Signal;
   April 1999

APPENDIX B

A CASE STUDY: IN-CLOSE APPROACH CHANGES (ICAC)

The following is an excerpt from the paper: Statler, I. C.; Morrison, R.; and Rosenthal, L. J.: Beyond Error Reporting Toward Risk Assessment. In Proceedings of the 12th Biennial International Symposium on Aviation Psychology, Dayton, Ohio, April 14–17, 2003.

During 2002, a milestone was achieved with the application of some ASMM computational tools and methods to a potentially hazardous scenario encountered in air-carrier operations. The approach involved using each of the ASMM tools in a set of (nearly) independent studies of the same operational scenario with the following goals:

- Demonstrate the kinds of information that each ASMM tool can contribute to gaining insight into the complete picture of an event;

- Show the methodology for utilizing each of the tools in a complementary and synergistic process of causal analysis and safety risk assessment.

The operational implications of the analysis are still under evaluation, and are presented at the end of this paper as preliminary observations. The primary purpose of this report is to present the process for identifying and evaluating event precursors using ASMM tools.

The scenario selected for the demonstration, In-close Approach Change (ICAC), involved changes to a landing runway assignment while on approach to the airport. ATC sometimes issues clearance changes to air-carrier pilots late in the approach to expedite traffic flows and resolve traffic conflicts. Pilots can usually accommodate these clearance amendments, but sometimes they experience unwanted consequences such as unstable approaches and hard landings. The research question was whether operationally significant risks were entailed in ICAC events, and if so, how could they be minimized. Figure B-1 illustrates the approach to addressing these questions using ASMM tools

In main text of this report, each of the ASMM tools is described individually as a stand-alone capability for extracting information from available data. The study of ICAC was intended to show how the ASMM tools could be used together in analyses of diverse data sources to provide further insight into safety from a systemic perspective. Each of the ASMM tools was used in (nearly) independent studies of the same scenario (ICAC) and demonstrated the feasibility and the value of merging information from heterogeneous databases to aid causal analyses and risk assessment. This study was intended solely to demonstrate the kinds of information that each data source can potentially contribute to gaining insight into the complete picture, and the method for utilizing each of the tools in a complementary and synergistic process of causal analysis and safety risk assessment. The results of this study were not intended to provide definitive conclusions regarding ICAC, the safety of its utilization, or mitigations should they be deemed necessary. The accuracy and reliability of the results of this demonstration was limited because of the small percentage of the general population of each data set that was available for the purpose. This is simply a description of the process illustrated in figure B-1 and its potential value.

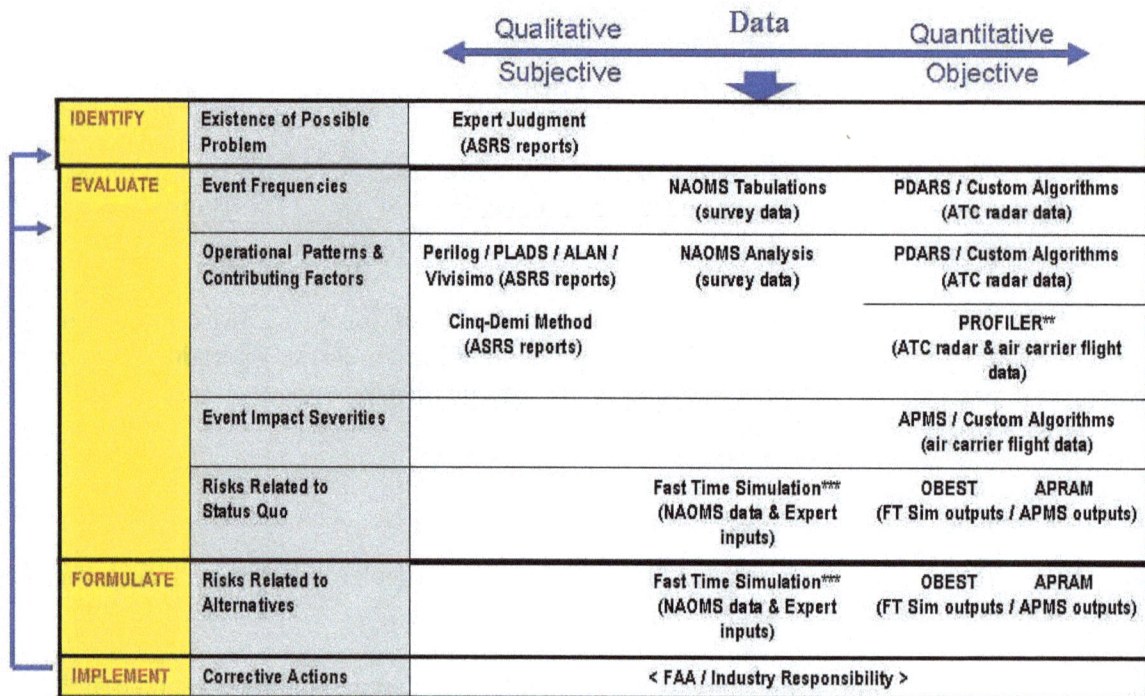

Figure B-1 – The Case Study Process

The pointer to a potentially hazardous aviation scenario could come from any data source. In this instance, it was ASRS analysts who first identified potential problems in pilots' accommodating changes to their runway assignment, altitude, or speed when close to an airport during approach. The ASRS report analyses set the stage for a more thorough examination of the problem.

Insights from the NAOMS Survey Tool

The first step was to incorporate specific questions regarding ICAC's in a NAOMS survey of air transport pilots. With sufficient data, this can provide reliable quantitative information on the frequency of occurrence of an event such as an ICAC.

In general, data obtained by NAOMS could not be published because of possible sensitivities. Also, care must be exercised in interpreting the results because the survey was mostly limited to commercial airline pilots and they may not represent the full population of the NATS on certain issues. Table B-1 is just a small representative sample of the results of the questions put to the commercial pilots regarding ICAC.

TABLE B-1 - PROBLEMS ASSOCIATED WITH ICAC

Type of ICAC Problem	Number Reported	Percentage of Itemized Problems Among Accepted ICACs[3]	Projected Number of Problems System-Wide in 2001[4]		Upper and Lower 95% Confidence Limit Estimates of Unwanted Approach Related Events
Unstabilized Approach	631	3.76	Estimated	17,045	15,740 to 18,349
Missed Approach	211	1.26	Estimated	5,700	4,935 to 6,464
Airborne Conflict	50	0.3	Estimated	1,350	977 to 1,724
Wake Turbulence	213	1.27	Estimated	5,754	4,985 to 6,521
Out-of-Limit Winds	33	0.20	Estimated	891	588 to 1,195
Wrong Runway Landing	0	0.00			
Long or Fast Landing	591	3.52	Estimated	15,964	14,700 to 17,228
Landing Without Clearance	7	0.04	Estimated	189	49 to 329
Ground Conflict	52	0.31	Estimated	1,405	1,210 to 1,599
Other	479	2.85	Estimated	12,939	11,796 to 14,081

NAOMS data available for this study suggested that somewhat less than 10 percent of all air carrier approaches involve ICAC's, but there is a great deal of variability among locations. A fraction of these ICAC's are followed by unwanted events such as unstabilized approaches, hard landings, and airborne/ground conflicts. The ICAC's do not necessarily cause all of these unwanted outcomes—they might have occurred even without the ICAC. However, a reasonable conclusion from the NAOMS data is that ICAC's contribute to many of these unwanted events. It appeared that the results of the NAOMS survey validated the concern identified by the ASRS analysts and further investigation was justified.

The NAOMS survey data provided input on the potential seriousness of the event and also contributed to the characterization of the contextual factors contributing to anomalous consequences of ICAC's that are reflected in the results section of this report.

Insights from the ALAN Tool

ALAN was used to cluster a subset of 179 ASRS reports related to ICAC's spanning the period of January 1988 to August 2000. An ICAC is only reported to the ASRS when it results in a reportable event such as a hard landing. Therefore, even the total number of ICAC reports in the ASRS database is likely to be much fewer than the number identified by the NAOMS interviews. The small subset of these was selected for this study in order to fulfill the purposes of the demonstration within the available resources. The purpose was to demonstrate the capability to identify groups of related events in these reports. Operational experts assisted in characterizing the clusters identified by ALAN. ALAN identified the following primary event clusters and sub-clusters:

- External factors causing approach difficulties

  - Distraction during approach leading to procedural lapse

  - Approach change to an ILS runway

  - Landings with visibility near legal minimums

- System providing information about approach problems
  - Communication with another aircraft on takeoff
  - TCAS advisories during approach
  - ATIS providing RVR
- Issues primarily relating to larger/newer aircraft
  - Issues with use of FMS
  - Interactions with a wide-body aircraft
  - Issues involving specific approach characteristics
- Issues resulting in approach/landing procedure problems
  - STAR procedures and restrictions
  - Winds at landing and landing speed
  - Problems with cockpit automation
  - Approach plates and briefings for changed runway

The selected subset of ASRS reports on ICAC (179) was small enough so that domain experts could read them all and correlate their evaluations of the factors entailed in the event with the automated analyses of ALAN. The experts used a structured analytical approach that confirmed the results of the automated analyses of the ALAN tool, and contributed to some of the observations in the Results section.

Insights from PDARS Tools

PDARS used radar-track data to quantify the traffic patterns during ICAC events at two metropolitan airports. One month's data for San Francisco International (SFO) and Los Angeles International (LAX) airports were used as representative samples for this study. Both airports have parallel runway configurations that are often used by ATC for "side-step" ICAC maneuvers.

PDARS was able to identify aircraft trajectory patterns during final approach that indicated the designated landing runway changed from one runway to another parallel runway within the last 15 miles to touchdown. PDARS tools identified the time and position of this side-step maneuver.

At SFO, an average of 11.3 ICAC's occurred each day. The average time from the side-step maneuver to the runway threshold was 93.3 seconds with a standard deviation of 50.7 seconds, and the average distance was 3.37 NM with a standard deviation of 1.83 NM.

At LAX, an average of 22 ICAC's occurred each day. The average time from the side-step maneuver to runway threshold was 95.8 seconds with a standard deviation of 62.0 seconds, and the average distance was 3.46 NM with a standard deviation of 2.24 NM.

The consistency in the average time of the maneuver before runway threshold is speculated to be due to the fact that controllers need to wait until they are certain of their ground situation before they approve a runway change. Experienced pilots seldom have a problem executing a side-step maneuver within about 95 seconds of threshold. However, the time is short enough so that a problem may occur if other adverse aircraft or environmental factors exist (Schade et al. (2002)).

Insights from APMS Tools

APMS was used to gain further insight into ICAC events by examining aircraft flight-recorded data. An objective was to demonstrate that an aircraft operator could use flight data and APMS tools to aid in the assessment of the risk associated with an identified issue. In this study, APMS tools were used to demonstrate how flight data might be a source of information about the frequency of occurrence and the severity of consequences associated with ICAC events. The referenced report (Chidester et al. (2002)) is for internal NASA use only because it includes references to the proprietary data to which NASA was granted access by the air carriers for the purpose of this study.

The APMS team developed search criteria to identify approach and runway assignment changes and applied existing APMS tools to examine their consequences. The definition of an In-close Approach Change used in this study was a change that occurred after being established on final approach and below 2,500 ft. afe.

The database for this study was extremely limited by the data available to be accessed for analysis. The study sample consisted of 2680 flights, including 577 arriving at LAX, 881 at SEA, and 283 at ANC. Of the 577 flights at LAX, 27 flights (4.7%) were found to have experienced close-in runway changes, of 881 flights at SEA, 65 flights (7.4%) experienced close-in changes, and of 283 flights at ANC, 3 flights (1%) experienced close-in changes.

The APMS team also demonstrated the capability to assess the consequences of ICAC using the Routine Events tool, which documents the distribution of key parameters relevant to standard operating procedures at specific points during a flight. This permits a comparison between the key parameter distributions at times during approach for all flights to that airport that have experienced an in-close change to the same set of parameter distributions for all flights landing at the same airport. If close-in runway assignment increases risk of incident or accident, these approaches may differ on average or be more variable on some Routine Event parameter distributions. These Routine Event parameter distributions are tied to stabilized approach criteria and reveal how closely a group of flights comply with those criteria by plotting the distributions of airspeed, localizer deviation, glide-slope deviation, vertical speed, and N1 (a measure of engine thrust) at 1,500, 1,000, 500, and 100 feet above the runway. Stabilized approaches will group together around the average on each distribution. Unstable approaches will gravitate towards the tails of the distribution. If in-close runway changes produce problems, they may differ from the general population on average or variance, showing greater spread towards the extremes of the distribution of each parameter.

Figures B-2 and B-3 are examples of the many results obtained. These two figures show a comparison of the Routine Event parameter distributions for flights arriving at LAX to contrast six parameters at four points on the approach. These distributions are amenable to standard inferential statistical tests. Differences on average from all flights landing at an airport were evaluated for significance using t tests. Differences in variance were evaluated using the F test.

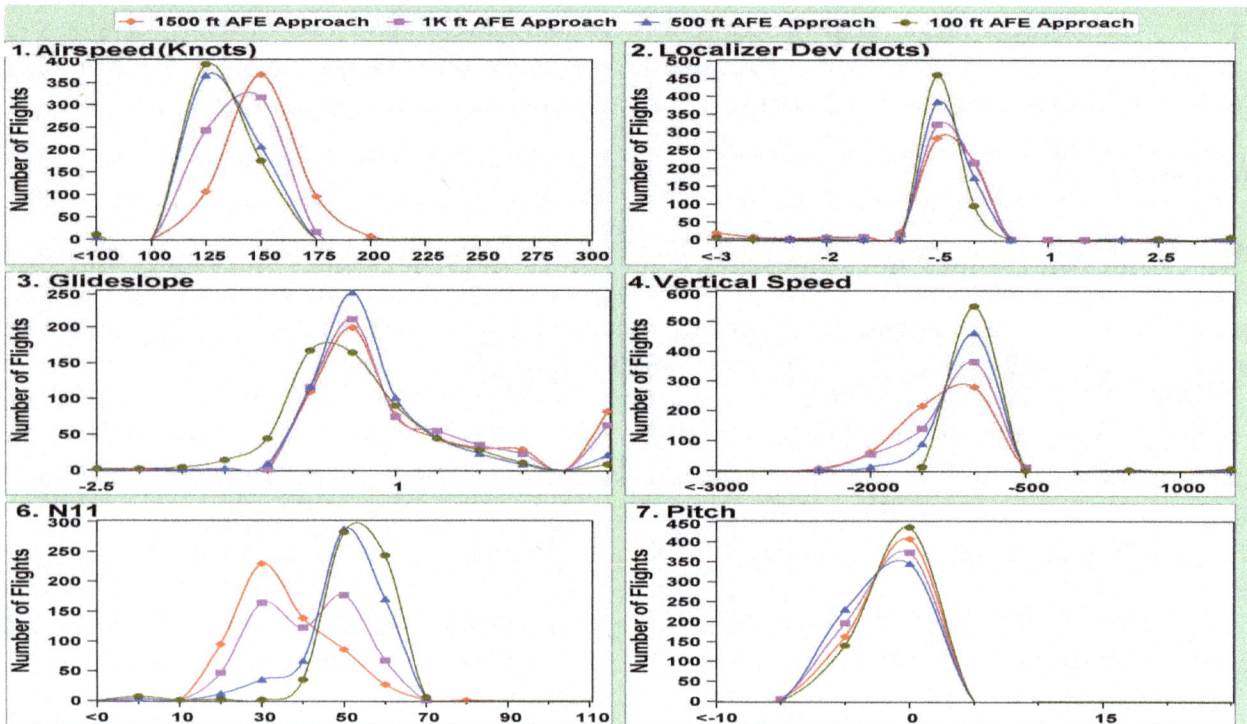

Figure B-2 – LAX – All Arrivals (557).

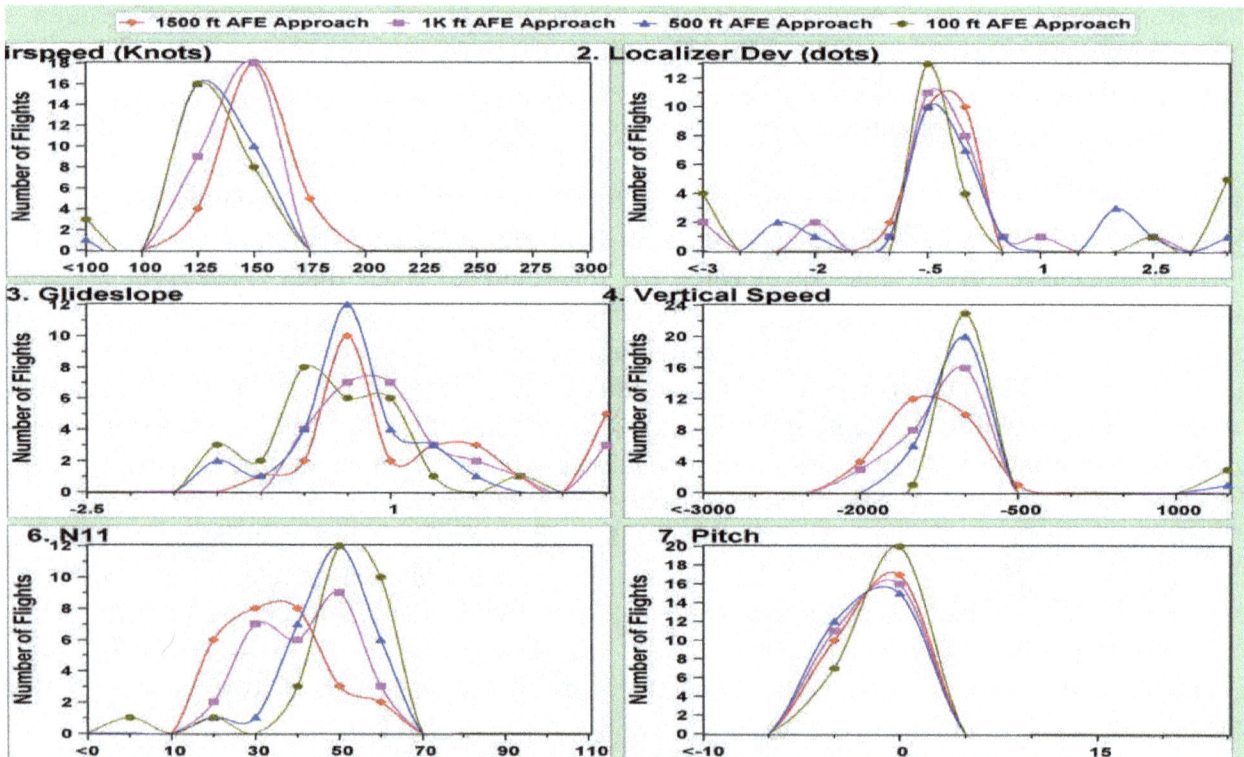

Figure B-3 – LAX – In-close Runway Changes.

When compared to distributions based on all flights landing at LAX, flights experiencing a close-in runway change showed greater localizer deviation at 1,500, 1,000, and 500 ft. above field elevation (afe) were higher on the glide slope at 1,000 ft. and lower at 500 and 100 ft., and had greater nose-down pitch at 500 and 100 ft. These flights were more variable on the localizer throughout the approach, in airspeed and vertical speed at 500 and 100 ft., and N1 at 100 ft. (All probabilities < .01).

In summary, while in-close runway assignment changes no more frequently resulted in unstable approaches, they differed on average and were more variable on a variety of dimensions making up a stable approach. Flights with in-close runway assignment changes become less stable, but not so extremely as to trigger unstable approach criteria based on current FOQA exceedances. Findings from this small study sample suggest that ICAC's frequently result in less stable approaches, implying greater risk. The small database was not necessarily representative of the full population, but this study demonstrated the kind of information that could be extracted from flight data using the APMS tools that complement the information from other data sources.

Insights from PROFILER Tools

PROFILER was applied to both APMS flight recorded data and PDARS radar track data to see whether it provided automated identification of patterns resembling runway changes, and "meaningful" clusters (singletons, atypical clusters) that correlated with the experts' analyses. PROFILER used APMS test data for a single carrier, and a limited number of flights, to examine recorded data parameters for the last 5 minutes of flight. PROFILER was able to identify flights that landed on a different runway than the majority in their cluster, and within this sub-cluster, to identify three flights that were ICAC candidates. Human experts agreed that two of these flights were ICACs. Using a month's worth of PDARS radar data, PROFILER examined 20,767 flights for an airport and identified 1,412 (7 percent) as potential ICAC candidates. These results were partially confirmed by a separate analysis.

Preliminary ICAC Case Study Observations

In this brief summary of this experiment, some examples of the information derived from the data were presented. . After integrating all of the information extracted from the various data sources (i.e., ASRS, NAOMS, flight and radar data) using the ASMM analysis tools (i.e., ALAN, PROFILER, Pattern Search, APMS, PDARS, and the Cinq-Demi Method), factors that could possibly contribute to the anomalous consequences of an ICAC were identified.

These factors are presented as hypotheses, because of the limited available data. . For example, as stated above, the NAOMS project surveyed only the commercial pilot community, the ATC perspective is represented to only a minor extent from ASRS reports as ASRS reports these are submitted primarily by the pilot community, and only a very small representative set of flight data were available for analyses with the APMS statistical tools..

Nevertheless, considerable insight was gained into the potential safety-risk of ICAC's drawn from the quantitative and qualitative data. Certainly, ICAC's contribute to a large number of unwanted consequences annually, and these unwanted consequences are likely associated with certain factors. Preliminary observations on changes that might mitigate anomalous consequences of ICAC's are:

- *ATC issuance of ICAC clearances may be problematic in relation to the following factors:*
  - Failure to appreciate the visual conditions from the perspective of the cockpit
  - Altitude at which clearance is given
  - Distance from airport at which clearance is given
  - Unfamiliarity with, or failure to consider correctly, the capabilities of the type of equipment being flown
  - Runway configuration
- *Air Carrier operating practices may be vulnerable to problems related to:*
  - Acceptance of in-close approach changes vis-à-vis go-around
  - Response to in-close approach changes (e.g., reprogramming the automatics vs. flying on raw data)

Whether and how these insights result in changes implemented to the system are the responsibilities of the FAA and the air carriers

Summary

The study was not able to arrive at definitive conclusions regarding the factors that may be causing pilots to have problems associated with ICAC or clear recommendations for interventions to mitigate them. However, the value of using the suite of ASMM tools to assist domain experts in gaining insight into an event was demonstrated.

Proactive management of safety risk starts with having in place a method for continuously monitoring the performance of the system, and a capability for comparing performance to expectations, to uncover and to understand potential risks of human error. Simply saying that one or more of the humans in a system may have made a mistake is not constructive. Analyses of the quantitative databases will help the domain experts understand exactly what happened. Analyses of textual databases and narrative reports are needed to understand why. That is the essence of an approach that will take the NATS beyond human error to proactive management of safety risk.

Appendix B References

Amidan, B.G., Cooley, S.K., and Ferryman, T.A.. Identifying In-Close-Approach-Changes in Aviation Performance Measuring System (APMS) Data. PNWD-3209. Battelle Pacific Northwest National Laboratory: Richland, WA. 2002

Amidan, B.G., Swickard, A.R., Allen, R.E. and Ferryman, T.A. Identifying In-Close-Approach-Changes in Air Traffic Control (ATC) Data. PNWD-3210. Battelle Pacific Northwest National Laboratory: Richland, WA. 2002.

Chidester, T.R. An Overview of the Enhanced Aviation Performance Measuring System. In Fifth GAIN World Conference Proceedings and Products, Miami, FL, December 5-6, 2001.

Chidester, T.R. Understanding Normal and Atypical Operations through Analysis of Flight Data. In Proceedings of the 12th International Symposium on Aviation Psychology, Dayton, Ohio. (This volume). 2003.

Chidester, T.R., Lynch, R.E., Lawrence, R.E., & Lowe, M.A. Applying the Aviation Performance Measuring System: Assessing risks associated with approach and runway assignment changes. NASA Technical Report. 2002

Ferryman, T.A. Cluster Analysis of Digital Flight Data for the Aviation Performance Measurement System. In Proceedings of American Statistics Association. Atlanta, GA, August 5-9, 2001.

Lecomte, P., Wanner J.C. and Wanner, N. Sécurité aérienne et erreurs humaines (Aviation safety and human errors). Aéronautique et Astronautique, 2 (153), 72-76. 1992.

Shade, J., Abkin, M., Davis, B., and den Braven, W. In-close approach change analysis using PDARS. NASA Contractor Report, ATAC Corp., Sunnyvale, CA 2002.

Wanner, J.C. Methodes d'analyse des incidents aériens (Methods for analyzing aviation incidents). In Proceedings of ANAE Forum 8 on "Relation Homme-Machine dans l'Aéronautique", January, 1999, Toulouse, 17-22. 1999.

Willse, A.R., Cooley, S. K., Chappell, A. R.; Daly, D.S., and Ferryman, T.A. Analysis of Rotary-wing Aircraft Incident and Accident Reports. PNWD-3070, Battelle Pacific Northwest National Laboratory: Richland, WA. 2002.

APPENDIX C

THE AVIATION SAFETY REPORTING SYSTEM (ASRS)

ASRS is a highly successful and trusted program that has served the needs of the aviation community for over 30 years, and is available to all participants in the national aviation system who wish to report safety incidents and situations. The ASRS was established in 1976 under an agreement between the Federal Aviation Administration (FAA) and NASA. This cooperative safety program invites pilots, air traffic controllers, flight attendants, maintenance personnel, and others to voluntarily report to NASA any actual or potential hazard to safe aviation operations. The FAA provides most of the program funding. NASA administers the program, ensures confidentiality, sets policies in consultation with the FAA and aviation community, and receives the reports submitted.

ASRS Data Are Used to Promote Safety

The ASRS collects and responds to voluntarily submitted incident reports to lessen the likelihood of aviation accidents (figure C-1). ASRS data are used to:

- *identify aviation system deficiencies for correction by appropriate authorities,*

- *support aviation system policy, planning, and improvements, and*

- *strengthen the foundation of aviation human-factors safety research.*

The ASRS analysts with expertise in all of the domains of aviation operations identify reports of events or circumstances that have implications of compromised safety and, after discussions with the FAA, issue Alert Bulletins or For Your Information letters to the appropriate authorities. A percentage of all of the reports are processed by the analysts and then entered into the database. The ASRS database has proven to be a very valuable resource over the years for many studies of human-factors issues.

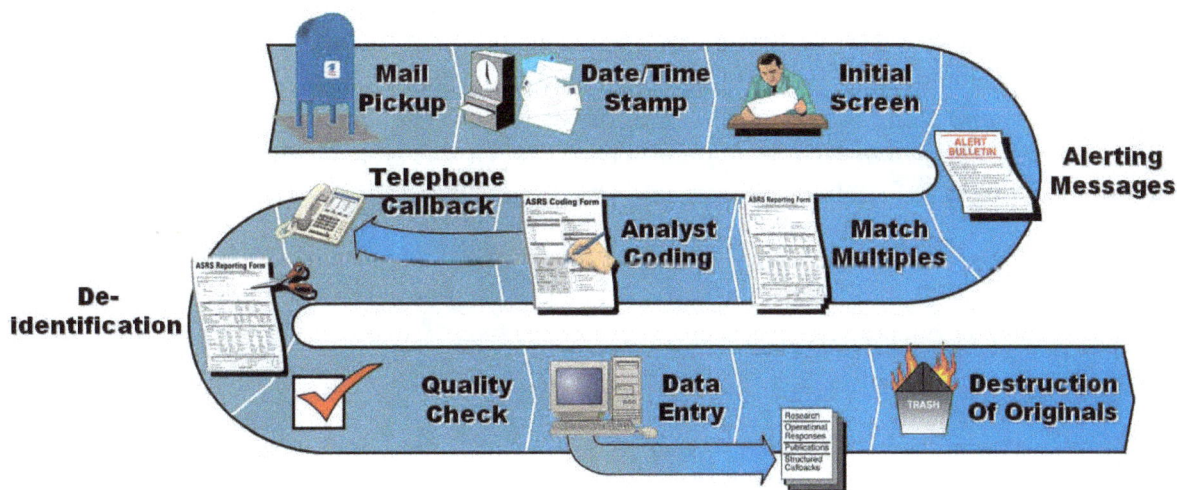

Figure C-1. Report processing.

ASRS routinely produces the following:

- *Alerting messages* are sent to individuals in positions of authority so that corrective action can be initiated when the ASRS receives a report describing a hazardous situation.

- *CALLBACK* is a monthly bulletin issued to more than 85,000 personnel in the aviation community. It contains extracts of the ASRS reports, summaries of research, and related safety information.

- *Database search responses* are issued when search requests are received from parties wishing to access information about particular subjects.

- *Operational support* is provided to the FAA and NTSB during rule-making, procedure and airspace design, and accident investigations by collating relevant information from the database.

- *Topical research* is carried out and the results of the research (mainly human-performance issues) are published.

ASRS Reporters Are Protected

The FAA offers those who report to the ASRS two important guarantees: *confidentiality* and *limited immunity*. The FAA offers these guarantees because of the value of the safety information obtained through the ASRS program.

Reports Are Held in Confidence

Reports sent to the ASRS are held in strict confidence. More than 600,000 reports have been submitted since the beginning of the program without a single reporter's identity being revealed. ASRS removes all personal names and other potentially identifying information before entering reports into its database. Reports submitted to ASRS may be amplified by further contact with the individual who submitted them, but the information provided by the reporter is not investigated further.

Reporters Receive Limited Immunity

ASRS reporters are also guaranteed limited immunity by the FAA, meaning that the FAA will not use information that has been filed with the ASRS in an enforcement action, and will waive fines and penalties for *unintentional* violations of federal aviation regulations that are reported, as long as violations are reported *within 10 days* of their occurrence. However, accidents and criminal activities are not protected from enforcement actions, and should not be submitted to the ASRS.

ASRS Benefits Reporters

In addition to immunity provisions associated with the ASRS program, reporters often mention other equally important motivations for using the program:

- *Satisfaction in knowing they are helping to improve the aviation system by giving safety information to the ASRS, and*

- *Increased understanding of the factors contributing to their safety incident.*

For the ASRS databases, a reasonable general rule is that each report occupies about 75 KB of storage, including all associated indices and, because it currently holds approximately 150,000 active records, an estimate of its size is approximately 10 GB.

Certain caveats apply to the use of ASRS statistical data. All ASRS reports are voluntarily submitted, and thus cannot be considered a measured random sample of the full population of like events. Moreover, not all pilots, controllers, air carriers, or other participants in the aviation system are equally aware of the ASRS or equally willing to report to it. Thus, the data reflect reporting biases. These biases, which are neither fully known nor measurable, may influence ASRS statistics. A safety problem such as near midair collisions (NMACs) may appear to be more highly concentrated in area "A" than area "B" simply because the airmen who operate in area "A" are more supportive of the ASRS program and more inclined to report to the ASRS when an NMAC occurs.

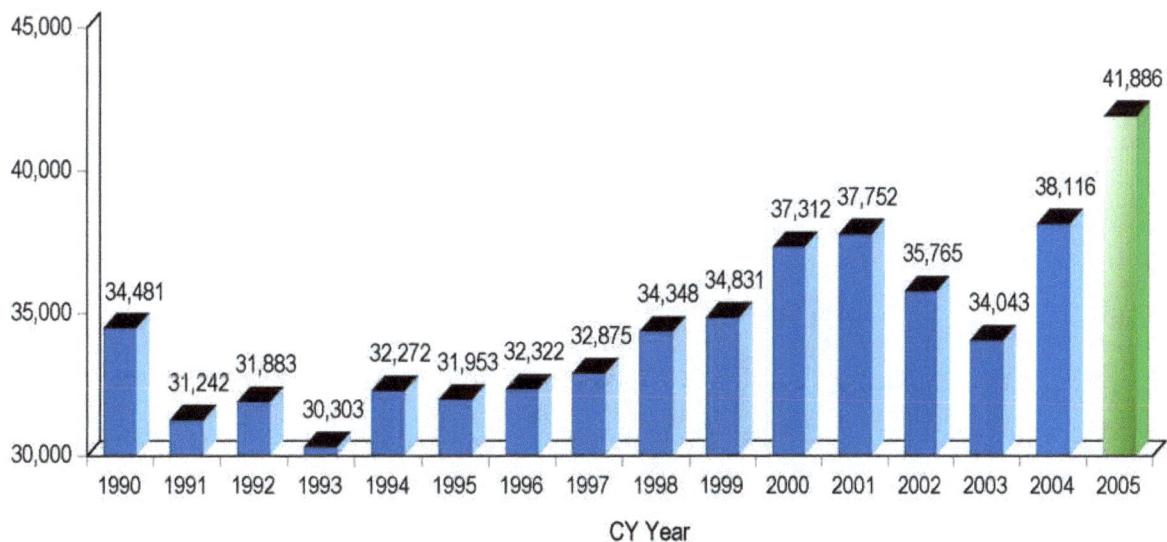

Figure C-2. ASRS reports received annually.

Appendix C Bibliography

1. FAA Regulations on ASRS:

   a. Advisory Circular 00-46C

   b. FAR 91.25

   c. Facility Operations and Administration Handbook (7210.3M) paragraph 2-38

APPENDIX D

COMMERCIAL AVIATION SAFETY INFORMATION SHARING PROCESS: CONCEPT PAPER AND ROADMAP

Synopsis

This document defines a future state and roadmap for safety information sharing that will help achieve an accident-free national air transportation system. Studies suggest that in many fatal accidents, information existed somewhere in the system that could have been used to prevent the accidents, had it been available to the right people at the right time. Currently there are many uncoordinated safety information collection and sharing efforts in the air transportation industry, which were typically initiated in response to specific needs. While industry is progressing in the development of many safety data sources, there has been no systemic effort to integrate this information across organizations and stakeholders in a process directed at improving overall system safety. Hazards to safe air carrier operations will always emerge in our rapidly changing and highly stressed national airspace system. New approaches to safety management and safety information sharing are required, to better detect these hazards and manage the risks before they cause air carrier accidents.

A key element of future safety management systems will be a dramatic change in the way we collect, manage, and share data and information. We envision a future state in which safety information from stakeholders is accessed through a safety information sharing process and is available to decision-makers for managing risks. Such a process would enable the industry to more effectively utilize its limited resources in maintaining and improving safety.

The aviation industry has developed several information sharing programs that show how beneficial this process can be. It is imperative that the Federal Aviation Administration (FAA) and commercial aviation industry provide the leadership, commit the resources, and allocate funding to achieve the envisioned future state. Success depends on several critical factors, including committed leadership that will consistently champion the cause and maintain a long term-commitment to this effort. Cultural changes and tool development for this new approach to safety information management will take time. Unless we begin immediately, we risk resurgence in fatal air transportation accidents as air traffic volume increases threefold by 2025.[18]

Background

FAA, industry, and the Commercial Aviation Safety Team (CAST) leadership identified a need to define an ideal future state for safety information sharing in the commercial aviation community, in order to better understand and manage current activities and to agree on a roadmap to implement the future state. CAST designated the Aviation Safety Information Sharing Task (ASIST) Team to

---

[18] *"Next Generation Air Transportation System, Integrated Plan,"* FAA JPDO, 2004.

address this need. The ASIST team includes representatives of various organizations within the FAA, the National Aeronautics and Space Administration (NASA), the air carrier industry, and manufacturers of large transport airplanes. Representatives from the air traffic community were not included on the team, but are identified as a major stakeholder for future work. ASIST developed this Concept Paper in response to its charter.

The ASIST team's products are also required to support the Joint Planning and Development Office (JPDO). Congress established the JPDO in December 2003 with the goal of transforming the current aviation system into the "Next Generation Air Transportation System" (NGATS) by the year 2025, working across six government agencies. To meet this goal, eight Integrated Product Teams (IPT) have been formed. FAA's Office of Regulation & Certification is leading the Safety Management IPT, which as part of its mission is defining steps to achieve a cross-agency information sharing and analysis capability. The ASIST team products will be provided to the IPT in order to support this mission from the aviation perspective, for inclusion in the inter-agency JPDO strategic plan for 2025.

Why Change is Urgent

Recent studies have suggested that in many fatal accidents, information existed somewhere in the system that could have been used to prevent the accidents, had it been available to the right people at the right time.[19] That is, existing information could have been used to identify the need to mitigate risks and thus prevent accidents. This is unacceptable – it means that we are not using existing safety data and information optimally in today's environment.

Meanwhile, the U.S. air transportation system is straining under rapidly growing demand, capacity limitations, and rapid technological change. This is not new – the commercial aviation industry has been in a constant state of change through its entire existence, and we must assume that change is a constant. But as the national airspace system (NAS) continues to increase in size and complexity, safety information proliferates across the system. Our static and reactive data-handling methods do not support change well. We must develop safety information sharing processes that can evolve to meet growth needs, and can ensure that safety information is consistently available to key decision-makers to mitigate risks, therefore preventing accidents.

---

[19] *Accident Precursor Analysis and Management, Reducing Technological Risk through Diligence*, National Academy of Engineering of the National Academies, The National Academies Press, Washington, D.C., 2004

Roelen, A.L.C, Wever, R., Verbeek, M.J., *Improving Aviation safety by better understanding and handling of interfaces, a pilot study*, NLR-CR-2004-025, National Aerospace Laboratory, Amsterdam, 2004.

Commercial Airplane Process Study, An Evaluation of Selected Aircraft Certification, Operations, and Maintenance Processes, The Report of the FAA Associate Administrator for Regulation and Certification, 2002.

"Continuing Airworthiness Risk Evaluation: An Exploratory Study," Flight Safety Foundation, *Flight Safety Digest* September-October, 1999.

Improving the Continued Airworthiness of Civil Aircraft, A Strategy for the FAA's Aircraft Certification Service, Committee on Aircraft Certification Safety Management, Aeronautics and Space Engineering Board, Commission on Engineering and Technical Systems, National Research Council, National Academy Press, 1998.

Commercial air carrier operations in the U.S. are very safe today, and accidents are prevented every day. We can expect the public to demand an increasing level of safety -- defined by freedom from accidents. However, because of the currently low accident rate, industry will rely more on safety management processes to systemically reduce the risk of accidents. Ironically, associated resources to improve processes may be difficult to secure when fewer accidents are occurring, due to increased public complacency – making implementation of safety management and information sharing processes more problematic.

Vision for the Commercial Aviation Safety Information Sharing Process

An integrated, over-arching process is required to optimize sharing of aviation safety information across the commercial air transport industry. We envision a future state in which safety information from stakeholders is rapidly accessed through a safety information sharing process and is easily available to decision-makers for managing risks. The process must support industry safety management processes in order to position the NAS for the tremendous change predicted from now through 2025.[20] We must make better use of valuable but frequently "stove piped" safety information systems, to support safety management processes[21] that identify hazards, assess their risk, assist in development of mitigation strategies, and measure effectiveness of corrective actions. We must end the era in which an accident occurs because an existing[22] hazard was not discovered and addressed.

The figure 'Future State' depicts the envisioned end state to which we strive, and the capabilities that safety management and safety information sharing must provide to achieve that future state. It should be noted that risk will still exist within the system, and accidents may still occur. This should in no way dilute the urgent need to develop safety management and safety information sharing processes in order to strive for this future accident-free environment.

---

[20] "Next Generation Air Transportation System, Integrated Plan," FAA JPDO, 2004.

[21] Among the excellent sources for discussion of risk management are Managing the Risks of Organizational Accidents, James T. Reason (Ashgate Publishing, Limited, January 1998), and System Safety Handbook: Practices and Guidelines for Conducting System Safety Engineering and Management (FAA, December 30, 2000). ASIST team acknowledges that risk will always exist within the system; the objective is to manage the risks in order to avoid accidents.

[22] The ASIST team notes that current data and information systems focus on detecting existing hazards in the aviation system. We foresee a need to improve our ability to anticipate future hazards and risks and eliminate, avoid, or mitigate them when we design the future aviation system; however, we have not addressed that prognostic capability in this analysis. The European Joint Safety Strategy Initiative (JSSI) is attempting to do this via their Future Aviation Safety Team (FAST). FAST and possible other hazard prognostication methods would ideally be developed to address future hazards.

Future State

The future air transportation system is <u>accident-free</u>

HOW?

| Hazards are discovered and risks are managed <u>before</u> they cause accidents | Through a safety management process, decision makers at all levels prevent accidents by managing risks | Managing risks requires <u>information</u> to discover unknown hazards and to validate risk mitigation |

Data and information from all stakeholders are accessed through a safety information sharing process and available to decision makers for managing risks

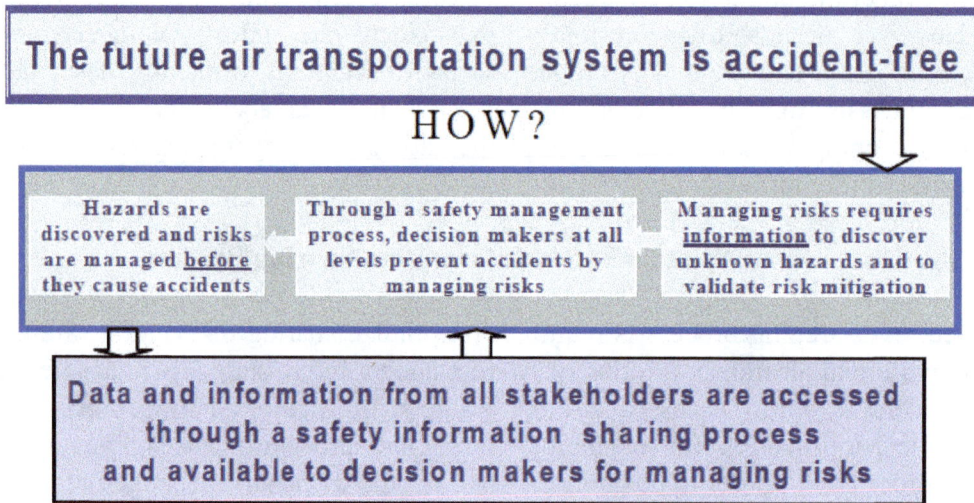

Key Attributes of Future State

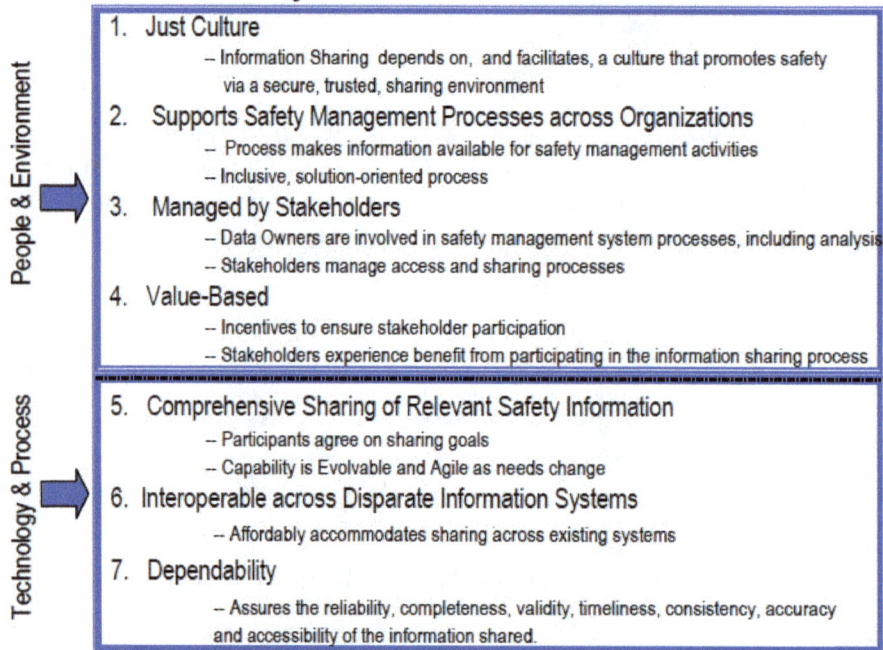

People & Environment

1. Just Culture
   -- Information Sharing depends on, and facilitates, a culture that promotes safety via a secure, trusted, sharing environment
2. Supports Safety Management Processes across Organizations
   -- Process makes information available for safety management activities
   -- Inclusive, solution-oriented process
3. Managed by Stakeholders
   -- Data Owners are involved in safety management system processes, including analysis
   -- Stakeholders manage access and sharing processes
4. Value-Based
   -- Incentives to ensure stakeholder participation
   -- Stakeholders experience benefit from participating in the information sharing process

Technology & Process

5. Comprehensive Sharing of Relevant Safety Information
   -- Participants agree on sharing goals
   -- Capability is Evolvable and Agile as needs change
6. Interoperable across Disparate Information Systems
   -- Affordably accommodates sharing across existing systems
7. Dependability
   -- Assures the reliability, completeness, validity, timeliness, consistency, accuracy and accessibility of the information shared.

The necessary characteristics, or requirements, of the envisioned safety information sharing process are shown in the figure 'Key Attributes of Future State.

The ASIST Team believes each Attribute identified here is necessary in order for safety management and safety information sharing processes to succeed in the commercial aviation industry.

The first four Attributes address "People & Environment" -- the required involvement of stakeholders across the community, as well as evolution toward a 'safety culture'. The last three Attributes describe the required characteristics of the eventual enabling safety information sharing technology and processes. Additional description of the Attributes is provided to insure the team's intent for each is clear.

1. *Just Culture*: The safety information sharing process requires, but also facilitates, a just culture that promotes information reporting and sharing within a trusted environment,[23] providing security for the data and withstanding challenges from litigants, media, and special interest groups to prevent misuse of information. The process entails emphasis on a "learning culture" rather than a blame culture, and the stakeholders are persuaded that there is "shared risk" in contributing information. That is, the stakeholder must have reasonable confidence that by contributing information, it will not be exposed to increased adverse consequences in relation to other stakeholders.

   As part of this cultural change, the regulatory authority would retain its responsibility for oversight of regulatory compliance, but also would recognize and foster the safety benefits of a systems approach to risk management and the industry's role in accomplishing this. This will require promotion of safety through sharing of data and information in an environment that removes barriers caused by fears of punitive response or misuse of the information.

2. *Supports Safety Management Processes across Organizations*: There must be a systematic safety management process to which safety information sharing is linked, to identify hazards, analyze causal factors, develop recommendations, implement mitigating strategies, and monitor implementation results. The process must be scalable to accommodate local issues up to national/international industry issues. Information to identify hazards and assess and manage risk, as well as to assess the health of the system, will be collected through the information sharing process. The process will enable information to be provided to decision makers.

   Additionally, methods must be defined to measure the health of the safety system in regard to information sharing, and insure processes are in place to measure and make improvements where needed. This is a component of any safety system, but will be especially important given the need to evolve and change quickly to meet changing needs.

3. *Managed by Stakeholders*: The process will provide for complete stakeholder communication, coordination, and cooperation as necessary. The process will support data owners being involved in the safety management processes, and aid broad assessment of systemwide risks involving more than one stakeholder. Stakeholders will appropriately manage access and sharing processes for safety needs. The process will not include non-safety interests, because information shared is limited to use for identifying, solving, or

---

23 *A Roadmap to a Just Culture: Enhancing the Safety Environment,* Global Aviation Information Network (GAIN) Working Group E: Flt Ops/ATC Safety Information Sharing, First Edition, 2004.

following up on safety or operational issues only. Based on this "need to know," the process will be mutually supportive, and secure, with success derived from an involvement in outcomes.

Although the flight operations community is leading current information sharing efforts, all stakeholders (manufacturers, repair stations, the Air Traffic Organization, accredited safety research activities, industry associations, etc.) must be included for comprehensive safety information sharing. The future state and attributes must be refined to meet all stakeholder needs and validated to a common end state as the goal. Interim steps will include increasing stakeholder access to de-identified, aggregate data as part of their participation in industry safety sharing processes.

4. *Value-Based*: The process will focus on data, information and knowledge that can be acted upon and provide incentives to participation: stakeholders will experience benefit from participating in the information sharing process, including the realization of intrinsic safety benefits and accreditation by the regulator. Examples include alternate methods of compliance to current reporting requirements for participants in information sharing prototypes; use of safety information to reduce and/or target regulatory oversight based on more risk-based oversight methods; and action by regulators to reduce data access needs based on better understanding of information needed for effective oversight.

   Participants may also find other safety and cost benefits as a result of sharing information. For example, operators may learn of adverse experiences from other operators with similar equipment and operating environments, and may be able to reduce costs by implementing preventive measures to avoid the same experiences.

5. *Comprehensive Sharing of Necessary Safety Information*: Information accessed will be only that which is necessary to manage safety. All contributors will provide this information under a common definition and understanding of its use. The process will be agile and flexible to allow for evolution of the aviation system. The future process will not just link existing data sources; that would simply produce a bigger haystack. Reaching our goals will require automated processes to identify issues of potential risk: monitoring of each database for potential safety issues and integrating events, incidents, and issues across sources of data to provide an understanding of possible contributing factors of these events.

   Additionally, safety information sharing must include a process for introducing new information providers to the trusted community. The process must provide protection for stakeholders of different types, guaranteeing anonymity while allowing data query and discussion, and allowing data owners to manage access. This would occur within a community of stakeholders and would facilitate data and information sharing while the envisioned 'just culture' of trust is continuing to develop.

6. *Interoperable across Disparate Information Systems*: Safety information consists of a variety of data sources distributed across organizations and stored in organization-specific formats. An effective safety information-sharing process requires routine analysis across data sources in their native format without imposing data structure across data systems,

stakeholders, and domains, so that risks identified in any given source of data may be examined from multiple perspectives represented in heterogeneous data. The process must accept information and enable analysis within and across all heterogeneous data types and systems and organizations. The process will facilitate a safety information sharing process that is low-cost, flexible, and scalable.

Multiple parallel research and application projects will be required to make these distributed data sources available for routine, collaborative analysis and response by stakeholders. The increasing complexity and sophistication of the NAS demands that we facilitate these multiple perspectives to better be able to "connect the dots" and prevent accidents by analyzing operational and other data rather than having to analyze accidents.

7. ***Provides Dependability***: Development of technology and processes must assure the reliability, completeness, validity, timeliness, consistency and accuracy of the information shared, and become a vital element of stakeholders' safety management systems.

Safety Information – Barriers to Sharing

Barriers exist that adversely affect the efficient sharing of safety data and information among the participants. These barriers are largely culturally induced, based on fears that the information may be used against the provider, through disciplinary action, litigation, enforcement, or public disclosure. Also, studies show that current data collection efforts often suffer from inefficiencies, lack of focus, and lack of coordination, duplication, or "stove piping", and declining quality. Stove piping is common -- an organization may have information that could be of great significance to another organization or discipline but is not shared because the structure of the current system has not widely institutionalized information sharing.

As an example, the first airline to implement a program under the FAA's Aviation Safety Action Program (ASAP, a voluntary reporting program further discussed below) was able to identify errors resulting from revisions in the Pre-Departure Clearances (PDC). In these cases, PDC revisions were not effectively communicated to flight crews, resulting in crews flying the original clearance route while ATC is expecting the revised route to be followed. The airline changed the PDC procedures and the events stopped. The airline communicated this information concerning the potential for PDC errors to other operators and by NASA's Aviation Safety Reporting System (ASRS) in the form of an Operations Bulletin.

Unfortunately, other operators did not follow this guidance. It was not until other airlines' ASAP were initiated following the FAA issuance of regulations allowing ASAP programs across airlines in 2001 that these same PDC errors were detected at the other airlines. The remaining operators then implemented the same resolutions used by the initial airline. Identifying this issue within one airline and alerting by our nation's safety reporting system did not result in correction across airlines. Only re-discovery within additional airlines' ASAP produced these needed changes.

This can be construed as a failure of information processes from several perspectives:

- At the beginning, only one airline had access to this information internally, though events were happening across airlines.

- Though the national-level reporting system alerted the problem as likely cross organizational, it was not tied to a change process.

- Though the identifying airline discussed its internal changes fairly publicly, other airlines did not recognize the need to act within their organizations.

- Organizations implemented internal change only when internal ASAP programs were established at multiple airlines.

All of these events represented some risk of in-flight collision. This is a specific example of why our industry needs an information-sharing process tied to inter-organizational safety management processes that ensure that identification of an issue by one stakeholder is sufficient to provoke evaluation throughout the national air transportation system and accomplish systemic change when required.

Safety Information – Experiences in Sharing

There are many data and information systems within the NAS and the various aviation communities. These provide a starting point for data and information collection, analysis, and sharing. In certain cases, the industry has implemented some limited programs to share information among certain stakeholders, addressing issues of trust and pro-active analysis with acceptable protections. These programs demonstrate the safety value of greater openness among appropriate stakeholders. They also demonstrate the wealth of data and information available in the industry, particularly within the air carrier industry, and how powerful these data and information can be when integrated across organizational lines and accessed by the right decision makers. Specific examples are relevant within air traffic, airframe manufacturing, and maintenance.

In the air traffic community, FAA's Office of Performance Management has implemented the Performance Data Analysis and Reporting System (PDARS). Using radar data, PDARS calculates a range of performance measures including traffic counts, travel times and distances, traffic flows, and in-trail separation. PDARS has been implemented at all US Air Route Traffic Control Centers (ARTCC) and the Command Center and is being extended to Regional Offices and Terminal Radar Approach Control (TRACON) facilities. Using these data, Air Traffic facilities have accomplished performance analysis, route and airspace design, noise abatement analysis, training, support for search and rescue, and research by FAA and NASA on future improvements. Similarly, the National Airspace Information Monitoring System (NAIMS) has been developed for sharing air traffic control safety event reports.

Aircraft and engine manufacturers generally have individually tailored systems for obtaining and sharing safety information. Input from operational events, customers, bulletin boards, and field representatives, is collected for internal analysis and action. Required reporting of events to FAA can be augmented through voluntary agreements such as the Boeing/FAA Continued Operational Safety Program (COSP) agreement.

Much like the airframe manufacturing community, the maintenance/repair/overhaul (MRO) community has several unique processes for obtaining and sharing relevant safety information. MRO events are reported and shared among several community stakeholders including operators, repair stations, manufacturers, and the FAA. One such sharing process is the Rotor Manufacturing Induced Anomaly Database (RoMAN). Within RoMAN, 11 engine manufacturers report and share high energy rotor events induced by manufacturing anomalies. The FAA's Service Difficulty Reporting System (SDRS) is yet another example of a MRO community sharing system. Operators and maintainers of aircraft report in-service events, such as defects, malfunctions, and failures.

Working across multiple communities, NASA has operated the Aviation Safety Reporting System (ASRS) on behalf of the FAA since 1976. It receives, processes, and alerts in response to safety reports from pilots, dispatchers, mechanics, and flight attendants.

In the airline flight operations community, Flight Operational Quality Assurance (FOQA) is a process that enables a systems safety manager to analyze continuous, pre-planned, flight data from all individual fleet aircraft across a common time interval from the aircraft's Digital Flight Data Acquisition Unit. ASAP is a pro-active safety process that encourages air carrier and repair station employees to voluntarily report safety information that may be critical to identifying potential precursors to accidents. FOQA and ASAP are examples of progress in developing a safety culture that places a higher value on reporting and learning, rather than enforcement and punitive action. Under the FAA's regulations, air carriers that implement approved FOQA programs must provide the FAA with aggregate FOQA data. Except for criminal or deliberate acts, the FAA will not use that FOQA data against the operator or its employees.

This sharing to date has been limited to sharing of issues in stakeholder meetings. The data-sharing process negotiations have been ongoing for several years, culminating in the Voluntary Aviation Safety Information-sharing Process (VASIP). VASIP is an activity that will integrate two key safety information systems, FOQA and ASAP. VASIP is intended to accomplish two complementary objectives. One objective is the development of a technical process to provide access to de-identified safety data from any participating airline's FOQA or ASAP program, aggregate it, and make it accessible to appropriate industry stakeholders for analysis. The other is the development of a comprehensive, structured process among all stakeholders that will permit them to analyze aggregate safety information, identify problem areas, develop and implement appropriate corrective action plans, measure the effectiveness of those actions, and share the conclusions with stakeholders.

We highlight VASIP here as a leading example for the direction safety information sharing should take, in collaboration among stakeholders and building the environment of trust and cooperation with regulators. Additionally, the CPS Response Aviation Rulemaking Committee (ARC) in 2004 prototyped a secure data sharing approach, validating the technological capability to guarantee anonymity to data providers while allowing for data query and follow-up investigation. The ASIST team recognizes VASIP and similar sharing approaches as building blocks in the roadmap toward the future state. Industry and government must expand these sharing concepts, building an environment of trust that provides incentives for timely and effective sharing of quality data and adequate protection for the information and the providers.

Conclusion

The leadership to attain the safety information sharing future state would be provided by a high-level leaders' group that would work with decision-makers to promote the understanding of the benefits of information sharing in a non-punitive environment, and allocate necessary resources. The group would draw from, and integrate as appropriate with, the substantial experience and history built by CAST, the ASAP and FOQA ARCs, and industry representative groups. The group would identify the stakeholders, enlist their participation, and work with these stakeholders throughout government and industry to drive long-term efforts necessary for effective safety management systems and safety information sharing processes within and across organizations.

Two roadmaps are provided in this document, a 20-year long-term broad view and an 18-month short-term detailed view. These roadmaps describe actions, decision points, and approximate timelines to progress from the current state of mostly disconnected data and information systems to a process in which safety information is consistently available to aviation decision makers for use in the management of risk and prevention of accidents. Each roadmap depicts the anticipated evolution of society's expectations of air transportation safety over the given timeframe, along with the evolution of the state of safety information sharing over the same period. It also shows that the broad activity areas that will be necessary to move the industry toward the future state and its attributes.

Implementing the future state safety information sharing process will require change leadership to develop the environment, culture, and people, and expert technical leadership to develop the enabling technology and processes. The team believes the single most important factor to success can be found within the commercial aviation community -- FAA and industry leadership to consistently champion the cause and maintain a long term-commitment to achieving the future state of safety information sharing.

Commercial Aviation Safety Information Sharing Process

ROADMAP - Long-Term

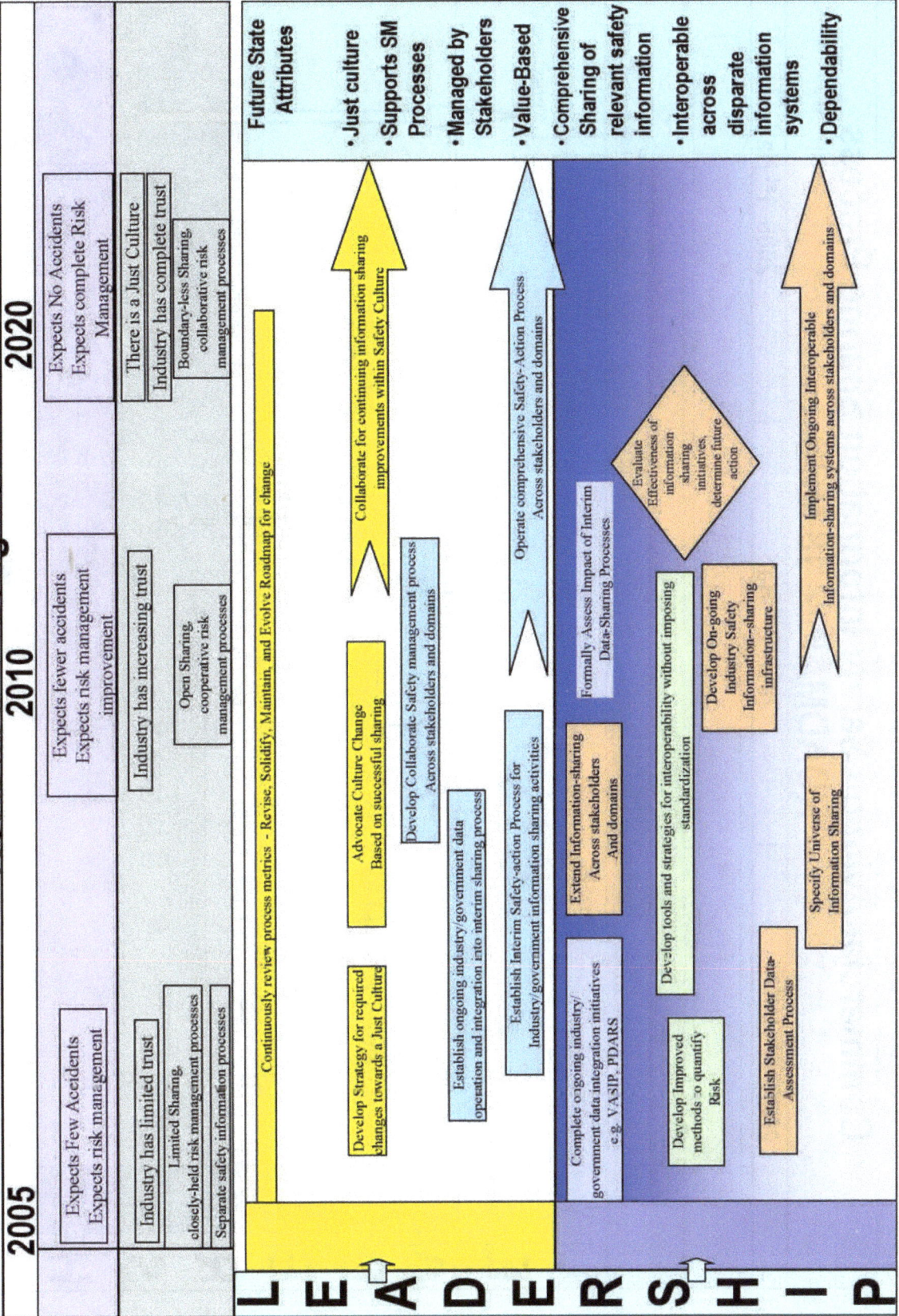

	2005	2010	2020	Future State Attributes

Expects Few Accidents
Expects risk management

Expects fewer accidents
Expects risk management improvement

Expects No Accidents
Expects complete Risk Management

Industry has limited trust

Industry has increasing trust

There is a Just Culture
Industry has complete trust

Limited Sharing, closely-held risk management processes
Separate safety information processes

Open Sharing, cooperative risk management processes

Boundary-less Sharing, collaborative risk management processes

Future State Attributes

- Just culture
- Supports SM Processes
- Managed by Stakeholders
- Value-Based
- Comprehensive Sharing of relevant safety information
- Interoperable across disparate information systems
- Dependability

LEADERSHIP

Continuously review process metrics - Revise, Solidify, Maintain, and Evolve Roadmap for change

Develop Strategy for required changes towards a Just Culture

Advocate Culture Change Based on successful sharing

Collaborate for continuing information sharing improvements within Safety Culture

Develop Collaborate Safety management process Across stakeholders and domains

Establish ongoing industry/government data operation and integration into interim sharing process

Establish Interim Safety-action Process for Industry/government information sharing activities

Operate comprehensive Safety-Action Process Across stakeholders and domains

Complete ongoing industry/ government data integration initiatives e.g. VASIP, PDARS

Extend Information-sharing Across stakeholders And domains

Formally Assess Impact of Interim Data-Sharing Processes

Evaluate Effectiveness of information sharing initiatives, determine future action

Develop Improved methods to quantify Risk

Develop tools and strategies for interoperability without imposing standardization

Develop On-going Industry Safety Information-sharing infrastructure

Establish Stakeholder Data-Assessment Process

Specify Universe of Information Sharing

Implement Ongoing Interoperable Information-sharing systems across stakeholders and domains

Commercial Aviation Safety Information Sharing Process
ROADMAP - 18 MOS

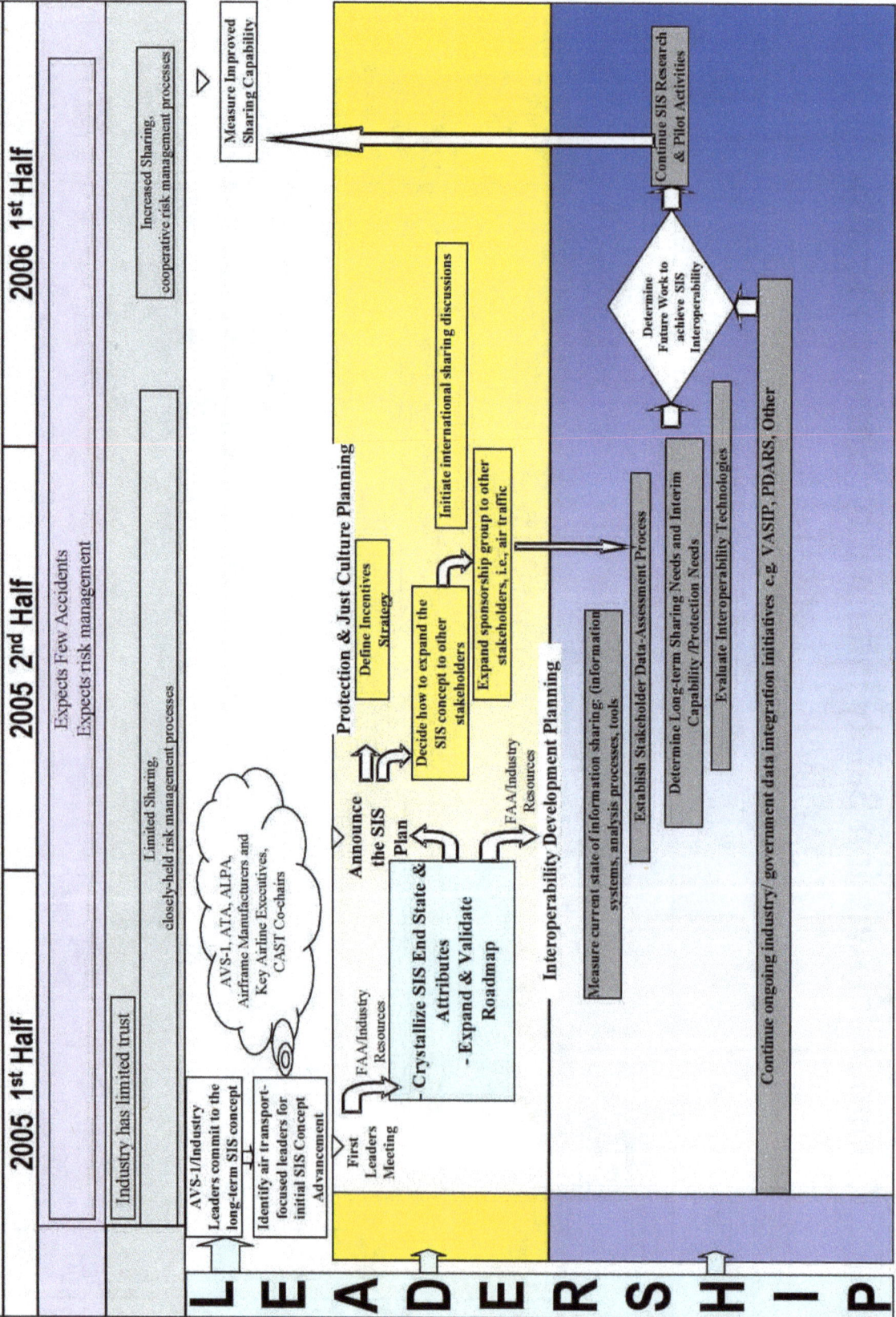

2005 1st Half	2005 2nd Half	2006 1st Half

Expects Few Accidents
Expects risk management

Industry has limited trust

Limited Sharing,
closely-held risk management processes

Increased Sharing,
cooperative risk management processes

Measure Improved
Sharing Capability

L

AVS-1/Industry
Leaders commit to the
long-term SIS concept

Identify air transport-
focused leaders for
initial SIS Concept
Advancement

E

AVS-1, ATA, ALPA,
Airframe Manufacturers and
Key Airline Executives,
CAST Co-chairs

A

First
Leaders
Meeting

FAA/Industry
Resources

Announce
the SIS
Plan

Protection & Just Culture Planning

Define Incentives
Strategy

Decide how to expand the
SIS concept to other
stakeholders

Initiate international sharing discussions

Expand sponsorship group to other
stakeholders, i.e., air traffic

D

Crystallize SIS End State &
Attributes
- Expand & Validate
Roadmap

E

FAA/Industry
Resources

Interoperability Development Planning

Measure current state of information sharing; (information
systems, analysis processes, tools

Establish Stakeholder Data-Assessment Process

Determine Long-term Sharing Needs and Interim
Capability /Protection Needs

Evaluate Interoperability Technologies

R

S

Continue ongoing industry/ government data integration initiatives e.g. VASIP, PDARS, Other

H

Determine
Future Work to
achieve SIS
Interoperability

Continue SIS Research
& Pilot Activities

I

P

APPENDIX E

HUMAN-PERFORMANCE MODELS

This appendix comprises two summary papers describing the work on developing human-performance models that was conducted under the System Wide Accident Prevention (SWAP) project of the Aviation Safety Program, and some summary comments from the meeting session at which they were presented.

The first paper (Foyle, Goodman, and Hooey, (2003)) is the introduction to a conference proceedings and provides an overview of the activities, the state of modeling human performance at the end of 2002, and a description of the future activities planned for 2003 and 2004. It appears in the proceedings of the NASA Aviation Safety Program Conference on Human Performance Modeling of Approach and Landing with Augmented Displays, *NASA Conference Proceedings NASA/CP-2003-212267*, 2003.

The second paper is a summary paper describing the modeling efforts that was presented as a panel presentation on the topic at the 2005 Human Factors and Ergonomics Society meeting. It appears in the *Proceedings of the Human Factors and Ergonomics Society 49th Annual Meeting*, Santa Monica, Calif., 2005.

AN OVERVIEW OF THE NASA AVIATION SAFETY PROGRAM (AVSP) SYSTEMWIDE ACCIDENT PREVENTION (SWAP) HUMAN PERFORMANCE MODELING (HPM) ELEMENT

David C. Foyle
NASA Ames Research Center
Allen Goodman
San Jose State University
Becky L. Hooey
Monterey Technologies, Inc.

Abstract

An overview is provided of the Human Performance Modeling (HPM) element within the NASA Aviation Safety Program (AvSP). Two separate model development tracks for performance modeling of real-world aviation environments are described: the first focuses on the advancement of cognitive modeling tools for system design, while the second centers on a prescriptive engineering model of activity tracking for error detection and analysis. A progressive implementation strategy for both tracks is discussed in which increasingly more complex, safety-relevant applications are undertaken to extend the state-of-the-art, as well as to reveal potential human-system vulnerabilities in the aviation domain. Of particular interest is the ability to predict the precursors to error and to assess potential mitigation strategies associated with the operational use of future flight deck technologies.

HPM Element Goals

This report provides a summary review of recent research activities conducted in support of the Human Performance Modeling (HPM) element within the Systemwide Accident Prevention (SWAP) Level 2 project of the NASA Aviation Safety Program (AvSP). In March 2003, a one-day conference was held at NASA Ames Research Center to present the interim results of the HPM element. Specifically, the 2003 NASA HPM conference was focused on scenarios related to approach and landing with synthetic vision systems (SVS).

The overall 5-year goal of the HPM element is to develop and advance the state of cognitive modeling while addressing real-world safety problems. To this end, the HPM element continues to develop and demonstrate cognitive models of human performance that will aid aviation product designers in developing equipment and procedures that support pilots' tasks, are easier to use, and are less susceptible to error. The modeling focus is on computational frameworks that facilitate the use of fast-time simulation for the predictive analysis of pilot behaviors in real-world aviation environments.

Rationale

More than two-thirds of all aircraft accidents are attributed to pilot error. Identifying when equipment and procedures do not fully support the operational needs of pilots is critical to reducing error and improving flight safety (Leiden, Keller & French, 2001). This becomes especially relevant in the development of new flight deck technologies, which have traditionally followed a design process more focused on component functionality and technical performance than pilot usage and operability. To help counter this bias and to better understand the potential for human error associated with the deployment of new and complex systems, advanced tools are needed for predicting pilot performance in real-world operational environments.

As noted in the literature on aviation safety, serious piloting errors and the resultant accidents are rare events (for a review, see Leiden, Keller & French (2001)). The low-probability of occurrence makes the study of serious pilot errors difficult to investigate in the field and in the laboratory. These errors characteristically result from a complex interaction between unusual circumstances, subtle "latent" flaws in system design and procedures, and limitations and biases in human performance. This can lead to the fielding of equipment, which puts flight safety at risk, particularly when operated in a manner or under circumstances that may not have been envisioned or tested.

When combined with nominal and off-nominal scenario human-in-the-loop real-time testing, human-performance modeling in non-real–time (usually, fast-time) simulations provides a complementary technique to develop systems and procedures that are tailored to the pilots' tasks, capabilities, and limitations. Because of its use in fast-time simulations, human-performance modeling is a powerful technique to uncover "latent design flaws" -- in which a system contains a design flaw that may induce pilot error only under some low-probability confluence of precursors, conditions and events

Human performance modeling using fast-time simulation offers a powerful technique to examine human interactions with existing and proposed aviation systems across an unlimited range of possible operating conditions. It provides a flexible and economical way to manipulate aspects of the task-environment, the equipment and procedures, and the human for simulation analyses. In particular, fast-time simulation analyses can suggest the nature of likely pilot errors, as well as highlight precursor conditions to error such as high levels of memory demand, mounting time pressure and workload, attentional tunneling or distraction, and deteriorating situational awareness. Fast-time simulation is the only practical way to generate the very large sample sizes that are needed to reveal low-rate-of-occurrence events. Human-in-the-loop real-time simulations are too costly to use for this purpose. Additionally, this can be done early in the design cycle, without the need to fabricate expensive prototype hardware.

HPM Models

The AvSP HPM element is organized along two model development tracks (see figure E-1). The first model development track is Predictive Human Performance Models, in which multiple predictive models of human performance simultaneously address several well-specified problems in aviation safety. The second model development track is the Prescriptive Engineering Human Performance Model, which consists of a model of error detection (specifically a prescriptive engineering model of operator performance in context). The six models comprising the AvSP HPM element are described below.

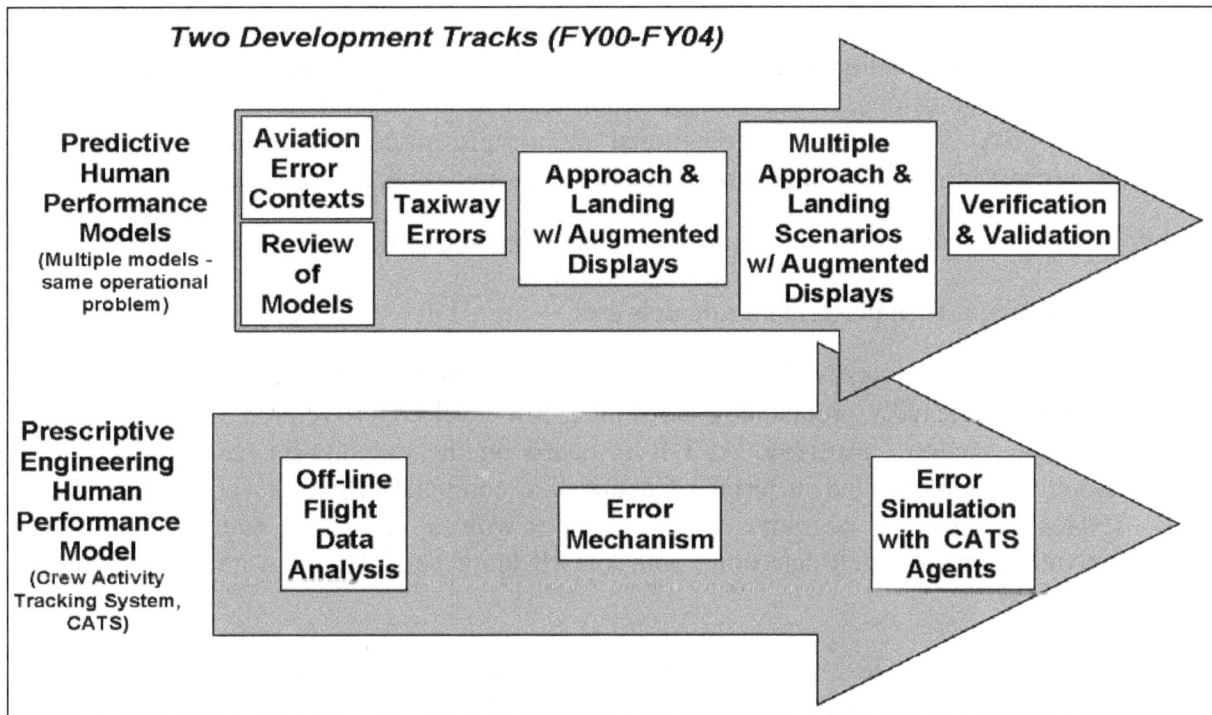

Figure E-1. Two model development tracks of the HPM element: Predictive Human Performance Models with multiple predictive models investigating a set of common problems; and, Prescriptive Engineering Human Performance Model.

Predictive Human Performance Models

From an initial review of past efforts in cognitive modeling, it was recognized that no single modeling architecture or framework had the scope to address the full range of interacting and competing factors driving human actions in dynamic, complex environments (Leiden, Laughery, Keller, French, Warwick & Wood, 2001). As a consequence, the HPM element sought to develop and extend multiple modeling efforts to further the current state of the art within a number of HPM tools. In 2001, five modeling frameworks were selected from a large group of responses to a proposal call for computational approaches for the investigation and prediction of operator behaviors associated with incidents and/or accidents in aviation. This was, in essence, a request for analytic techniques that employed cognitive modeling and simulation. The proposals were peer-reviewed with selection criteria including model theory, scope, maturity, and validation as well as the background and expertise of the respective research team.

All five of the predictive human performance modeling frameworks share common, important human characteristics. The models are:

(1) Generative -- Output results from the flow of internal model processes and is not "scripted";

(2) Have stochastic elements -- Simulation runs are not identical, even when all parameters defining the environment external to the human operator are held constant;

(3) Context sensitive -- Changes in the task-environment effect changes in simulation output.

Four of the five selected modeling frameworks were based on mature, validated, and integrative architectures which linked together embedded component processes of cognition with capabilities to construct representations of the task-environment and for simulations. The fifth modeling framework (A-SA) is a set of computational algorithms, more limited-in-scope, focused on attentional processes and the assessment of situational awareness that will be described below.

Additional characteristics of these five models are summarized in figure E-2.

The five predictive human performance models are:

ACT-R (Rice University; University of Illinois). *Atomic Components of Thought-Rational* is an experimentally grounded, open-source, low-level cognitive architecture developed at Carnegie Mellon University. ACT-R is based on the assumption that human cognition should be implemented in terms of neural-like computations on a very small time scale (50 ms–200 ms). A cognitive layer interacts with a perceptual-motor layer to create activation levels which determine both knowledge accessibility and goal-oriented conflict resolution.

Air MIDAS (San Jose State University). Air MIDAS is a version of *the Man-machine Integration Design and Analysis System* (MIDAS) developed as a joint Army-NASA program to explore computational representations of human-machine performance. Air MIDAS is driven by a set of user inputs specifying operator goals, procedures for achieving those goals, and declarative knowledge appropriate to a given simulation. These asserted knowledge structures interact with, and are moderated by, embedded models of cognition for managing resources, memory, and action.

228

A-SA (University of Illinois). *Attention-Situational Awareness* is a computational model developed at the University of Illinois. The underlying theoretical structure of the A-SA model is contained in two modules, one governing the allocation of attention to events and channels in the environment, and the second drawing an inference or understanding of the current and future state of the aircraft within that environment. Four factors are used to compute attention allocation within a dynamic environment; salience, effort, expectancy, and value. In turn, attentional allocation modulates situational awareness.

D-OMAR (BBN Technologies). The *Distributed Operator Model Architecture* was originally developed by BBN Technologies under sponsorship from the Air Force Research Laboratory. D-OMAR supports the notion of an agent whose actions are driven not only by actively seeking to achieve one or more goals, but also by reacting to the input and events of the world. It was designed to facilitate the modeling of human multi-tasking behaviors of team members interacting with complex equipment.

IMPRINT/ACT-R (Micro Analysis and Design, Inc.; Carnegie Mellon University; Army Research Laboratory). This hybrid framework integrates *Improved Performance Research Integration Tool (IMPRINT)*, a task network-based simulation tool developed by Micro Analysis and Design and *Atomic Components of Thought-Rational* (ACT-R), a low-level cognitive architecture developed at Carnegie Mellon University. This approach is meant to exploit the advantages of top-down control with the emergent aspects of bottom-up behavior for evaluating human performance in complex systems.

Model	Type	Research Team	Demonstrated Sources of Pilot Error
ACT-R	Low-level Cognitive with Statistical Environment Representation	Rice University University of Illinois	* Time pressure * Misplaced expectations * Memory retrieval problems
Air MIDAS	Integrative Multi-component Cognitive	San Jose State University	* Workload * Memory interference * Misperception
A-SA	Component Model of Attention & Situational Awareness	University of Illinois	* Misplaced attention * Lowered situation awareness
D-OMAR	Integrative Multi-component Cognitive	BBN Technologies	* Communications errors * Interruption & distraction * Misplaced expectation
IMPRINT / ACT-R	Hybrid: Task Network with Low-level Cognitive	Micro Analysis and Design, Inc. Carnegie Mellon University Army Research Laboratory	* Time pressure * Perceptual errors * Memory retrieval * Inadequate knowledge

Figure E-2. The five predictive human performance models, type of model, and demonstrated sources of pilot error.

Prescriptive Engineering Human Performance Model

CATS (San Jose State University/NASA Ames Research Center). The Crew Activity Tracking System (CATS) is a prescriptive engineering model which provides a representation of the task that the user is attempting to complete, a representation of how the task should be completed, and the capability to track and compare actual performance against prescribed performance. The model allows for error-detection and is being expanded to include mechanisms that produce observed operator errors.

HPM Element Approach and Scope

The 2003 HPM Conference and the resulting Conference Proceedings focused on the particular aviation safety-related problem of approach and landing with and without augmented displays. By plan, only the predictive human performance models addressed this problem. For this reason, the Prescriptive Engineering Human Performance Model, CATS, is not included in these proceedings. For more information, the reader is referred to articles on that topic (e.g., Callantine, 2002). The problem and approach described below refers only to the five predictive models of human performance.

The approach used in the AvSP HPM element involves applying different cognitive modeling frameworks to the analysis of a well-specified operational problem for which there is available empirical data of pilot performance in the task (see figure E-3). In 2001, the five different modeling frameworks were used to analyze a series of land-and-taxi-to-gate scenarios taken from a high-fidelity full mission simulation study that produced an extensive data-set of pilot performance. This completed 2001 effort is represented by the left-most panel of figure E-3. Overall, this approach enables the HPM Element to assess and contrast the predictive ability of a diverse range of human performance modeling frameworks while encouraging the advancement of these frameworks. For 2002-2003 (figure E-3 center panel), the five predictive modeling frameworks have been extended to the more complex problem of modeling pilot behaviors during approach and landing operations with and without the availability of a synthetic vision system (SVS)). This is in accord with the HPM Element's 2002 milestone objective requiring the development of cognitive models of an approach/ landing scenario with an augmented display. In 2003-2004 (figure E-3 right panel), these five models will focus on other specific approach and landing scenarios. A schematic representing the multiple off-nominal conditions (e.g., late runway reassignment; SVS display malfunction; and "go-arounds" because of cloud cover and runway traffic) to be investigated in the last years of the program is shown in the rightmost panel of figure E-3.

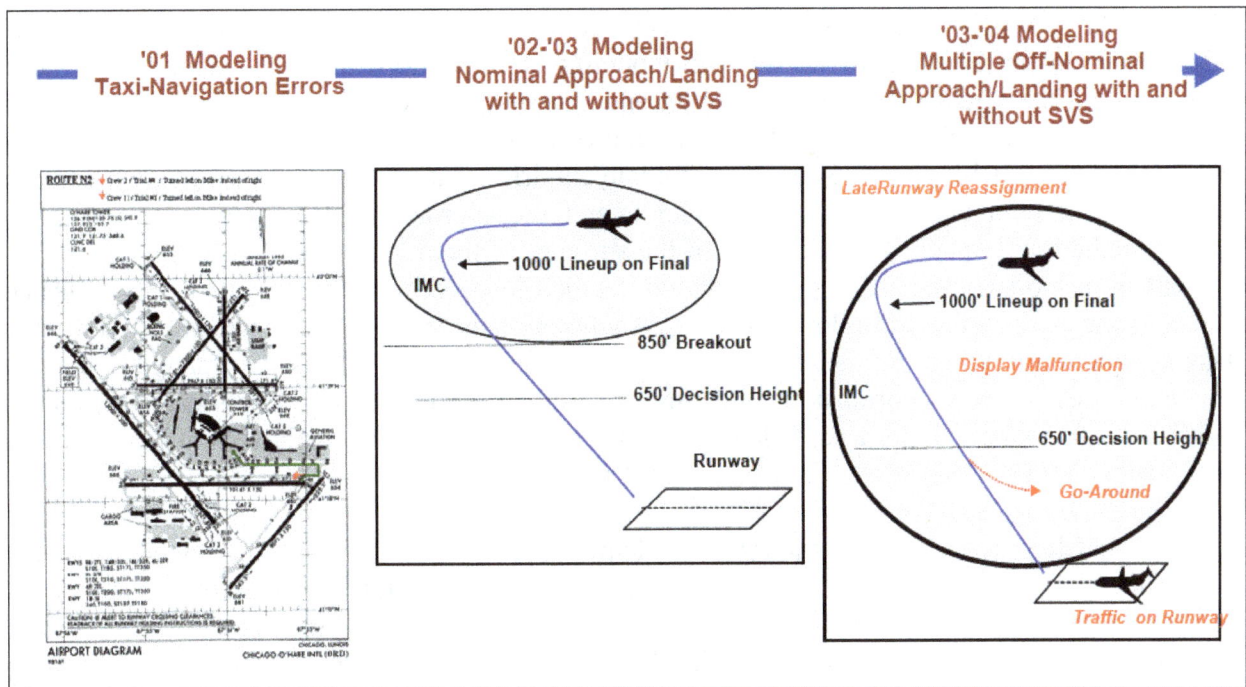

The figure contains the following labels:

'01 Modeling
Taxi-Navigation Errors

'02-'03 Modeling
Nominal Approach/Landing
with and without SVS

'03-'04 Modeling
Multiple Off-Nominal
Approach/Landing with and
without SVS

ROUTE N2

AIRPORT DIAGRAM   CHICAGO-O'HARE INTL (ORD)

IMC   1000' Lineup on Final
850' Breakout
650' Decision Height
Runway

LateRunway Reassignment
1000' Lineup on Final
Display Malfunction
IMC
650' Decision Height
Go-Around
Traffic on Runway

Figure E-3. Three phases of aviation safety-related problems addressed by the five Human Performance Modeling predictive models during 2001-2004.

Current HPM Efforts and Findings

In these conference proceedings, papers describing current accomplishments of the predictive human performance models of the HPM element are presented. The first two papers serve to set the stage for the modeling efforts which follow. The first paper in these proceedings describes a cognitive task analysis of the approach and landing phase of flight conducted by Keller, Leiden and Small. Next, is a discussion of a part-task human-in-the-loop simulation, the tested scenarios, and the data supplied to the five modeling teams by Goodman, Hooey, Foyle and Wilson. Following these two papers in these proceedings are descriptions of the modeling efforts and their results to date. Summaries of these five predictive human-performance-modeling efforts are given below.

In the first modeler's report in these proceedings, Byrne and Kirlik describe three central principles which guided their modeling approach: 1) the desire to create a dynamic, close-loop model of pilot cognition in interaction with the cockpit, aircraft, and environment; 2) the presumption that pilots are knowledgeable and adapted operators; and, 3) a focus on the allocation of visual attention as crucial to yielding important design and training-related insights. Their model, implemented in the ACT-R/PM cognitive architecture and referencing a statistical description of the environment, produced high-level predictions of gaze time that fit well with human-in-the-loop simulation data. Additionally, the model proved sensitive to the local properties of the SVS display, demonstrating that the type and format of presented flight symbology is a strong determinant of SVS usage. This suggests one line of focused investigation in which model predictions are used to assess a range of small variations to the symbology set of the SVS display in order to optimize pilot performance.

231

Next, in the report by Corker, Gore, Guneratne, Jadhav and Verma, the authors document their efforts to augment the standard Air MIDAS modeling architecture with an advanced vision model incorporating the affects of contrast legibility and visual search/reading time to better account for performance using a SVS display. To gain additional accuracy, the visual sampling model was calibrated and verified with an extensive, alternate empirical data set. The revised model generated predictions of pilot visual scanning behavior over three approach and landing scenarios. These model predictions explained 31% to 77% of the variance of the human-in-the-loop simulation data. Output from model simulations also permitted detailed inspection of the executed task sequences for both the pilot flying and pilot not flying. Analyses of these sequences indicate differences in task completion ordering, timing, and success between scenario conditions. This suggests possible vulnerabilities in crew coordination and timing resulting from specific situational demands. In another finding, the authors acknowledge that a better understanding of how flight crews select from redundant information sources is needed to improve fidelity.

Deutsch and Pew describe their efforts to implement a dedicated model of approach and landing within the D-OMAR simulation framework. The cognitive architecture evolved for this application focuses on multi-task behavior, the role of vision, and working memory. In simulation, the resultant models of the Captain, First Officer, and Air Traffic Controller working in concert demonstrated a commendable robustness by executing successful landings across five different scenarios circumstances. The model's prediction that the availability of the SVS display would reduce time devoted to HSI display is matched in the human data. This finding supports the implication that information redundancy on the SVS display may reduce workload. The authors also note that additional scenario complexity can lead to better models by teasing out flaws. This was the case when certain D-OMAR model shortcomings only became apparent when a distracter aircraft was added to the scenario.

The paper by Lebiere, Archer, Schunk, Biefeld, Kelly and Allender details the unique integration of the low-level cognitive architecture, ACT-R, with the task network simulation tool, IMPRINT, to provide a viable approach for modeling complex domains. Functionality of the resulting model of approach and landing operations permitted sensitivity analyses of mission success rates to global parameters regarding latency of procedural, visual, motor, and auditory actions, as well as stochastic manipulations of decision-making times. These analyses provided important inferences regarding effective design objectives for both information display and procedures. Among other findings, the model found that pilot performance is very sensitive to the speed of visual shifts between widely separated information sources. Similarly, pilot performance proved highly sensitive to the overhead of communications with increases in the number and/or duration of communications acts rapidly deteriorating performance. Noteworthy in these modeling analyses was the apparent "performance tipping point" in which near-perfect mission success rates would suddenly plummet with only the slightest increase in parameter latency.

Wickens, McCarley and Thomas describe the modification of their algorithmic SEEV Model of attentional allocation in dynamic environments to the prediction of visual scanning during approach and landing operations. The refined algorithm for this application is based on the parameters of effort, expectancy, and value. This revised model, the Attention-Situation Awareness (A-SA) Model, accounted for roughly 30% - 80% of the variance in the scanning behavior seen in the human data. Surprisingly, the effort parameter added no predictive power to the model beyond expectancy and value in this application. The authors do make a qualitative distinction between "good" and "poor" SA pilots based on latency to execute go-around maneuvers. They find that deviation from model predicted dwell times to the outside world clearly discriminated between these two categories of pilots. This is seen as supporting the model's ability to infer pilot SA. The authors also note that the large observed variance between individual pilot scanning behaviors (and resulting impact on model fit) may be attributable to one of two causes: 1) different pilot strategies of accessing information from redundant displays; or, 2) less-than optimal scanning behavior from some pilots. Again, it is asserted that the model can make that discrimination.

As will be seen in the following papers in these conference proceedings, the HPM predictive modeling efforts resulted in both design solutions and procedural recommendations to enhance the safety of SVS systems. The models identified potential problems that merit further investigation through human-in-the-loop simulations. Significant advancements to the state of human performance modeling were achieved by broadening the scope of the five models to include the aviation domain, and through the augmentation and expansion of specific modeling capabilities.

References

All these publications are available for download at the "publications link" at: http://humanfactors.arc.nasa.gov/ihi/hcsl/

Byrne, M.D. and Kirlik, A. (2003). Integrated modeling of cognition and the information environment: A closed-loop, ACT-R approach to modeling approach and landing with and without synthetic vision system (SVS) technology. In Conference Proceedings of the 2003 NASA Aviation Safety Program Conference on Human Performance Modeling of Approach and Landing with Augmented Displays, (David C. Foyle, Allen Goodman & Becky L. Hooey, Eds.). NASA Conference Proceedings NASA/CP-2003-212267.

Callantine, T. (2002). Activity tracking for pilot error detection from flight data. NASA Contractor Report 2002-211406. Moffett Field, CA: NASA Ames Research Center.

Corker, K.M., Gore, B.F., Guneratne, E., Jadhav, A. and Verma, S. (2003). Human Performance Modeling Predictions in Reduced Visibility Operation with and without the use of Synthetic Vision System Operations. In Conference Proceedings of the 2003 NASA Aviation Safety Program Conference on Human Performance Modeling of Approach and Landing with Augmented Displays, (David C. Foyle, Allen Goodman & Becky L. Hooey, Eds.). NASA Conference Proceedings NASA/CP-2003-212267.

Deutsch, S. & Pew, R. (2003). Modeling the NASA baseline and SVS-equipped approach and landing scenarios in D-OMAR. In Conference Proceedings of the 2003 NASA Aviation Safety Program Conference on Human Performance Modeling of Approach and Landing with Augmented Displays, (David C. Foyle, Allen Goodman & Becky L. Hooey, Eds.). NASA Conference Proceedings NASA/CP-2003-212267.

Goodman, A., Hooey, B.L., Foyle, D.C. and Wilson, J.R. (2003). Characterizing visual performance during approach and landing with and without a synthetic vision display: A part-task study. In Conference Proceedings of the 2003 NASA Aviation Safety Program Conference on Human Performance Modeling of Approach and Landing with Augmented Displays, (David C. Foyle, Allen Goodman & Becky L. Hooey, Eds.). NASA Conference Proceedings NASA/CP-2003-212267.

Keller, J., Leiden, K. and Small, R. (2003). Cognitive Task Analysis of Commercial Jet Aircraft Pilots during Instrument Approaches for Baseline and Synthetic Vision Displays. In Conference Proceedings of the 2003 NASA Aviation Safety Program Conference on Human Performance Modeling of Approach and Landing with Augmented Displays, (David C. Foyle, Allen Goodman & Becky L. Hooey, Eds.). NASA Conference Proceedings NASA/CP-2003-212267.

Lebiere, C., Archer, R., Schunk, D., Biefeld, E., Kelly, T. and Allender, L. (2003). Using an integrated task network model with a cognitive architecture to assess the impact of technological aids on pilot performance. In Conference Proceedings of the 2003 NASA Aviation Safety Program Conference on Human Performance Modeling of Approach and Landing with Augmented Displays, (David C. Foyle, Allen Goodman & Becky L. Hooey, Eds.). NASA Conference Proceedings NASA/CP-2003-212267.

Leiden, K., Keller, J. W., and French, J.W. (2001). Context of Human Error in Commercial Aviation. Contractor Report.

Leiden, K., Laughery, K.R., Keller, J. W., French, J.W., Warwick, W. and Wood, S.D. (2001). A Review of Human Performance Models for the Prediction of Human Error. Contractor Report.

Wickens, C., McCarley, J. and Thomas, L. (2003). Attention-Situation Awareness (A-SA) Model. In Conference Proceedings of the 2003 NASA Aviation Safety Program Conference on Human Performance Modeling of Approach and Landing with Augmented Displays, (David C. Foyle, Allen Goodman & Becky L. Hooey, Eds.). NASA Conference Proceedings NASA/CP-2003-212267.

HUMAN PERFORMANCE MODELS OF PILOT BEHAVIOR

David C. Foyle, NASA Ames Research Center; Moffett Field, CA
Becky L. Hooey, San Jose State University/NASA ARC; Moffett Field, CA
Michael D. Byrne, Rice University; Houston, TX
Kevin M. Corker, San Jose State University; San Jose, CA
Stephen Deutsch, BBN Technologies; Cambridge, MA
Christian Lebiere, Micro Analysis and Design; Boulder, CO
Ken Leiden, Micro Analysis and Design; Boulder, CO
Christopher D. Wickens, University of Illinois; Savoy, IL

Five modeling teams from industry and academia were chosen by the NASA Aviation Safety and Security Program to develop human performance models (HPM) of pilots performing taxi operations and runway instrument approaches with and without advanced displays. One representative from each team will serve as a panelist to discuss their team's model architecture, augmentations and advancements to HPMs, and aviation-safety related lessons learned. Panelists will discuss how modeling results are influenced by a model's architecture and structure, the role of the external environment, specific modeling advances and future directions and challenges for human performance modeling in aviation.

INTRODUCTION

More than two-thirds of all aircraft accidents are attributed to pilot error. Identifying when equipment and procedures do not fully support the operational needs of pilots is critical to reducing error and improving flight safety (Leiden, Keller, & French, 2001). This becomes especially relevant in the development of new flight deck technologies that have traditionally followed a design process more focused on component functionality and technical performance than pilot usage and operability. To help counter this bias and to better understand the potential for human error associated with the deployment of new and complex systems, advanced tools are needed for predicting pilot performance in real-world operational environments. Serious piloting errors and accidents are rare events and the low-probability of occurrence makes the study of pilot error difficult to investigate in the field and in the laboratory. These errors characteristically result from a complex interaction between unusual circumstances, subtle "latent" flaws in system design and procedures, and limitations and biases in human performance. This can lead to the fielding of equipment that puts flight safety at risk, particularly when operated in a manner or under circumstances that may not have been envisioned or tested.

Human performance modeling, when combined with nominal and off-nominal scenario human-in-the-loop testing, provides a complementary technique to develop systems and procedures that are tailored to the pilot's tasks, capabilities, and limitations (Leiden, Laughery, Keller, French, Warwick, & Wood, 2001). Because of its fast-time nature, human performance modeling is a powerful technique to uncover "latent design flaws" -- in which a system contains a design flaw that may induce pilot error only under some low-probability confluence of precursors, conditions and events. Human performance modeling also offers a powerful technique to examine human interactions with existing and proposed aviation systems across an unlimited range of possible operating conditions. It provides a flexible and economical way to manipulate aspects of the task-environment, the equipment and procedures, and the human for simulation analyses. In particular, modeling and simulation analyses can suggest the nature of likely pilot errors, as well as highlight precursor conditions to error such as high levels of memory demand, mounting time pressure and workload, attentional tunneling or distraction, and deteriorating situation awareness. Fast-time simulation permits the generation of very large sample sizes from which low-rate-of-occurrence events are more likely to be revealed. Additionally, this can be done early in the design cycle, without the need to fabricate expensive prototype hardware.

Five modeling teams from industry and academia were chosen by the NASA Aviation Safety and Security Program to develop human performance models (HPMs) that address two problems in the aviation domain that have significant implications for aviation safety, and that are representative of general classes of problems faced by the aviation industry today. First, modeling teams addressed a current-day aviation problem within the realm of surface operations safety, by identifying causal factors of navigation errors and potential error mitigations (procedural, technical or operational). The notion of understanding causal factors of human error, and the importance of being able to predict when human operators might be vulnerable to error, and predicting which potential mitigating strategies might be successful is pervasive throughout every phase of flight. Second, the modeling teams modeled pilot performance during the approach and landing phases of flight for both a baseline configuration representing today's glass cockpit and a configuration that also included a Synthetic Vision System (SVS). The research issues inherent in this problem are common to the design, development and integration of any advanced cockpit display technology.

Human-In-The-Loop Studies

Effective HPMs require extensive understanding of the task and the domain environment in order to produce valid and meaningful results. HPMs are most informative when supported by empirical data derived from laboratory studies, human-in-the-loop (HITL) simulations, and field studies. To enable model development, information was provided to the modeling teams including task analyses and objective data and subjective ratings from two HITL simulations that were conducted at NASA Ames Research Center. The HITL data were used in two different ways in this project. In some cases the data were used by the modeling teams to populate and develop their models, and in other cases, the data were used to validate the model output.

Airport Surface Operations. A high-fidelity surface operations HITL simulation was conducted to understand the factors that contribute to taxiway navigation errors and potential mitigating solutions (Hooey, Foyle, & Andre, 2001). The simulation, conducted in NASA Ames Research Center's high-fidelity, glass cockpit Advanced Concepts Flight Simulator (ACFS), compared taxi performance under current-day baseline operations with a prototypical cockpit display system called the Taxiway Navigation and Situation Awareness (T-NASA) system which is comprised of an electronic moving map (EMM), a head-up display (HUD), and auditory alerts and warnings. The simulation trials required pilots to land and taxi to the gate following an ATC-issued taxi clearance. All trials were conducted in low visibility at a high-fidelity rendering of Chicago O'Hare airport. The study included common taxi scenarios including hold short instructions and route amendments as well as off-nominal taxi events that represented failures or errors in the system. In current-day operation (baseline) trials, pilots made navigation errors on approximately 20% of the trials, while these errors were eliminated with T-NASA. NASA provided the HPM teams with data including taxi speed, navigation errors, intra-cockpit communications, pilot-ATC communications, workload, and situation awareness to enable the teams to develop models of the pilot tasks and taxi scenarios necessary to predict taxiway navigation errors.

Synthetic Vision Systems (SVS). A part-task HITL simulation study was conducted to investigate the effect of new synthetic vision systems (SVS) on pilot performance, visual attention, and crew roles and procedures during low-visibility instrument approaches (Goodman, Hooey, Foyle, & Wilson, 2003). The HPM teams were provided with a cognitive task analysis of the approach phase of flight and human performance data including eye movements, communications, and control panel responses from the NASA part-task simulation of instrument approaches with and without the SVS. Events such as no visibility at decision altitude (forcing a missed approach/go-around), a misalignment between instruments and the out-the-window view, and a late runway reassignment, were included in the scenarios to yield a robust set of human performance data.

OVERVIEW OF HPM APPROACHES

From an initial review of past efforts in cognitive modeling, it was recognized that no single modeling architecture or framework had the scope to address the full range of interacting and competing factors driving human actions in dynamic, complex environments (Leiden, Laughery, Keller, French, Warwick, & Wood, 2001). As a consequence, the decision was made to develop and expand multiple modeling efforts to extend the current state of the art within a number of HPM tools. Five modeling frameworks were selected based on a peer-reviewed process with selection criteria including model theory, scope, maturity, and validation as well as the background and expertise of the respective research team.

In Phase 1, each team modeled some aspect of the airport surface operations domain problem with an emphasis on replicating or predicting pilot error. In Phase 2, each team built on their existing model capabilities to address issues relating to SVS design and integration. The approach and specific research questions were left to the discretion of the modeling teams yielding diverse models with a demonstrated capability of answering a variety of important aviation domain questions. In the sections that follow, each modeling team will briefly describe their model architecture, augmentations to the models, and significant findings for the aviation community.

Attention-Situation Awareness (A-SA)
C.D. Wickens (Panelist) and J. McCarley

The Attention-Situation Awareness (A-SA) model has two components. The first (attention, A) describes the way in which three factors of the visual environment – the *salience*, *expectancy* and **value** of events – drive attention allocation, as this allocation is inhibited by a fourth factor, the **effort** required to scan between information sources within that environment. All four factors can be quantified, to make ordinal predictions of the degree to which areas of interest will be fixated as assessed by visual scanning (Wickens, Goh, Helleberg, Horrey, & Talleur, 2003). Such visual input supports the second component: situation awareness (SA), or understanding of the current and future state of the aircraft. SA is based on a belief updating module, such that SA is updated any time a new piece of information is encountered. The pieces are weighted based on the value of the information to the task such that the current value of SA is increased or decreased. SA decays when no new pieces of information are encountered. The model has both an analytic and a real-time dynamic version. The model augments classical optimal scanning models to include the design-layout related factors of salience and effort, to accommodate auditory channels, to consider circumstances related to task priority and those in which there is a one-to-many mapping between fixation areas and tasks.

The surface operations simulation data were used to exercise the model, which predicted both intersections where errors were likely because of degraded SA, and the benefits of the T-NASA display in mitigating those errors. The SVS simulation data from NASA Ames, as well as SVS data collected at the University of Illinois were used to validate the attention allocation component of the A-SA model as inferred from visual scanning measures. The latter data, involving a low altitude flight through a terrain-challenged environment, revealed that the model accounted for an average of 85% of the variance in scanning behavior across the five areas of interest in the cockpit, for eight pilots (Wickens, McCarley, Alexander, Thomas, Ambinder, & Zheng, 2005). Model fit was not improved by considering the inhibiting role of effort, above and beyond that of expectancy and value, suggesting that pilots were quite optimal. Furthermore, greatest deviations from model predictions were shown by those who suffered greater decrements in flight path tracking and hazard detection performance, thereby suggesting that variations in optimality of attention allocation translated to variations in performance.

ACT-R Version 5.0
M.D. Byrne (Panelist), A. Kirlik and M.D. Fleetwood

Our models are detailed closed-loop, pilot-displays-aircraft system models. That is, for both tasks, the ACT-R model of the pilot was connected to an executable model of the aircraft and relevant visual environment and both models run in real time. ACT-R provides a platform to support modeling at a fine temporal detail; the output of an ACT-R model is a time-stamped series of fairly primitive behaviors, down to the individual saccade. ACT-R models are also knowledge-intensive and require extensive task analysis and consultation with subject matter experts to supply the model with the knowledge necessary to actually execute the task. Taxiing and autopilot-guided instrument approach are, in fact, substantially different tasks, and the two models overlap in the general approach taken and the kind of outputs provided but less so elsewhere.

The ACT-R cognitive architecture was originally designed to model the results of laboratory psychology experiments, which typically consist of simple tasks requiring little knowledge and performed in restricted environments. While ACT-R has an extensive track record of being successful in such domains, how ACT-R would "scale up" to aviation-relevant tasks was not entirely clear for both technical (e.g., software integration) and theoretical (e.g., could we represent environmentally-based constraints appropriately in this formalism) reasons. We thus conclude that ACT-R is indeed a viable option for serious HPM research (Byrne & Kirlik, 2005).

Cognitive modeling efforts have traditionally focused on "in the head" cognition, but one of the important lessons learned in these modeling projects is the importance of a high-fidelity

238

representation of the environment; both of our models are highly sensitive to the structure of the environment. The taxiing model, for example, is very sensitive to the fact that the physical layout of Chicago's O'Hare airport generates taxi routes that are systematically different from typical taxi routes at other airports. This, in turn, contributes to errors in taxiing. The SVS modeling work revealed how a new display with information that is redundant with other cockpit instrumentation has a substantial impact on pilot scanning behavior, even in phases of flight where such changes were not the system designers' intent.

Air-MIDAS
K.M. Corker (Panelist)

Human performance models were developed and applied to surface and flight operations in order to predict errors, and evaluate the impact of new information technologies and new procedures on flight crew performance. The human performance model component of these studies was developed using Air-MIDAS (Man Machine Integrated Design and Analysis System). The model was used to represent the flight deck crew responding to information systems and ATC.

The first study on surface operations represented flight crew responses to ground control commands for post landing roll out and taxi. The model used working memory limits, interference processes, and heuristics to successfully predict errors observed in the HITL simulation. The second round of studies concentrated on the use of SVS technologies to allow pilots to continue approach under visual minima. The perceptual model (visual sampling of information) was both statistically verified and validated against calibration data and HITL simulation data. With the validated model in place, we analyzed approach and go-around performance under standard and SVS technologies, and under conditions of approach and go-around decisions based on flight crew decisions and based on air traffic controller performance. The conclusions of this examination of SVS were that:

- SVS would not adversely affect the flight safety in approach, landing, and go-around phases regardless of the decision altitude and go-around triggers including the pilot-flying's intention at decision altitude and ATC's command, while it would allow approach and landing in conditions that would otherwise be unattainable;

- Small delays of action initiation in flight control were observed in the approach phase with SVS operations. This occurred because the chances of fixation on each display was decreased by adding the SVS to the conventional display configuration; and,

- No human performance degradation and no delay of task initiation were observed in the landing and go-around phases, although there were time shifts in the approach phase.

D-OMAR
S. Deutsch (Panelist) and R.W. Pew

The Distributed Operator Model Architecture (D-OMAR) provides an event-based simulator and a suite of representation language -- a frame language, a procedure language and a rule language -- that we have used to instantiate a cognitive architecture that is the basis for our aircrew and air traffic controller models. Using the procedural language, we have constructed a set of basic person procedures -- perceptual, cognitive, and motor processes that are the building blocks for the expertise exhibited by aircrews and controllers.

As a general purpose simulator, D-OMAR has enabled us to also construct models for the essential elements of the commercial airspace: aircraft and their flight decks, ATC workplaces, airports with their runways, taxiways, and concourses, and the airways and navigation aides.

Our approach to examining aircrew error has been to build models that exhibit the robust behaviors of aircrews and then probe the models for the seams along which error can intrude (Deutsch & Pew, 2004). The infrequent errors in which aircrews mistakenly turned away for their designated concourse attracted our attention. Our analysis suggested that this was a point at which habit might intrude and lead to such an error. And indeed, one of the modeled competing sources for the action to take at an intersection was grounded in habit. Subsequent review of the human subject data validated the models prediction of the source of the error. In a similar manner, we modeled a case in which an aircrew, contrary to the ground controller's directive, turned toward their destination gate. To open the window for these errors, we constructed situations that prevented the first officer from prompting the captain on the correct turn to take at the intersection based on notes taken when the ground controller provided the taxi routing.

One aspect of our modeling of the use of the SVS, focused on our observation that it might be utilized as a second primary flight display (PFD) leading to an inefficient scan pattern. A review of the human subject data suggested that this might well be the case. To counter this potential problem, we designed a single attitude instrument combining PFD and SVS functionality. Model trials then predicted that the more efficient baseline scan pattern would be restored when using the combined PFD-SVS.

In each problem area addressed, the model and its underlying theory led to important new insights into the sources of aircrew behaviors.

IMPRINT / ACT-R
C. Lebiere (Panelist) and R. Archer

The approach that was used by our team to perform the approach and landing modeling task was an integration of the Improved Performance Research Integration Tool (IMPRINT) and the Adaptive Control of Thought – Rational (ACT-R) cognitive architecture. We found that a natural integration of the two tools had IMPRINT assuming the role of implementing the simulation and ACT-R assuming the role of the cognitive agents, i.e., the pilots. The IMPRINT model represents the state of the aircraft, its controls, and the environment. The ACT-R model represents the cognitive state of the pilots and their decision-making process. The two models communicate through a general, scalable, reusable interface called Link IMPRINT/ACT-R (LIA) that reduces the burden of integration from weeks or months of effort down to days. This approach of combining the strengths of task network and cognitive modeling for different portions of the same modeling scenario makes an important contribution to the capabilities of Human Performance Modeling: It allows complex simulations to be assembled in a modular fashion that makes explicit all communications requirements and enables high-fidelity cognitive modeling to be deployed where needed most in a tractable, affordable manner.

The cognitive models were built to emphasize the mechanisms and constraints of the cognitive architecture. Errors result from inherent limitations in the architecture's cognitive and perceptual abilities. Conversely, the model also exploits the architecture's powerful learning mechanisms to adapt to the introduction of new technology and other changes in its environment. We varied a number of parameters representing both variations in individual cognitive, perceptual and motor abilities as well as changes in the composition of the environment (e.g., by the addition of an SVS system) that affect the cognitive, perceptual and motor operations of the human operator (and cognitive model). This sensitivity analysis provides an indication of where the primary benefits of technological aids are likely to reside, as well as the most likely sources of error (Best, Lebiere, Schunk, Johnson, & Archer, 2004).

ADVANCES AND LESSONS LEARNED

K. Leiden (Panelist), D. C. Foyle and B. L. Hooey

Advances in Human Performance Modeling

Each of the HPMs discussed in this paper has extended their capabilities significantly to support the surface operations and SVS display modeling efforts. Considerable effort has been expended for the development of the external models to represent the aircraft flight dynamics, flight deck displays, and the communication link between external environment and HPM tool. (Some teams expended between 50-70% of their effort to represent these functions.) The ACT-R 5.0 and Air-MIDAS teams connected their models to higher-fidelity flight simulators and thus are poised to tackle future problems in which the closed-loop behavior between the pilot's action and the aircraft's response (or vice versa) is a key factor.

The usefulness of HPMs to the design and evaluation of new technology is determined to a significant extent by the core capabilities – visual attention allocation, workload, crew interactions, procedures, situation awareness, and error prediction. For example, the A-SA and ACT-R models focused specifically on what drives visual attention from a bottom-up (e.g., effort to move the eyes) as well as top-down perspective. Hence, if visual attention allocation needs to be understood for a particular technology, the ACT-R and A-SA modeling frameworks would more easily facilitate the analysis. In contrast, multiple operator models (e.g., pilot-flying and pilot-not-flying) are more easily accommodated by Air-MIDAS and D-OMAR. Thus, if flight crew or pilot/ATC interactions are expected to be significant drivers to a future modeling effort then the Air-MIDAS and D-OMAR frameworks would be more straightforward to apply. Of course, the respective HPMs and capabilities are dynamic. Each successive modeling effort in a complex environment such as aviation most likely adds to a framework's repertoire of capabilities.

Lessons Learned for the Aerospace Community

The modeling efforts revealed that HPMs, even those cognitive architectures that have traditionally been used in the context of psychological laboratory experiments, can indeed be useful tools for complex, context-dependent, domains such as aviation. Specifically, the tools can be used to address the design and evaluations of aviation displays, procedures, and operations.

Error Prediction and Mitigation. Across all of the modeling efforts, the tools were able to predict errors, or error vulnerabilities, that occurred because of situation awareness degradation, memory degradation and interference, airport layout, pilot expectation and habit, distraction, and workload. Further, it was shown that HPMs can be used to identify and evaluate the effectiveness of various technologies in the mitigation of such errors.

Display Design and Information Allocation. The models proved useful as tools to estimate the impact of new display technology on pilot scanning behavior. As such, HPMs can be used to inform display design and the allocation of information so as to optimize efficient scan patterns and increase the uptake of relevant information in a timely manner. As a salient example, the models showed that when redundant information is overlaid on an SVS display (e.g., altitude, heading, speed on both traditional and SVS displays), pilots altered their baseline scan pattern to attend to the more easily acquired redundant information (based on saccade latency). Thus, the time spent attending to traditional displays under the SVS configuration was reduced significantly compared to the baseline configuration.

DISCUSSION

Under NASA's Human Performance Modeling Project, five models of human performance have been applied to a specific set of aviation problems. As a result, we are in a unique position to characterize some of the different ways in which specific models interact with the problem, and to note similarities and differences in how model representations affect the characterization of the modeling problem. The present panel discussion addresses this topic, and, in the near future, this will be documented as part of the HPM effort.

Some of the specific model characterizations include: 1) Model architecture and structures - To what extent do the specific architectures and structures in the various models impact: the user's choice of a modeling tool; the ability to describe/predict the data; and, the validation of results?; 2) Role of the external environment - How is the external environment captured in the model; how does the model interact with the external environment?; 3) Model predictive ability - To what extent do the specific modeling tools accurately model/predict behavior, produce emergent behavior, are predictive vs. simulation in nature, and, allow for the extrapolation to other non-tested display/procedural conditions?; and, 4) Usefulness/implications of the modeling results - What specific implications do the models make regarding procedures, communication, ATC - pilot interactions, intra-cockpit interactions, and, SVS display issues? Additionally, the future directions and challenges for human performance modeling in aviation are discussed and addressed.

ACKNOWLEDGMENTS

The work presented in this paper was supported by NASA's Aviation Safety and Security Program, Systemwide Accident Prevention, Human Performance Modeling element (21-078-20-10). For more information and for publication downloads see:

http://human-factors.arc.nasa.gov/ihi/hcsl/publications.html

REFERENCES

Best, B., Lebiere, C., Schunk, D., Johnson, I., and Archer, R. (2004). Validating a Cognitive Model of Approach Based on the ACT-R Architecture. (Technical Report). Micro Analysis and Design, Inc., Boulder, CO.

Byrne, M. D., and Kirlik, A. (2005). Using computational cognitive modeling to diagnose possible sources of aviation error. International Journal of Aviation Psychology, vol. 15, pp. 135-155.

Deutsch, S., and Pew, R. (2004). Examining new flight deck technology using human performance modeling. Proceedings of the Human Factors and Ergonomics Society 48th Annual Meeting. Santa Monica, CA, pp. 108-112.

Goodman, A., Hooey, B. L., Foyle, D. C., and Wilson, J. R. (2003). Characterizing Visual Performance During Approach and Landing With and Without a Synthetic Vision Display: A Part-Task Study. In D. C. Foyle, A. Goodman, and B. L. Hooey (Eds.), Proceedings of the 2003 NASA Aviation Safety Program Conference on Human Performance Modeling of Approach and Landing with Augmented Displays, NASA Ames Research Center.

Hooey, B. L., Foyle, D. C., and Andre, A. D. (2000). Integration of cockpit displays for surface operations: The final stage of a human-centered design approach. SAE Transactions, Journal of Aerospace, vol. 109, pp. 1053-1065.

Leien, K., Keller, J. W., and French, J. W. (2001). Context of Human Error in Commercial Aviation. (Technical Report), Micro Analysis and Design, Inc., Boulder, CO.

Leiden, K., Laughery, K. R., Keller, J. W., French, J. W., Warwick, W., and Wood, S. D. (2001). A Review of Human Performance Models for the Prediction of Human Error. (Technical Report), Micro Analysis and Design, Inc., Boulder, CO.

Wickens, C. D., Goh, J., Helleberg, J., Horrey, W., and Talleur, D. A. (2003). Attentional models of multi-task pilot performance using advanced display technology. Human Factors, vol. 45, pp. 360-380.

Wickens, C. D., McCarley, J. S., Alexander, A. L., Thomas, L. C., Ambinder, M., and Zheng, S. (2005). Attention-Situation Awareness (A-SA) Model of Pilot Error. (Technical Report AHFD-04-15, NASA-04-5), University of Illinois at Urbana-Champaign, Institute of Aviation Human Factors Division, Savoy, Illinois.

AVSP SWAP HUMAN PERFORMANCE MODELING (HPM) PUBLICATIONS

POC: David C. Foyle, Ph.D.
David.C.Foyle@nasa.gov
650-604-3053

Many of these are available for download at the "publications link" at:
http://humanfactors.arc.nasa.gov/ihi/hcsl/

As of September, 2005

Journals, Books, Conference Proceedings

Byrne, M.: Cognitive Architectures in HCI: Present Work and Future Directions. In Proceedings of the 11th International Conference on Human Computer Interaction, Las Vegas, NV, 2005.

Byrne, M. D.; and Kirlik, A.: Using computational cognitive modeling to diagnose possible sources of aviation error. International Journal of Aviation Psychology, vol. 15, no. 2, 2005, pp. 135–155.

Byrne, M. D.; Kirlik, A.; Fleetwood, M. D.; Huss, D. G.; Kosorukoff, A.; Lin, R.; and Fick, C. S.: A closed-loop, ACT-R approach to modeling approach and landing with and without synthetic vision systems (SVS) technology. Proceedings of the Human Factors and Ergonomics Society 48th Annual Meeting, Santa Monica, CA, 2004, pp. 2111–2115.

Byrne, M. D.; and Kirlik, A.: Integrated modeling of cognition and the information environment: A closed-loop, ACT-R approach to modeling approach and landing with and without synthetic vision system (SVS) technology. In Conference Proceedings of the 2003 NASA Aviation Safety Program Conference on Human Performance Modeling of Approach and Landing with Augmented Displays, David C. Foyle, Allen Goodman, and Becky L. Hooey (Eds.), NASA/CP-2003-212267, 2003.

Callantine, T. J.: Performance Evaluation of a Computational Model of En Route Air Traffic Control. Proceedings of the 13th International Symposium on Aviation Psychology, Oklahoma City, April 18–21, 2005.

Callantine, T.: Detecting and simulating pilot errors for safety enhancement. SAE Technical Paper 2003-01-2998, SAE International, Warrendale, PA, 2003.

Callantine, T.: A representation of air traffic control clearance constraints for intelligent agents. In Proceedings of the 2002 IEEE International Conference on Systems, Man, and Cybernetics, A. El Kamel, K. Mellouli, and P. Bourne (Eds.), no. WA1C2, (CD-ROM), 2002.

Callantine, T.: Activity tracking for pilot error detection from flight data. Proceedings of the 21st European Annual Conference on Human Decision Making and Control, Glasgow, 2002, pp. 16–26.

Callantine, T.: Agents for analysis and design of complex systems. Proceedings of the 2001 IEEE International Conference on Systems, Man, and Cybernetics, 2001, pp. 567–573.

Callantine, T.: Analysis of flight operational quality assurance data using model-based activity tracking. SAE Technical Paper 2001-01-2640, SAE International, Warrendale, PA, 2001.

Callantine, T.: The crew activity tracking system: Leveraging flight data for aiding, training, and analysis. Proceedings of the 20th Digital Avionics Systems Conference, 5.C.3-1–5.C.3-12 (CD-ROM), 2001.

Corker, K. M.; Gore, B. F.; Guneratne, E.; Jadhav, A.; and Verma, S.: Human Performance Modeling Predictions in Reduced Visibility Operation with and without the use of Synthetic Vision System Operations. In Conference Proceedings of the 2003 NASA Aviation Safety Program Conference on Human Performance Modeling of Approach and Landing with Augmented Displays, David C. Foyle, Allen Goodman, and Becky L. Hooey, (Eds.), NASA/CP-2003-212267, 2003.

Deutsch, S.: Reconceptualizing expertise: Explaining an expert's error. Proceedings of the 7th Naturalistic Decision Making Conference, Amsterdam, 2005.

Deutsch, S.; and Pew, R.: Examining new flight deck technology using human performance modeling. Proceedings of the Human Factors and Ergonomics Society 48th Annual Meeting, Santa Monica, CA, 2004, pp. 108–112.

Deutsch, S.; and Pew, R.: Modeling the NASA baseline and SVS-equipped approach and landing scenarios in D-OMAR. In Conference Proceedings of the 2003 NASA Aviation Safety Program Conference on Human Performance Modeling of Approach and Landing with Augmented Displays, David C. Foyle, Allen Goodman, and Becky L. Hooey (Eds.), NASA/CP-2003-212267, 2003.

Deutsch, S.; and Pew, R.: Modeling human error in a real-world teamwork environment. Proceedings of the 24th Annual Meeting of the Cognitive Science Society, Fairfax, VA, 2002, pp. 274–279.

Foyle, D. C.; and Hooey, B. L. (Eds.): Human Performance Modeling in Aviation. Lawrence Erlbaum Associates, Inc., 2005.

Foyle, D. C.; Hooey, B. L.; Byrne, M. D.; Corker, K. M.; Deutsch, S.; Lebiere, C.; Leiden, K.; and Wickens, C. D.: Human performance models of pilot behavior. Proceedings of the Human Factors and Ergonomics Society 49th Annual Meeting, Santa Monica, CA, 2005.

Foyle, D. C.; Goodman, A.; and Hooey, B. L.: Conference Proceedings of the 2003 NASA Aviation Safety Program Conference on Human Performance Modeling of Approach and Landing with Augmented Displays, David C. Foyle, Allen Goodman, and Becky L. Hooey (Eds.), NASA/CP-2003-212267, 2003.

Foyle, D. C.; Goodman, A.; and Hooey, B. L.: An overview of the NASA Aviation Safety Program (AvSP) Systemwide Accident Prevention (SWAP) Human Performance Modeling (HPM) element. In Conference Proceedings of the 2003 NASA Aviation Safety Program Conference on Human Performance Modeling of Approach and Landing with Augmented Displays, David C. Foyle, Allen Goodman, and Becky L. Hooey (Eds.), NASA/CP-2003-212267, 2003.

Goodman, A.; Hooey, B. L.; Foyle, D. C.; and Wilson, J. R.: Characterizing visual performance during approach and landing with and without a synthetic vision display: A part-task study. In Conference Proceedings of the 2003 NASA Aviation Safety Program Conference on Human Performance Modeling of Approach and Landing with Augmented Displays, David C. Foyle, Allen Goodman, and Becky L. Hooey (Eds.), NASA/CP-2003-212267, 2003.

Gore, B. F.; and Corker, K. M.: Increasing aviation safety using human performance modeling tools: An Air Man-machine Integration Design and Analysis System application. In 2002 Military, Government and Aerospace Simulation, M. J. Chinni, (Ed.), Society for Modeling and Simulation International, San Diego, CA, vol. 34, no. 3, 2002, pp. 183–188.

Gore, B. F.: Human performance cognitive-behavioral modeling: A benefit for occupational safety. In International Journal of Occupational Safety and Ergonomics (JOSE), B. Chase and W. Karwowski (Eds.), vol. 8, no. 3, 2002, pp. 339–351.

Gore, B. F.: An emergent behavior model of complex human-system performance: An aviation surface related application. VDI Bericht 1675, vol. 1, no. 1, Düsseldorf, Germany: VDI Verl, 2002, pp. 313–328.

Gore, B. F.; and Corker, K. M.: Human error modeling predictions: Increasing occupational safety using human performance modeling tools. In Computer-Aided Ergonomics and Safety (CAES) 2001 Conference Proceedings, B. Das, W. Karwowski, P. Modelo, and M. Mattila (Eds.), Maui, Hawaii, July 28–August 4, 2001.

Keller, J.; Leiden, K.; and Small, R.: Cognitive Task Analysis of Commercial Jet Aircraft Pilots during Instrument Approaches for Baseline and Synthetic Vision Displays. In Conference Proceedings of the 2003 NASA Aviation Safety Program Conference on Human Performance Modeling of Approach and Landing with Augmented Displays, David C. Foyle, Allen Goodman, and Becky L. Hooey (Eds.), NASA/CP-2003-212267, 2003.

Kirlik, A.; and Byrne, M.: Time Design Panel – Timely and Accurate Decisions. Proceedings of the Human Factors and Ergonomics Society 48th Annual Meeting, HFES, Santa Monica, CA, 2004.

Lebiere, C.; Archer, R.; Warwick, W.; and Schunk, D.: Integrating Modeling and Simulation into a General-Purpose Tool. In Proceedings of the 11th International Conference on Human Computer Interaction, Las Vegas, NV, 2005.

Lebiere, C.; Biefeld, E.; Archer, R.; Archer, S.; Allender, L.; and Kelly, T. D.: Imprint/ACT-R: Integration of a task network modeling architecture with a cognitive architecture and its application to human error modeling. In 2002 Military, Government and Aerospace Simulation, M. J. Chinni (Ed.), Society for Modeling and Simulation International, San Diego, CA, vol. 34, no. 3, 2002, pp. 13–19.

Lebiere, C.; Archer, R.; Schunk, D.; Biefeld, E.; Kelly, T.; and Allender, L.: Using an integrated task network model with a cognitive architecture to assess the impact of technological aids on pilot performance. In Conference Proceedings of the 2003 NASA Aviation Safety Program Conference on Human Performance Modeling of Approach and Landing with Augmented Displays, David C. Foyle, Allen Goodman, and Becky L. Hooey (Eds.), NASA/CP-2003-212267, 2003.

McCarley, J. S.; Wickens, C. D.; Goh, J.; and Horrey, W. J.: A computational model of attention/situation awareness. Proceedings of the 46th Annual Meeting of the Human Factors and Ergonomics Society, Santa Monica, CA, 2002, pp. 1669–1673.

Muthard, E. K.; and Wickens, C. D.: Display Size Contamination of Attentional and Spatial Tasks: An Evaluation of Display Minification and Axis Compression. (AHFD-05-12/NASA-05-3), University of Illinois at Urbana-Champaign, Aviation Human Factors Division, Savoy, IL, 2005.

Newman, R. L.; and Foyle, D. C.: Test scenarios for rare events. Proceedings of the 12th International Symposium on Aviation Psychology, Wright State University, Dayton, OH, 2003, pp. 873–882.

Thomas, L. C.; and Wickens, C. D.: Eye-tracking and individual differences in off-normal event detection when flying with a Synthetic Vision System Display. Proceedings of the Human Factors and Ergonomics Society 48th Annual Meeting, Santa Monica, CA, 2004, pp. 223–227.

Wickens, C. D.; Alexander, A. L.; Horrey, W. J.; Nunes, A.; and Hardy, T.: Traffic and flight guidance depiction on a Synthetic Vision System display: The effects of clutter on performance and visual attention allocation. Proceedings of the Human Factors and Ergonomics Society 48th Annual Meeting. Santa Monica, CA, 2004.

Wickens, C.; McCarley, J.; and Thomas, L.: Attention-Situation Awareness (A-SA) Model. In Conference Proceedings of the 2003 NASA Aviation Safety Program Conference on Human Performance Modeling of Approach and Landing with Augmented Displays, David C. Foyle, Allen Goodman, and Becky L. Hooey (Eds.), NASA/CP-2003-212267, 2003.

Technical Reports

Best, B.; Lebiere, C.; Schunk, D.; Johnson, I.; and Archer, R.: Validating a Cognitive Model of Approach Based on the ACT-R Architecture. Contractor Report, Micro Analysis & Design, Inc., Boulder, CO, July, 2004.

Byrne, M. D.; and Kirlik, A.: Integrated Modeling of Cognition and the Information Environment: A Closed-loop, ACT-R Approach to Modeling Approach and Landing with and without Synthetic Vision Systems (SVS) Technology. Rice University Technical Report, July, 2004.

Byrne, M. D.; and Kirlik, A.: Integrated Modeling of Cognition and the Information Environment: A Closed-Loop, ACT-R Approach to Modeling Approach and Landing with and without Synthetic Vision System (SVS) Technology. Technical Report AHFD-03-4/NASA-03-3, Institute of Aviation, University of Illinois at Urbana-Champaign, 2003.

Byrne, M. D.; and Kirlik, A.: Integrated Modeling of Cognition and the Information Environment: Closed-Loop, ACT-R Modeling of Aviation Taxi Errors and Performance. Technical Report AHFD-02-19/NASA-02-10, Institute of Aviation, University of Illinois at Urbana-Champaign, 2002.

Callantine, T.: Catalog/Classification of Errors Detectable with the Crew Activity Tracking System (CATS). San Jose State University Milestone Report, July, 2004.

Callantine, T.: Error Generation in CATS-based Agents. NASA/CR–2003-212263, 2003.

Callantine, T.: CATS-based agents that err. NASA/CR–2002-211858, 2002.

Callantine, T.: CATS-based air traffic controller agents. NASA/CR–2002-211856, 2002.

Callantine, T.: Activity tracking for pilot error detection from flight data. NASA/CR–2002-211406, 2002.

Corker, K. M.; Gore, B. F.; Guneratne, E.; Jadhav, A.; and Verma, S.: SJSU/NASA coordination of Air MIDAS safety development human error modeling: NASA aviation safety program. Integration of Air MIDAS human visual model requirement and validation of human performance model for assessment of safety risk reduction through the implementation of SVS technologies, NASA Contract Task Order No. NCC2-1307, Interim Report and Deliverable, Ames Research Center, 2003.

Deutsch, S.; and Pew, R.: Modeling the NASA SVS Part-task Scenarios in D-OMAR. BBN Report No. 8399, BBN Technologies, Cambridge, MA, 2004.

Deutsch, S.; and Pew, R.: Modeling the NASA baseline and SVS-equipped approach and landing scenarios in D-OMAR. BBN Report No. 8364, Contractor Report, 2003.

Deutsch, S.; and Pew, R.: Modeling human error in D-OMAR. BBN Report No. 8328, Contractor Report, 2001.

Goodman, A.; Hooey, B. L.; and Foyle, D. C.: Developing Cognitive Models of Approach and Landing with Augmented Displays, NASA Milestone Report, 2003.

Gore, B. F.; Verma, S.; Jadhav, A.; Delnegro, R.; and Corker, K. M.: Human error modeling predictions: Air MIDAS human performance modeling of T-NASA. NASA Ames Research Center, Contract No. 21-1307-2344, CY01 Final Report, 2002.

Keller, J. W.; and Leiden, K.: Information to Support the Human Performance Modeling of a B757 Flight Crew during Approach and Landing: RNAV. Contractor Report, 2002.

Keller, J. W.; and Leiden, K.: Information to Support the Human Performance Modeling of a B757 Flight Crew during Approach and Landing SVS Addendum. Contractor Report, 2002.

Lebiere, C.; Biefeld, E.; and Archer, R.: Cognitive models of approach and landing. Contractor Report, 2003.

Leiden, K.; and Best, B.: A Cross-Model Comparison of Human Performance Modeling Tools Applied to Aviation Safety (Technical Report). Micro Analysis & Design, Inc., Boulder, CO, 2005.

Leiden, K.; Keller, J. W.; and French, J.: Information to Support the Human Performance Modeling of a B757 Flight Crew during Approach and Landing. Contractor Report, 2002.

Leiden, K.; Laughery, K. R.; Keller, J. W.; French, J. W.; Warwick, W.; and Wood, S. D.: A Review of Human Performance Models for the Prediction of Human Error. Contractor Report, 2001.

Leiden, K.; Keller, J. W.; and French, J. W.: Context of Human Error in Commercial Aviation. Contractor Report, 2001.

Miller, S.; Kirlik, A.; Kosorukoff, A.; and Byrne, M. D.: Ecological Validity as a Mediator of Visual Attention Allocation in Human-Machine Systems. Technical Report AHFD-04-17/NASA-04-6. University of Illinois at Urbana-Champaign, Aviation Human Factors Division, Savoy, IL, 2004.

Muraoka, K.; Verma, S.; Jadhav, A.; Corker, K. M.; and Gore, B. F.: Human Performance Modeling of Synthetic Vision System Use. Technical Report, San Jose State University, San Jose, CA, July, 2004.

Newman, R. L.: Scenarios for "rare event" simulation and flight testing. Crew Systems TR-02-07A, Monterey Technologies Inc., 2002.

Srivastava, A. N.; Akella, R.; Diev,V.; Kumaresan, S. P.; McIntosh, D. M.; Pontikakis, E. D.; Xu, Z.; and Zhang, Y.: Enabling the Discovery of Recurring Anomalies in Aerospace Problem Reports using High-Dimensional Clustering Techniques.

Uhlarik, J.; and Prey, C. M.: Functional Allocation Issues and Tradeoffs (FAIT) Analysis of Synthetic Vision Systems (SVS). Contractor Report, 2002.

Wickens, C. D.; McCarley, J. S.; Alexander, A. L.; Thomas, L. C.; Ambinder, M.; and Zheng, S.: Attention-Situation Awareness (A-SA) Model of Pilot Error. Technical Report AHFD-04-15/NASA-04-5, University of Illinois at Urbana-Champaign, Savoy, IL, 2005.

Wickens, C. D.; Alexander, A. L.; Thomas, L. C.; Horrey, W. J.; Nunes, A.; Hardy, T. J.; and Zheng, X.: Traffic and Flight Guidance Depiction on a Synthetic Vision System Display: The Effects of Clutter on Performance and Visual Attention Allocation. Technical Report AHFD-04-10/NASA(HPM)-04-1, University of Illinois at Urbana-Champaign, Aviation Human Factors Division, Savoy, IL, 2004.

Wickens, C. D.; and McCarley, J. S.: Attention-Situation Awareness (A-SA) Model of Pilot Error. Final Technical Report ARL-01-13/NASA-01-6, University of Illinois, Aviation Research Lab, Savoy, IL, 2001.

Wickens, C. D.: Spatial Awareness Biases. Technical Report ARL-02-6/NASA-02-4, University of Illinois, Aviation Research Lab, Savoy, IL, 2002.

Wickens, C. D.; McCarley, J. S.; and Thomas, L.: Attention-Situation Awareness (A-SA) Model. Contractor Report, 2003.